Genome Editing in Plants

Genome Editing in Plants
Principles and Applications

Edited by

Om Prakash Gupta and Suhas Gorakh Karkute

CRC Press
Taylor & Francis Group
Boca Raton London New York

CRC Press is an imprint of the
Taylor & Francis Group, an **informa** business

First edition published 2022
by CRC Press
6000 Broken Sound Parkway NW, Suite 300, Boca Raton, FL 33487-2742

and by CRC Press
2 Park Square, Milton Park, Abingdon, Oxon, OX14 4RN

CRC Press is an imprint of Taylor & Francis Group, LLC

Library of Congress Cataloging-in-Publication Data
Names: Gupta, Om Prakash (Plant Biochemistry), editor. | Karkute, Suhas Gorakh, editor.
Title: Genome editing in plants : principles and applications /
edited by Om Prakash Gupta, Suhas Gorakh Karkute.
Description: First edition. | Boca Raton, FL : CRC Press, 2021. | Includes bibliographical references and index.
Identifiers: LCCN 2021005809 | ISBN 9780367415907 (hardback) |
ISBN 9781032039343 (paperback) | ISBN 9780367815370 (ebook)
Subjects: LCSH: Gene editing. | Plant biotechnology. | Molecular biology—Technique.
Classification: LCC QH447 .G465123 2021 | DDC 572.8/2—dc23
LC record available at https://lccn.loc.gov/2021005809

ISBN: 9780367415907 (hbk)
ISBN: 9781032039343 (pbk)
ISBN: 9780367815370 (ebk)

Typeset in Times
by codeMantra

Affectionately Dedicated to
the Natural Vehicle of Genome Editing

Contents

List of Figures

List of Tables

Preface

Since the dawn of time, crop improvement programmes have been the essential component of global agriculture for ensuring food as well as nutritional security of the ever-growing population. The early scientific crop improvement programmes were principally focused on enhancing the yield only to feed the global human population, however; owing to the significant buffer stock of food grains, the nutritional aspects have also been targeted in the recent past. Despite several efforts, various environmental factors, including biotic and abiotic stresses, remain major challenges for sustainable agriculture, and therefore breeders have to develop crops that are tolerant to detrimental effects of such environmental stresses. Besides, the global warming resulting in frequent changes in climatic conditions also poses a major threat to crop production. Thus, future crop improvement programmes should be dedicated towards integrating all these aspects by utilizing modern breeding tools and techniques. The different methods of crop improvement such as traditional, mutation, molecular breeding and transgenic approaches have taken place due to their greater potential. More recently, genome editing techniques have revolutionized the whole crop improvement programmes due to their high efficiency and specificity. The era of genome editing started with meganucleases followed by sequence-specific nucleases such as ZFNs and TALENs, which fine-tuned the process of genome editing with greater specificity. ZFNs and TALENs were the techniques of choice for genome editing and were largely used in animal and plant systems until the development of CRISPR/Cas tools. CRISPR/Cas technique has replaced the older genome editing tools due to its simple designing and precise DNA manipulation. Various advanced versions of CRISPR tools have several applications such as gene knockout, gene knock-in, base editing, chromosome engineering, chromatin imaging, gene repression, gene activation, RNA editing, nucleic acid detection, etc. The versatility of this tool has made it the most important tool for crop improvement and is being widely implemented in several crops for imparting biotic and abiotic stress tolerance, increasing yield, enhancing nutritional quality and also for metabolic engineering. Thus, the Norwegian Nobel Committee has rightly selected E. Charpentier and J. A. Doudna for the 2020 Nobel Prize in chemistry for the discovery of CRISPR/Cas system, which is the game-changing technology in molecular biology and genome engineering.

This book compiles all the information of genome editing, starting from principles and historical aspects to the latest advancements in the field. The initial chapters describe different genome editing tools, and their principles and applications. Other chapters are dedicated to the present status and future applications of genome editing techniques in various crop improvement programmes. Some of the advanced applications of CRISPR/Cas tools like base editing and RNA detection along with regulatory aspects of genome-edited crops are described in detail. This book serves as a valuable resource to the researchers in the field of crop improvement, graduate and postgraduate students engaged in plant molecular biology and biotechnology, academicians and policy makers.

Dr. Om Prakash Gupta
Scientist (Plant Biochemistry), Division of Quality & Basic Sciences,
ICAR-Indian Institute of Wheat & Barley Research (IIWBR),
Karnal-132001, Haryana, India

Dr. Suhas Gorakh Karkute
Scientist (Plant Biotechnology), Division of Crop Improvement,
ICAR-Indian Institute of Vegetable Research,
Varanasi-221 305, Uttar Pradesh, India

Editors

Om Prakash Gupta received his M.Sc. and Ph.D. from Chandra Shekhar Azad University of Agriculture and Technology, Kanpur, India, and ICAR-Indian Agricultural Research Institute (IARI), New Delhi, India, respectively. He is presently working as a scientist (Plant Biochemistry) at ICAR-Indian Institute of Wheat and Barley Research (ICAR-IIWBR), Karnal, India. Dr. Gupta has more than 10 years of research experience in the area of plant biochemistry and molecular biology, and has significantly contributed in deciphering the role of small RNA during various biotic and abiotic stresses, biofortification, marker development, nutritional and processing quality, etc. He has co-developed three genetic stocks and one bread wheat variety suitable for various quality traits.

Dr. Gupta has been bestowed with more than 80 publications to his credit, which includes 30 research and review papers, 20 book chapters, 4 books and 30 popular articles. He has presented his research papers in several national and international symposia/workshops/conferences. He has immensely contributed in resource generations through several externally funded projects funded by Indian agencies like DBT & DST and international agency, including CIMMYT. Dr. Gupta is also involved in teaching fundamentals of molecular biology course to Ph.D. students. Due to his immense research contributions, Dr. Gupta has been serving as an editorial board member and a reviewer of many international journals. Due to his outstanding contribution in research and meritorious profile, Dr. Gupta has the distinction of receiving numerous honours, fellowships and awards in recognition to his excellent academic and research contributions. He is a fellow of the Society for Advancement of Wheat and Barley Research (SAWBAR), and received several awards, including Jawaharlal Nehru Award for outstanding doctoral thesis by ICAR, University Silver Medal, ASPEE Gold Medal and Dr. Kirtikar Memorial Gold Medal, and Chaudhary Charan Singh Memorial Award during his bachelor degree programme.

Suhas Gorakh Karkute is a scientist in Indian Council of Agricultural Research, India. He is a graduate in Agricultural Sciences from Dr. Vasantrao Naik Marathwada Agricultural University, Parbhani, Maharashtra, India. He has been awarded with M.Sc. in Molecular Biology and Biotechnology from a prestigious ICAR-Indian Agricultural Research Institute, New Delhi, in 2010. He is the recipient of several fellowships and awards in recognition to his performance in academics and scientific career. He has also qualified for a number of national-level competitive examinations such as ICAR-JRF, ARS, CSIR-NET, IARI-Ph.D., JNU-Biotech, IISC-Ph.D. etc.

Dr. Karkute has gained more than 6 years of research experience since he joined as an ARS scientist (Agricultural Biotechnology) at ICAR-Indian Institute of Vegetable Research, Varanasi, India. He is also working as an Institute Biosafety Officer for transgenic projects in the institute since 2017. Besides, he has the experience of working as an Examiner of Patents and Designs in Patent Office, Government of India, for nearly 2 years.

During his scientific career at ICAR-IIVR, he is actively involved in several research projects of vegetable improvement. He is working in the area of functional genomics, transgenic development and genome editing with the aim to understand the abiotic stress tolerance in plants. He is also associated with studies on plant bioactive compounds. He has published many original research and review articles in peer-reviewed international journals and several book chapters and popular articles. He has participated and presented his research work in different national and international conferences/workshops/seminars. He is also contributing as a reviewer of many national and international journals.

Contributors

Israr Ahmad
Division of Crop Improvement and Biotechnology
ICAR-Central Institute for Subtropical Horticulture
Lucknow, India

Sivalingam Anandhan
ICAR-Directorate of Onion and Garlic Research
Rajgurunagar, India

Antim
ICAR-Indian Institute of Wheat and Barley Research
Karnal, India

and

Chaudhary Charan Singh Haryana Agricultural University
Hisar, India

Ashish
ICAR-Central Soil Salinity Research Institute
Karnal, India

and

Chaudhary Charan Singh Haryana Agricultural University
Hisar, India

Nabaneeta Basak
Crop Physiology and Biochemistry Division
Central Genomics and Quality Laboratory
ICAR-National Rice Research Institute
Cuttack, India

Alok Das
Division of Plant Biotechnology
ICAR-Indian Institute of Pulses Research
Kanpur, India

P. T. K. Jagannadham
Division of Crop Improvement
ICAR-Central Citrus Research Institute
Nagpur, India

Sandeep Jaiswal
ICAR-Research Complex for North-Eastern Hill Region
Umiam, India

Gitanjali Jiwani
ICAR-National Institute for Plant Biotechnology
New Delhi, India

Ravindra Ramrao Kale
Crop Improvement Division
ICAR-Indian Institute of Rice Research
Rajendranagar
Hyderabad, India

Taruneet Kaur
ICAR-National Dairy Research Institute
Karnal, India

Anil Kumar
College of Horticulture and Forestry
Rani Lakshmi Bai Central Agricultural University
Jhansi, India

Ashok Kumar
ICAR-Directorate of Onion and Garlic Research
Rajgurunagar, India

Awadhesh Kumar
Division of Crop Physiology and Biochemistry
ICAR-National Rice Research Institute
Cuttack, India

Gaurav Kumar
Crop Physiology and Biochemistry Division
Central Genomics and Quality Laboratory
ICAR-National Rice Research Institute
Cuttack, India

Jitendra Kumar
ICAR-National Dairy Research Institute
Karnal, India

Manoj Kumar
ICAR-Central Institute for Research on Cotton Technology
Mumbai, India

Rajeev Kumar
ICAR-Indian Institute Sugarcane Research
Lucknow, India

Sudhir Kumar
Division of Plant Biotechnology
ICAR-Indian Institute of Pulses Research
Kanpur, India

Vikas Kumar
Department of Biotechnology
Maharishi Markandeshwar (Deemed to be University)
Ambala, India

Neetu Kushwah
Division of Plant Biotechnology
ICAR-Indian Institute of Pulses Research
Kanpur, India

Mayuri D. Mahalle
ICAR-National Institute for Plant Biotechnology
New Delhi, India

H. M. Mamrutha
ICAR-Indian Institute of Wheat and Barley Research
Karnal, India

Vikas Mangal
ICAR-Central Research Institute for Jute and Allied Fibres
Barrackpore, India

N. Manjunatha
Division of Seed Technology
ICAR-Indian Grassland and Fodder Research Institute
Jhansi, India

Nand Lal Meena
Division of Germplasm Evaluation
ICAR-National Bureau of Plant Genetic Resources
New Delhi, India

Shriram Jaganath Mirajkar
Department of Biotechnology
Dr. D. Y. Patil Arts, Commerce and Science College
Pimpri, India

Mangesh Pradeep Moharil
Biotechnology Centre
Department of Agricultural Botany
Dr. D. Y. Patil Arts, Commerce and Science College
Akola, India

Senthilkumar K. Muthusamy
Division of Crop Improvement
ICAR-Central Tuber Crops Research Institute
Thiruvananthapuram, India

K. Bhojaraja Naik
ICAR-Indian Institute of Seed Science
Bengaluru, India

Lopamudra Nayak
Division of Crop Physiology and Biochemistry
ICAR-National Rice Research Institute
Cuttack, India

Umakanta Ngangkham
ICAR-Research Complex for North-Eastern Hill Region
Umiam, India

Bhupendra Singh Panwar
Karnavati Seeds Pvt Ltd
Aslali, India

Amit Pareek
ICAR-National Institute for Plant Biotechnology
New Delhi, India

Jyoti Priya
Division of Plant Physiology
ICAR-Indian Agricultural Research Institute
New Delhi, India

Ritesh Kumar Raipuria
ICAR-National Institute for Plant Biotechnology
New Delhi, India

Kavita Rani
ICAR-National Dairy Research Institute
Karnal, India

Darasing Ramsing Rathod
Biotechnology Centre
Department of Agricultural Botany
Dr. D. Y. Patil Arts, Commerce and Science College
Akola, India

Meenal Rathore
Division of Plant Biotechnology
ICAR-Indian Institute of Pulses Research
Kanpur, India

Chandrasekhar Sahu
MS Swaminathan School of Agriculture
Centurion University of Technology and Management
Paralakhemundi, India

Ravi Prakash Saini
Division of Seed Technology
ICAR-Indian Grassland and Fodder Research Institute
Jhansi, India

Siddanna Savadi
ICAR-Directorate of Cashew Research
Puttur, India

Sonia Sheoran
ICAR-Indian Institute of Wheat and Barley Research
Karnal, India

Prashant Raghunath Shingote
Vasantrao Naik College of Agricultural Biotechnology
Yavatmal, India

and

Biotechnology Centre
Department of Agricultural Botany
Dr. Panjabrao Deshmukh Krishi Vidyapeeth
Akola, India

Narendra Pratap Singh
Division of Plant Biotechnology
ICAR-Indian Institute of Pulses Research
Kanpur, India

Prabha Singh
Department of Seed Science
ICAR-Indian Grassland and Fodder Research Institute
Jhansi, India

Prateek Singh
Division of Plant Biotechnology
ICAR-Indian Institute of Pulses Research
Kanpur, India

Rupesh Kumar Singh
Centro de Química de Vila Real (CQ-VR)
Universidade de Trás-os-Montes e Alto Douro (UTAD)
Vila Real, Portugal

and

Departamento de Agronomia
Universidade *de* Trás-os-Montes e Alto Douro (UTAD)
Vila Real, Portugal

Shweta Singh
Centre for the Research and Technology of Agro-Environment and Biological Sciences (CITAB)
University of Minho (UMINHO)
Braga, Portugal

and

Biosystems and Integrative Sciences Institute (BioISI)
Plant Functional Biology Center
University of Minho (UMINHO)
Braga, Portugal

Amolkumar U. Solanke
ICAR-National Institute for Plant Biotechnology
New Delhi, India

Mohd Tasleem
ICAR-National Institute for Plant Biotechnology
New Delhi, India

Shallu Thakur
Division of Plant Biotechnology
ICAR-Indian Institute of Pulses Research
Kanpur, India

Sushma Tiwari
Department of Plant Molecular Biology and Biotechnology
Rajmata Vijayaraje Scindia Krishi Vishwa Vidyalaya
Gwalior, India

Ram Sewak Singh Tomar
College of Horticulture and Forestry
Rani Lakshmi Bai Central Agricultural University
Jhansi, India

Bhudev Singh Tyagi
ICAR-Indian Institute of Wheat and Barley Research
Karnal, India

Vivek Chandra Verma
Vegetable Division, Soil Science Lab
Defense Institute of High Altitude Research - DRDO
Chandigarh, India

Pawar Deepak Viswanath
ICAR-Directorate of Weed Research
Jabalpur, India

Dhiraj Lalji Wasule
Vasantrao Naik College of Agricultural Biotechnology
Yavatmal, India

Anshul Watts
ICAR-National Institute for Plant Biotechnology
New Delhi, India

Archana Watts
Division of Plant Physiology
ICAR-Indian Agricultural Research Institute
New Delhi, India

1 Historical Developments of Genome Editing in Plants

Ravindra Ramrao Kale
ICAR-Indian Institute of Rice Research

Prashant Raghunath Shingote
Vasantrao Naik College of Agricultural Biotechnology
Dr. Panjabrao Deshmukh Krishi Vidyapeeth

Dhiraj Lalji Wasule
Vasantrao Naik College of Agricultural Biotechnology

Shriram Jagannath Mirajkar
Dr. D. Y. Patil Arts, Commerce and Science College

Darasing Ramsing Rathod and Mangesh Pradeep Moharil
Dr. Panjabrao Deshmukh Krishi Vidyapeeth

CONTENTS

1.1 GENESIS

An increase in population and the limitation of available resources for food production have caused food security problems globally. Crop improvement through conventional and molecular breeding approaches can help to tackle the problem of food security through increased food production. Conventional breeding is a very widely used approach for crop improvement and found to be promising for the establishment and increased crop production. Major constraints with conventional breeding are time required for crop improvement; laborious; huge population size; less accurate and reliable; phenotypic selection (PS)-dependent, which is affected by environment; less successful for low-heritable multigenic quantitative traits and the problem with distant crosses, etc. (Tester and Langridge, 2010). Thus, to fasten the crop improvement programme, various molecular breeding approaches such as transgenic, RNAi technology, marker-assisted breeding, gene editing and so on were proven to be effective in terms of the current scenario (Limbalkar et al., 2019). The transgenic approach was found to be effective as it helps to transfer genes from distant species or genera and reduces the time for improvement of crops for a particular trait, but due to societal distrust, biosafety and regulatory issues, the use of transgenic is limited to few countries (Limbalkar et al., 2019). Genome editing has the potential to alter an organism's genetic makeup through precise changes in the DNA sequence. Genome editing is done using enzymes, especially nucleases to make cuts into the DNA strands, allowing the removal or insertion of the target DNA. It is one of the leading approaches for crop improvement that allows for the modification of a specific region of the genome and allows for the removal, alteration and insertion of DNA sequences at a particular location in the genome with very high precision. Genome editing is based on natural enzymes, including site-directed recombinase (SDR) and site-specific nuclease (SSN) repair enzymes. Genome editing has many benefits, such as site-driven mutagenesis rather than random mutation, site-driven insertion of transgenes with small copy numbers, analysis of functional genomics research, virus or pathogen gene disruption and lesser biosafety issues as compared to transgenics (Pohare et al., 2019).

In the 1960s, scientists started to search for ways to alter genomes. Working in test tubes, researchers in the University of California, San Francisco and Stanford, bombarded DNA with complex variations of molecular gadgets, all borrowed from bacteria. For the first time in 1972, it had shown that genes could be mixed and matched to establish hybrid sequences called recombinant DNA, which had never existed before (Jackson et al., 1972). The initial experiments in the direction of genome editing were carried out by Scherer and Devis (1979), where they demonstrated *in vitro* replacement of target DNA, which involved the double-stranded break (DSB) repair and recombination mechanisms. There are four main subtypes of SSNs developed to make cuts in genomic sequences, *viz.* meganucleases, zinc finger nucleases (ZFNs), transcriptional activator-like effector nucleases (TALENs) and clustered regularly interspaced short palindromic repeat/CRISPR-associated proteins (CRISPR/Cas) nuclease system. The timeline of major historical discoveries in genome editing is summarized in Table 1.1.

TABLE 1.1
Timeline of Major Scientific Breakthroughs in Genome Editing

Year	Major Discovery
1978	Ishino and colleagues for the first time observed an unusual repetitive sequence activity in bacteria
1979	*In vitro* genome editing with the replacement of chromosome segment with other DNA sequence
1985	Identification of zinc-binding domain in protein transcription factor
1987	CRISPRs, first described as a bacterial immune system in *Escherichia coli*
1989	Studied the double-stranded break repair and recombination mechanism in yeast
1993	Francisco Mojica was the first to characterize what is now called a CRISPR locus
1996	Use of hybrid restriction enzyme including zinc finger and FokI cleavage domain
2001	Spanish researchers at University of Alicante proposed the name CRISPR
2005	Discovered the involvement of CRISPR as a form of adaptive immunity in bacteria and the discovery of Cas9 and PAM sequences
2007	Experimental demonstration of adaptive immunity of CRISP-Cas9 in bacteria
2008	Spacer sequences are transcribed into guide RNAs (crRNAs) and CRISP target DNA
2009	Identified action of TAL-type III effecters in *Xanthomonas*
2010	Role of type II, CRISPR-Cas identified
2011	Discovery of tracrRNA for Cas9 system
2012	Duodna and Charpentier use the CRISPR-Cas9 system as a tool for genome editing, by employing genetically engineered CRISPR-Cas system
2013	First-time demonstration of use of the CRISPR/Cas system for genome editing in plants, humans and other eukaryotic
2015	DNA-free genome editing technique in plants

1.2 TECHNIQUES FOR GENOME EDITING

1.2.1 Cre-Lox System

Cre-lox system has many applications involving controlled removals of DNA fragments and the targeted insertion of DNA into specific genome sites, whereas recent advances, including controlled creative expression and innovative use of wild-type and modified lox sites, have enhanced its applicability. Cre-lox is a member of a broad family of recombinases of tyrosine that includes many other common recombinases, including Flp and Int that are derived from the bacteriophage P1. Cyclization recombination (Cre) gene encodes a 38-kDa protein, which identifies 34-base pair (bp) sites known as lox (locus of crossing over), which includes 13-bp inverted repeats at both sites of 8-bp non-palindromic core region and catalyses recombination between two lox sites (Abremski et al., 1983). The role of such a system involves a precise recombination, removal of a selectable gene (transgene) and site-specific gene transfer for crop improvement (Abremski et al., 1983; Gilbertson, 2003). It is observed that prokaryotic recombinase can be used in the eukaryotes where it enters into the nucleus and Cre recombinase is capable of performing successful recombination events on the chromosomes (Sauer, 1987). A Cre-recombinase protein from microbes may be useful to study the genome modification in other higher organisms (Sauer and Henderson, 1988; Orban et al., 1992).

1.2.2 Meganucleases

Meganucleases are a class of endonucleases characterized by a large recognition site and act like 'molecular DNA scissors' that can be used to remove, delete or alter sequences in a highly targeted way. Meganucleases possess recognition sites of 12–40 bp, consequently occur only once in any given genome and are known to be the most common naturally occurring restriction enzymes (Figure 1.1a). By altering their recognition sequence through protein engineering, the enzymes can be specified to the specific desired sequence. Meganucleases can be expressed in different compartments of the cell such as mitochondria or chloroplasts of archaebacteria, bacteria, phages, fungi, yeast, algae and some plants, and are mainly represented by the two main enzyme families, namely intron endonucleases and intein endonucleases. Meganucleases are most commonly found in *Saccharomyces cerevisiae* (I-SceI), *Chlamydomonas reinhardtii* and *Desulfurococcus mobilis*, named as I-CreI and Dmo, respectively (Choulika et al., 1995; Roberts et al., 2003; Marcaida et al., 2008). Chimeric meganucleases, derived from homing endonucleases that can induce recombination and cleave a selected new DNA target site, have been used in genome modification (Epinat et al., 2003). Sylvain et al. (2007) used an engineered homing endonuclease approach for the first time to modify the I-CreI meganuclease DNA-binding interface to target sequences of a chromosomal locus. In another study, Grizot et al. (2009) customized the chimeric meganuclease, DmoCre, by successfully assembling mutants with locally altered specificities, and these newly engineered variants of DmoCre were found selective with low toxicity level, which help to induce successful homologous recombination events. Further, meganuclease engineering was used to make a frameshift mutation to recover the normal reading frame of a dystrophin gene (Chapdelaine et al., 2010). An artificial I-SceI recognition sequence incorporated in genomic DNA was analysed for maize mutations which were transmitted faithfully to the next generation, which is a valuable molecular method for the introduction of targeted mutations in plants (Yang et al., 2009; Gao et al., 2010).

1.2.3 Zinc Finger Nucleases

Perhaps ZFN system is the first method for highly targeted genome engineering, which improves the effectiveness of gene targeting in several ways. Zinc fingers help to recognize DNA, making it possible to design ZFNs for a variety of different genomic targets varying throughout many different types of cells and organisms. A repetitive zinc-binding domain discovered in the protein TFIIIA from *Xenopus laevis* (Miller et al., 1985) fused into a cleavage domain FokI (*Flavobacterium*

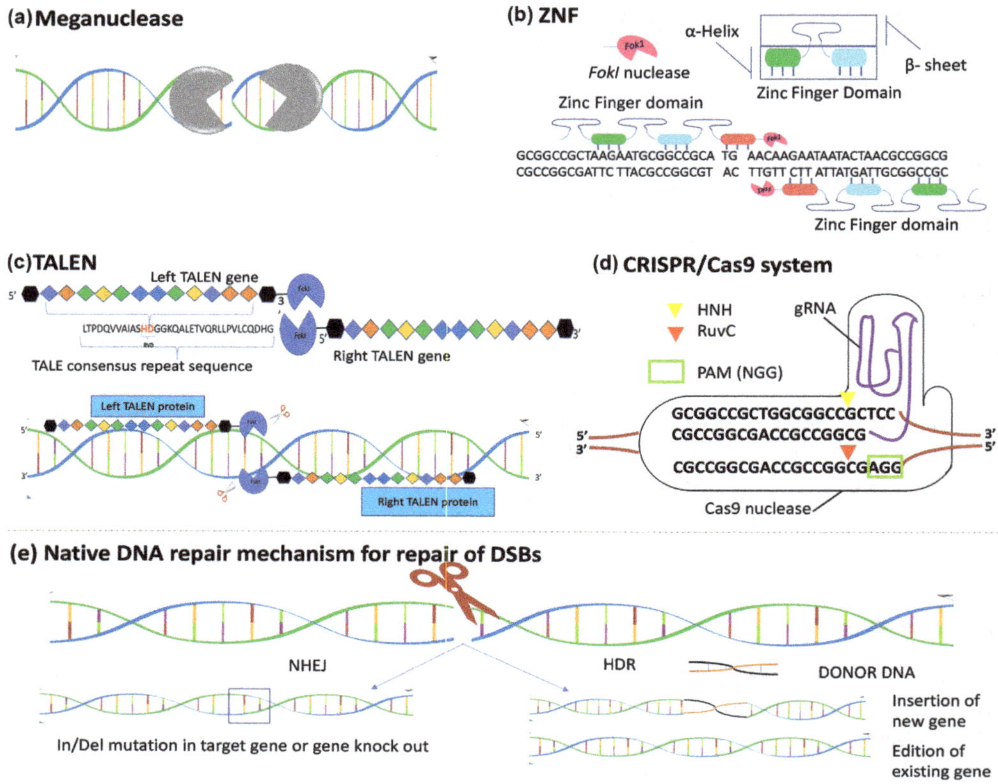

FIGURE 1.1 Overview of current genome editing methods. (a) Meganuclease recognizes long (20-bp) DNA sequences; (b) an engineered ZF domain consists of ~30 amino acids composed of αββα-fold; each domain contacts 3 or 4 bp in the target DNA. The ZFN target sites consist of two ZF-binding sites; (c) TALEN causes DSB that result in DNA repair; (d) Cas9 has two nuclease domains (RuvC and HNH) to digest different DNA strands to create DSBs at the target sites via direction of the gRNA to the PAM sequence. (*Source*: Lee et al. (2016).) (e) Native DNA repair mechanism for the repair of DSBs generated by engineered nucleases. The DSB in DNA is either repaired by non-homologous end-joining (NHEJ) causing small In/Del mutations or through homology-directed repair (HDR) for the insertion of new gene/correction of gene; RVD, repeat variable diresidue; PAM, protospacer adjacent motif; DSB, double-stranded breakage; NHEJ, non-homologous end-joining; HDR, homologous DNA recombination.

okeanokoites) has made it possible for zinc finger motifs to recognize different sites (Kim et al., 1996). Many proteins have been shown to contain tandem repeats of amino acid, which form structural domains around bound zinc ions termed as ZFNs (Berg, 1988). ZFNs consist of FokI that cleaves DNA fused with a modified zinc finger DNA-targeting domain (Figure 1.1b) that makes a DNA DSB at a specific site. The cells repair the break either through non-homologous recombination with small In/Del mutations or through homologous recombination, where the cell incorporates external DNA specifically at a breakpoint (Figure 1.1e).

Single ZFN binds precisely to a set of three nucleotides and can be collectively designed to bind longer DNA sequences. Amino acid residues at 1, 3, 6 on the alpha helix of zinc finger molecule interact with DNA and act as a recognition site (Roy et al., 2012), so mutating these main residues at three positions will generate the desired zinc finger that can recognize the site of our interest. Researchers have fused the restriction enzyme FokI to zinc finger to make a DSB at a specific site, and the cleavage domain is active at the dimeric state to have a precise cut in DNA (Kim et al., 1996). Zinc finger nickase classes of ZNFs (ZFNickases) created by inactivating one ZFN monomer's catalytic activity in the ZFN dimer are needed for DNA cleavage (Ramirez et al., 2012).

This technique encourages the idea of reverse genetics over Mendelian genetics: with Mendelian genetics, scientists generate random mutations looking for something significant and then characterize the gene they believe that is responsible for the mutation. With reverse genetics, a scientist first mutates the well-characterized gene to identify the specific effect on the phenotype. ZFNs provide advantages over other methods in terms of performance, high specificity and limited non-target impact. Current efforts are centred on further improving design and delivery and extending their applications in various crops (Kamburova et al., 2017).

1.2.4 Transcription Activator-Like Effector Nucleases (TALENs)

A plant pathogen bacteria of genus *Xanthomonas* produces proteins that act like transcription activator and changes gene expression by binding to the specific DNA sequence of host and thus, binding proteins are called as transcription activator-like effectors (TALEs), and it uses type III secretion system to introduce TALE proteins into plant cells and modify the cellular processes of the host (Figure 1.2) (Boch et al., 2009; Bogdanove and Voytas, 2011; Lahaye and Bonas, 2001). The DNA-binding domain includes a highly conserved sequence of 33–34 amino acids with variations at the 12th and 13th place, called repeat variable diresidues (RVD), a factor that decides the specificity of DNA binding and is used to modify in association with site-directed nucleases (SDNs) when TALE is involved in genome editing techniques (Boch et al., 2009; Moscou and Bogdanove, 2009). The plasmid with TALEN construct uses the sequences of RVDs that determine the target nucleotide sequences, and frequently associated nucleotide and RVD pairs are NI that binds to nucleotide A, HD that binds to nucleotide C, NN that binds to nucleotide G or A, NK that binds to

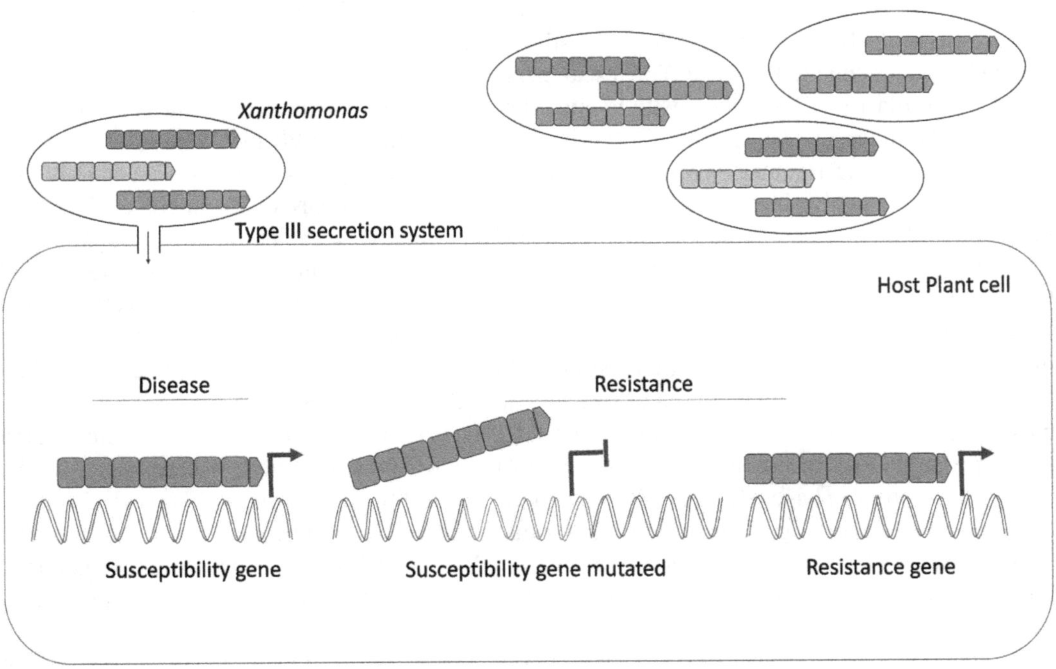

FIGURE 1.2 Transcription activators like effector proteins from *Xanthomonas* as virulence factors in plant disease. Following the entry into the host plant cytoplasm via the bacterial type III secretion system and translocation to the nucleus, TAL effectors bind to RVD-specified sequences in host DNA. Binding causes the activation of host susceptibility (S) genes, which leads to disease. Resistant plants may have an S gene mutation that blocks binding and activation by the TAL effector, or they may harbour an R gene that has a TAL effector binding site in its promoter but encodes a protein that mediates the localized cell death and limits the infection when activated.

nucleotide G and NG that binds to nucleotide T (Cermak et al., 2011). Various online tools such as TALEN Targeter, TAL Effector Targeter and Target Finder are available to design TAL effectors for TAL effector nucleases (TALENs), to identify binding sites in a DNA sequence for a given TAL effector and to identify the best-scoring sites in a DNA sequence for a specified RVD sequence, respectively (Cermak et al., 2011; Doyle et al., 2012).

The FokI endonuclease that functions as a dimer can be used to construct hybrid nucleases by combining them with TALE and forming TALENs to develop DSB at specific locations in the genome (Boch et al., 2009; Cermak et al., 2011) (Figure 1.1c). Proper orientation and spacing of the FokI domain is needed for its action on sites in the target genome. The important factors for high activity of TALENs are the number of residues of amino acids between the binding and cleavage domain, and the number of bases between the two TALEN-binding sites.

1.2.5 Clustered Regularly Interspaced Short Palindromic Repeats (CRISPRs)

The region CRISPR contains alternately repeated elements (CRISPR repeats) and variable sequences (CRISPR spacers) with regular intervals (Barrangou and Marraffini, 2014). CRISPR-systems have different origins such as bacteria and archaea and are ubiquitously present. These systems were initially discovered as unusual repeat sequences in bacteria separated by non-repeating sequences in a nearly palindromic pattern in *E. coli* (Ishino et al., 1987). Mojica also spotted the similar structure in an archaeal microbe *Haloferax mediterranei*, although there was no sequence similarity, with the *E. coli* repeats (Mojica et al., 1993). These discoveries elucidated that the presence of such similar repeats in distant microbes must signal an important function in prokaryotes. Mojica used the sequence of each spacer from an *E. coli* strain to search for similarity with CRISPR locus using BLAST, and found one of the spacers matching with a P1 phage sequence that infects many strains of *E. coli* (Mojica et al., 2005). Interestingly, the particular spacer-carrying strain was recognized to be resistant to P1 phage. The researchers searched the thousands of different spacers and found the similarity of 88 spacers with identified sequences from viruses or conjugative plasmids related to the microbe. Mojica et al. (2005) interpreted that the CRISPR loci were responsible for the specific adaptive immune system of microbes. Later, these repeat structures were found to be involved in the adaptive immune system of bacteria against foreign DNA (Barrangou et al., 2007). Further, similar sequences were identified in *M. tuberculosis* (Groenen et al., 1993). Major developments were made in understanding the CRISPR system, with the identification of genes coding CRISPR-associated system (Cas) protein (Jansen et al., 2002) followed by the identification of the non-repeating CRISPR spacer sequences derived from non-native DNA that lead to a hypothesis of involvement of CRISPRs in the immune system (Mojica et al., 2005; Pourcel et al., 2005). While bacteria contain several innate immune systems, the CRISPR-Cas system has been identified as a most effective adaptive immune system (Figure 1.3). The first biological evidence was demonstrated in *S. thermophilus* where resistance specificity in the CRISPR-Cas system is determined by the similarity of spacer-phage sequence. Barrangou et al., (2007) used *S. thermophilus* strain along with two phages for the genetic selection experiment to identify the phage-resistant bacteria. The resistant bacterial strains were observed to acquire CRISPR loci derived from the phage genome. Furthermore, they correlated the insertion of multiple spacers with increased resistance, suggesting it as an acquired immunity.

Brouns et al. (2008) transformed the CRISPR system from one type of *E. coli* to another type of strain that was deficient in its own CRISPR system. Through that system, they recognize a complex of five Cas proteins biochemically, called cascade (Brouns et al., 2008). By mutating each component at a time, they recognized that the whole cascade is essential for the cleavage of CRISPR locus, which transcribes the long precursor RNA into CRISPR RNAs (crRNAs) of 61 nucleotides. Sequencing of cascade-co-purified crRNAs revealed that all crRNAs had a repeat sequence of eight bases at the start, thereafter the whole spacer and again followed by the start of the subsequent repeat region. Their results were in line with the earlier suggestions of Sorek et al. (2008) that the

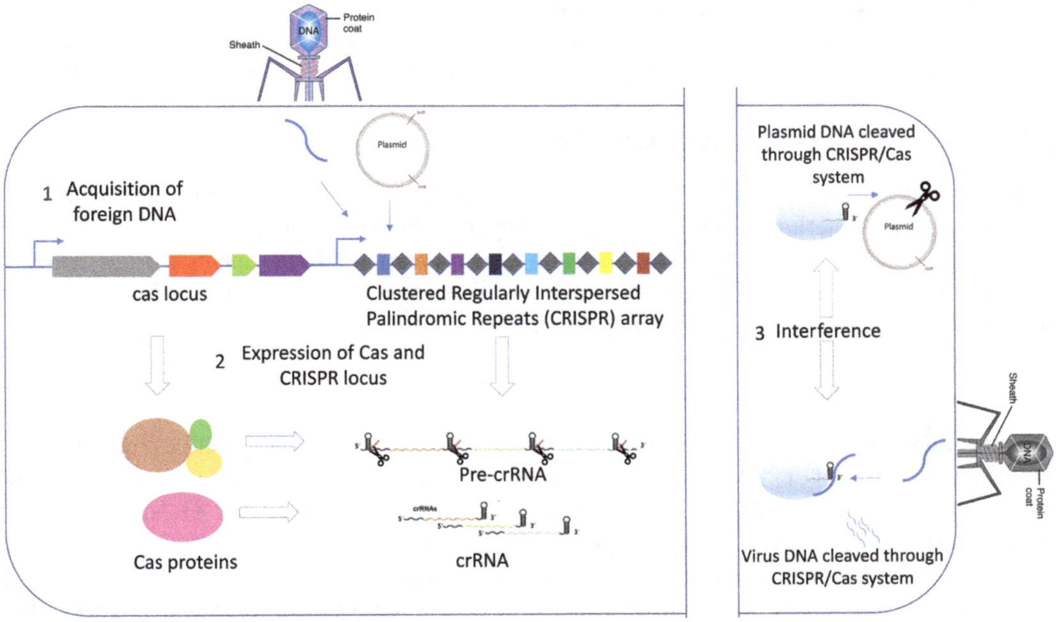

FIGURE 1.3 Features of the CRISPR-Cas adaptive immune system. (1) Acquisition of foreign DNA as CRISPR spacer: Protospacers (shown as ▮ different colours with an alternative protospacer-associated motif ◆) of double-stranded DNA from invading virus or pathogen are adapted at the leader end of a CRISPR array on bacterial genome (host). A CRISPR array involves unique spacers ▮ interspaced between repeats ◆. Acquisition ensued by Casl and Cas2 proteins. (2) Expression of CRISPR locus: RNA polymerase synthesized Pre-CRISPR RNA (Pre-crRNA) on the leader region further cleaved by Cas proteins into smaller crRNAs that contain a single spacer and a partial repeat. (3) Interference: crRNA comprising a spacer with a perfect complementarity to incoming foreign DNA initiates a cleavage mechanism through Cas proteins. This cleavage event interferes with virus replication or plasmid activity, and governs immunity to the host bacteria.

palindromic nature of the repeats would lead to crRNA secondary structure formation. Subsequent studies showed that the system involves a sequence-specific, RNA-mediated targeting of mainly DNA and sometimes RNA (Barrangou et al., 2007; Brouns et al., 2008; Marraffini and Sontheimer, 2008; Hale et al., 2009). Generally, this system follows the work in three steps: adaptation, transcription and interference (Horvath and Barrangou, 2010; Barrangou and Marraffini, 2014). Based on the genetic content and variations at the structural and functional level, CRISPR-Cas systems have been categorized into three main groups (Type I to Type III) and 12 subtypes. Cas3, cas9 and cas10 were classified as Type I, Type II and Type III signature genes, respectively (Makarova et al., 2011, 2013). Further progress in the CRISPR-Cas study found that it uses RNA-DNA binding, rather than protein-DNA binding, to guide nuclease activity (Jinek et al., 2012). CRISPR-Cas9 system was preferable over ZFNs and TALENs due to the simplicity of construct designing (Figure 1.1d), as a complicated fusion protein designing is needed for using ZFNs or TALEN. Cpf-1 nuclease from microbes is found to be more specific and allows the double-stranded (rather than blunt) DNA cuts, and thus Cpf-1 nuclease is preferred over Cas9 (Zetsche et al., 2015). CRISPR-Cas9 has been used in a variety of ways: it has been applied to early embryos to create genetically modified organisms, and to modify the genomes of crop plants, farm animals and laboratory model organisms (including mice, rats and non-human primates) (Pohare et al., 2019).

In addition to the classic *Streptococcus pyogenes* Cas9 system, multiple Cas variants have been used for plant genome editing (Endo et al., 2014). The classical Cas9 system consists of a dual RNA complex of CRISPR (cr) RNA and trans-activating CRISPR (tracr) RNA. Cas13, a CRISPR variant, has recently been exploited for RNA editing and tracking, and needs more study to make it available for commercial applications (Ali et al., 2020). Recent studies performed in plants and mammals with CRISPR-Cas system have not observed any non-target, collateral RNA degradation effects (Aman et al., 2018).

1.3 APPLICATIONS OF GENOME EDITING

Compared to the traditional approach of crop breeding, targetable nucleases are found to be advantageous due to the inheritance of mutations created by them from one generation to another and ease of generating the mutations. Genome editing has a widespread role in enhancing the resistance to viruses, fungal and bacterial diseases, as well as nutritional and other traits, and has been proven to be an excellent approach for sustainable agriculture.

1.3.1 Applications for Developing Biotic Stress Tolerance

TALENs were used to establish gene modifications of a particular rice susceptibility gene to prevent bacterial attacks (Li et al., 2012). In case of wheat, resistance to powdery mildew has been developed by introducing TALEN and CRISPR-mediated mutations in genes that encode MILDEW-RESISTANCE LOCUS (MLO) proteins (Wang et al., 2014). A CRISPR/Cas9 SSN targeting both the *OsERF922* gene and multiple sites within *OsERF922* was used to enhance rice blast resistance (Wang et al., 2016). Zhou et al. (2015) demonstrated that *OsSWEET13* gene, which acts as the disease-susceptibility gene for *PthXo2*, can be mutated using CRISPR/Cas9 technology to control disease attack. A specific gene *SlDMR6-1* orthologue associated with disease resistance in tomato has been mutated using the CRISPR-Cas9 system and showed a good level of resistance to the number of pathogens (de Toledo Thomazella et al., 2016). In another study with tomato, *SlJAZ2* gene was edited using CRISPR/Cas9 that provided resistance to the bacterial speck (Ortigosa et al., 2019). The first report of the development of non-transgenic cucumber resistance to the virus using Cas9/subgenomic RNA (sgRNA) has demonstrated success with targeting the recessive *eIF4E* (eukaryotic translation initiation factor 4E) gene (Chandrasekaran et al., 2016).

1.3.2 Applications for Developing Abiotic Stress Tolerance and Nutritional Quality

Truncated gRNA (tru-gRNA)/Cas9 combination that is used to modify *OST2* gene in *Arabidopsis* leads to the identification of the important role of the gene in water stress conditions by changing the stomatal closing pattern under stress (Osakabe et al., 2016). Application of ZFNs in maize for gene editing to produce variants showed a good level of herbicide tolerance as well as alteration of the inositol phosphate expression in developing seeds (Shukla et al., 2009). The 9-*cis*-epoxycarotenoid dioxygenase (NCED), a key enzyme of the ABA synthesis pathway mutated using CRISPR/Cas9 in rice, showed multistress tolerance, which may provide a new strategy for rice breeding (Huang et al., 2018). A CRISPR-Cas used to generate novel variants of *ARGOS8* gene, which is a negative regulator of ethylene responses in maize, results in an increased grain yield under stress (Shi et al., 2017). SELF-PRUNING 5G (SP5G) is the florigen paralog and flowering repressor expressed more in the wild-type compared to the cultivated tomato species, and modification of this gene results in the development of the compact determinate growth habit of cultivated tomatoes that translates to an early yield (Soyk et al., 2017). In case of rice, CRISPR-Cas9 was used for the targeted insertion of genes involved in carotenoid biosynthesis, and plants with no off-target effect and free from marker genes were recovered (Dong et al., 2020).

1.4 CONCLUSION

Traditional breeding is one of the prime methods that has contributed to agriculture development, and it mainly depends on the variability in plant populations that is mostly derived from mutations that are random and happen rarely. It is very difficult and time-consuming with conventional breeding to introgress a desirable variation from wild-type cultivars to cultivated cultivars. Contrariwise, an advanced molecular biology technique of genome editing produces modifications in targeted regions in the plant genome with precision. In addition to the availability of so many tools for genome editing, different systems need to be defined according to the purpose and target species. With the progress already made in the development of genome editing, it promises to play a key role in speeding up crop breeding approaches to achieve global food requirements.

REFERENCES

Abremski, K., R. Hoess, and N. Sternberg. "Studies on the properties of P1 site-specific recombination: Evidence for topologically unlinked products following recombination." *Cell* 32 (1983): 1301–11.

Ali, Z., A. Shami, K. Sedeek, et al. "Fusion of the Cas9 endonuclease and the VirD2 relaxase facilitates homology-directed repair for precise genome engineering in rice." *Communications Biology* 3 (2020): 44.

Aman, R., Z. Ali, H. Butt, et al. "RNA virus interference via CRISPR/Cas13a system in plants." *Genome Biology* 19 (2018): 1.

Barrangou, R. and L. A. Marraffini. "CRISPR-Cas systems: Prokaryotes upgrade to adaptive immunity." *Molecular Cell* 54, no. 2 (2014): 234–44.

Barrangou, R., C. Fremaux, H. Deveau, et al. "CRISPR provides acquired resistance against viruses in prokaryotes." *Science* 315 (2007): 1709–12.

Berg, J. M. "Proposed structure for the zinc-binding domains from transcription factor IIIA and related proteins." *Proceedings of the National Academy of Sciences of the United States of America* 85 (1988): 99–102.

Boch, J., H. Scholze, S. Schornack, et al. "Breaking the code of DNA binding specificity of TAL-type III effectors." *Science* 326 (2009):1509–12.

Bogdanove, A. J. and D. F. Voytas. "TAL effectors: Customizable proteins for DNA targeting." *Science* 333 (2011): 1843–6.

Brouns, S. J., M. M. Jore, M. Lundgren, et al. "Small CRISPR RNAs guide antiviral defense in prokaryotes." *Science* 321 (2008): 960–4.

Cermak, T., E. L. Doyle, M. Christian, et al. "Efficient design and assembly of custom TALEN and other TAL effector-based constructs for DNA targeting." *Nucleic Acids Research* 39, no. 12 (2011): e82.

Chandrasekaran, J., M. Brumin, D. Wolf, et al. "Development of broad virus resistance in non-transgenic cucumber using CRISPR/Cas9 technology." *Molecular Plant Pathology* 17, no. 7 (2016): 1140–53.

Chapdelaine, P., C. Pichavant, J. Rousseau, et al. "Meganucleases can restore the reading frame of a mutated dystrophin." *Gene Therapy* 17, no. 7 (2010): 846–58.

Choulika, A., A. Perrin, B. Dujon, and J. F. Nicolas. "Induction of homologous recombination in mammalian chromosomes by using the IScel system of *Saccharomyces cerevisiae*." *Molecular Cell Biology* 15 (1995): 1968–73.

de Toledo Thomazella, D. P., Q. Brail, and D. Dahlbeck. "CRISPR-Cas9 mediated mutagenesis of a DMR6 ortholog in tomato confers broad-spectrum disease resistance." BioRxiv (2016). doi: 10.1101/064824.

Dong, O. X., S. Yu, R. Jain, et al. "Marker-free carotenoid-enriched rice generated through targeted gene insertion using CRISPR-Cas9." *Nature Communications* 11, no. 1 (2020): 1178.

Doyle, E. L., N. J. Booher, D. S. Standage, et al. "TAL effector-nucleotide targeter (TALE-NT) 2.0: Tools for TAL effector design and target prediction." *Nucleic Acids Research* 40 (2012): W117–22.

Endo, M., M. Mikami, and S. Toki. "Multigene knockout utilizing off-target mutations of the CRISPR/Cas9 system in rice." *Plant and Cell Physiology* 56, no. 1 (2014): 41–7.

Epinat, J. C., S. Arnould, P. Chames, et al. "A novel engineered meganuclease induces homologous recombination in yeast and mammalian cells." *Nucleic Acids Research* 31, no. 11 (2003): 2952–62.

Gao, H., J. Smith, M. Yang, et al. "Heritable targeted mutagenesis in maize using a designed endonuclease." *Plant Journal* 61 (2010): 176–87.

Gilbertson, L. "Cre–lox recombination: Creative tools for plant biotechnology." *Trends in Biotechnology* 21, no. 12 (2003): 550–5.

Grizot, S., J. C. Epinat, S. Thomas, et al. "Generation of redesigned homing endonucleases comprising DNA-binding domains derived from two different scaffolds." *Nucleic Acids Research* 38, no. 6 (2009): 2006–18.

Groenen, P. M., A. E. Bunschoten, D. van Soolingen, et al. "Nature of DNA polymorphism in the direct repeat cluster of mycobacterium tuberculosis; application for strain differentiation by a novel typing method." *Molecular Microbiology* 10 (1993): 1057–65.

Hale, C. R., P. Zhao, S. Olson, et al. "RNA-guided RNA cleavage by a CRISPR RNA-Cas protein complex." *Cell* 139 (2009): 945–56.

Horvath, P. and R. Barrangou. "CRISPR/Cas, the immune system of bacteria and archaea." *Science* 327, no. 5962 (2010): 167–70.

Huang, Y., Y. Guo, Y. Liu, et al. "9-cis-epoxycarotenoid dioxygenase 3 regulates plant growth and enhances multi-abiotic stress tolerance in rice." *Frontiers in Plant Science* 9 (2018): 162.

Ishino, Y., H. Shinagawa, K. Makino, et al. "Nucleotide sequence of the iap gene, responsible for alkaline phosphatase isozyme conversion in *Escherichia coli*, and identification of the gene product." *Journal of Bacteriology* 169 (1987): 5429–33.

Jackson D. A., R. H. Symons, and P. Berg. "Biochemical method for inserting new genetic information into DNA of Simian Virus 40: Circular SV40 DNA molecules containing lambda phage genes and the galactose operon of *Escherichia coli*." *Proceedings of the National Academy of Sciences of the United States of America* 69, no. 10 (1972): 2904–9.

Jansen, R., J. D. Embden, W. Gaastra, et al. "Identification of genes that are associated with DNA repeats in prokaryotes." *Molecular Microbiology* 43 (2002): 1565–75.

Jinek, M., K. Chylinski, I. Fonfara, et al. "A programmable dual-RNA-guided DNA endonuclease in adaptive bacterial immunity." *Science* 337 (2012): 816–21.

Kamburova, V. S., E. V. Nikitina, S. E. Shermatov, et al. "Genome editing in plants: An overview of tools and applications." *International Journal of Agronomy* (2017). doi: 10.1155/2017/7315351.

Kim, Y. G., J. Cha, and S. Chandrasegaran. "Hybrid restriction enzymes: Zinc finger fusions to FokI cleavage domain." *Proceedings of the National Academy of Sciences of the United States of America* 93 (1996): 1156–60.

Lahaye, T. and U. Bonas. "Molecular secrets of bacterial type III effector proteins." *Trends in Plant Science* 6 (2001): 479–85.

Lee, H. B., B. N. Sundberg, A. N. Sigafoos, et al. "Genome engineering with TALE and CRISPR systems in neuroscience." *Frontiers in Genetics* (2016). doi: 10.3389/fgene.2016.00047.

Li, T., B. Liu, M. H. Spalding, et al. "High-efficiency TALEN-based gene editing produces disease-resistant rice." *Nature Biotechnology* 30, no. 5 (2012): 390–2.

Limbalkar, O. M., D. R. Dhumale, and P. R. Shingote. "Current understanding of genome editing and its application in plant breeding." Proceedings of Current Trends in Plant Science and Molecular Biology for Food Security and Climate Resilient Agriculture, Gwalior, (2019) pp. 151–8.

Makarova, K. S., L. Aravind., Y. I. Wolf, and E. V. Koonin. "Unification of Cas protein families and a simple scenario for the origin and evolution of CRISPR-Cas systems." *Biology Direct* 6 (2011): 38.

Marcaida, M. J., J. Prieto., P. Redondo, et al. "Crystal structure of I-DmoI in complex with its target DNA provides new insights into meganuclease engineering." *Proceedings of the National Academy of Sciences of the United States of America* 105 (2008): 16888–93.

Marraffini, L. A. and E. J. Sontheimer. "CRISPR interference limits horizontal gene transfer in staphylococci by targeting DNA." *Science* 322, no. 5909 (2008): 1843–5.

Miller, J., A. D. McLachlan, and A. Klug. "Repetitive zinc-binding domains in the protein transcription factor IIIA from Xenopus oocytes." *EMBO Journal* 4 (1985): 1609–14.

Mojica, F. J., G. Juez, and F. Rodriguez-Valera. "Transcription at different salinities of *Haloferax mediterranei* sequences adjacent to partially modified PstI sites." *Molecular Microbiology* 9 (1993): 613–21.

Mojica, F. J., C. Diez-Villasenor, J. Garcia-Martinez, et al. "Intervening sequences of regularly spaced prokaryotic repeats derive from foreign genetic elements." *Journal of Molecular Evolution* 60 (2005): 174–82.

Moscou, M. J. and A. J. Bogdanove. "A simple cipher governs DNA recognition by TAL effectors." *Science* 326 (2009): 1501.

Orban, P. C., D. Chui, and J. D. Marth. "Tissue- and site-specific DNA recombination in transgenic mice." *Proceedings of the National Academy of Sciences of the United States of America* 89, no. 15 (1992): 6861–5.

Ortigosa, A., S. Gimenez-Ibanez, N. Leonhardt, et al. "Design of a bacterial speck resistant tomato by CRISPR/Cas9-mediated editing of SlJAZ2." *Plant Biotechnology Journal* 17 (2019): 665–73.

Osakabe, Y., T. Watanabe, S. S. Sugano, et al. "Optimization of CRISPR/Cas9 genome editing to modify abiotic stress responses in plants." *Scientific Report* 6 (2016): 26685.

Pohare, M. B., M. Sharma, and S. G. Wagh. "CRISPR/Cas9 genome editing and its medical potential." In Kumar, S. (Eds), *Advances in Biotechnology and Biosciences*, vol. 3, pp. 69–90. Pusa, India: NavNik Publications (2019).

Pourcel, C., G. Salvignol, and G. Vergnaud. "CRISPR elements in Yersinia pestis acquire new repeats by preferential uptake of bacteriophage DNA, and provide additional tools for evolutionary studies." *Microbiology* 151 (2005): 653–63.

Ramirez, C. L., M. T. Certo, C. Mussolino, et al. "Engineered zinc finger nickases induce homology-directed repair with reduced mutagenic effects." *Nucleic Acids Research* 40, no. 7 (2012): 5560–8.

Roberts, R. J., M. Belfort, T. Bestor, et al. "A nomenclature for restriction enzymes, DNA methyltransferases, homing endonucleases and their genes." *Nucleic Acids Research* 31, no. 7 (2003): 1805–12.

Roy, S., S. Dutta, K. Khanna, et al. "Prediction of DNA-binding specificity in zinc finger proteins." *Journal of Biosciences* 37, no. 3 (2012): 483–91.

Sauer, B. "Functional expression of the cre-lox site-specific recombination system in the yeast *Saccharomyces cerevisiae*." *Molecular and Cellular Biology* 7, no. 6 (1987): 2087–96.

Sauer, B. and N. Henderson. "Site-specific DNA recombination in mammalian cells by the Cre recombinase of bacteriophage P1." *Proceedings of the National Academy of Sciences of the United States of America* 85, no. 14 (1988): 5166–70.

Scherer, S. and R. W. Davis. "Replacement of chromosome segments with altered DNA sequences constructed in vitro." *Proceedings of the National Academy of Sciences of the United States of America* 76, no. 10 (1979): 4951–5.

Shi, J., H. Gao, H. Wang, et al. "ARGOS8 variants generated by CRISPR-Cas9 improve maize grain yield under field drought stress conditions." *Plant Biotechnology Journal* 15 (2017): 207–16.

Shukla, V. K., Y. Doyon, J. C. Miller, et al. "Precise genome modification in the crop species Zea mays using zinc-finger nucleases." *Nature* 459, no. 7245 (2009): 437–41.

Sorek, R., Kunin, V., and P. Hugenholtz. "CRISPR–a widespread system that provides acquired resistance against phages in bacteria and archaea." *Nature Reviews Microbiology* 6 (2008): 181–6.

Soyk, S., N. A. Müller, S. J. Park, et al. "Variation in the flowering gene SELF PRUNING 5G promotes day-neutrality and early yield in tomato." *Nature Genetics* 49 (2017): 162–8.

Sylvain, A., P. Christophe, C. Jean-Pierre, et al. "Engineered I-CreI derivatives cleaving sequences from the human XPC gene can induce highly efficient gene correction in mammalian cells." *Journal of Molecular Biology* 371, no. 1 (2007): 49–65.

Tester, M. and P. Langridge. "Breeding technologies to increase crop production in a changing world." *Science* 327 (2010): 818–22.

Wang, Y., X. Cheng, Q. Shan, et al. "Simultaneous editing of three homoeoalleles in hexaploid bread wheat confers heritable resistance to powdery mildew." *Nature Biotechnology* 32 (2014): 947–51.

Wang, F., C. Wang, P. Liu, et al. "Enhanced rice blast resistance by CRISPR/Cas9-targeted mutagenesis of the ERF transcription factor gene OsERF922." *PLoS One* 11, no. 4 (2016): 1–18.

Yang, M., V. Djukanovic, J. Stagg, et al. "Targeted mutagenesis in the progeny of maize transgenic plants." *Plant Molecular Biology* 70 (2009): 669–79.

Zetsche, B., J. S. Gootenberg, O. O. Abudayyeh, et al. "Cpf1 is a single RNA-guided endonuclease of a class 2 CRISPR-Cas system." *Cell* 163, no. 3 (2015): 759–71.

Zhou, J., Z. Peng, J. Long, et al. "Gene targeting by the TAL effector PthXo2 reveals cryptic resistance gene for bacterial blight of rice." *Plant Journal* 82, no. 4 (2015): 632–43.

2 Mechanism of ZFN-Mediated Genome Editing

Scope and Opportunities

Nabaneeta Basak
ICAR-National Rice Research Institute

Vivek Chandra Verma
Defense Institute of High Altitude Research-DRDO

Rajeev Kumar
ICAR-Indian Institute of Sugarcane Research

Gaurav Kumar
ICAR-National Rice Research Institute

CONTENTS

2.1 INTRODUCTION

Since the time of domestication, improvement of crop traits has been progressively made with the utmost interest. Conventional techniques of plant breeding involve backcrossing of desirable traits obtained from a pool of natural genetic variation into an elite background. The incapacity of introducing trait characteristics selectively constitutes the major limitation of this process (Bilichak and

Eudes, 2016). The rise of mutation breeding in the 1920s (Stadler, 1928a,b) led to the artificial introduction of new traits heralding an era of forward genetics where screening of the mutagenized plant population for the precise identification of specific phenotypes was carried out. The recessive nature of the randomly generated mutations combined with the masking of phenotypic traits arising from sequence mutation due to polyploid nature of plants presents a drawback to this method (Petolino, 2015). The advancement in the field of gene cloning and DNA sequencing paved the way for the utilization of reverse genetics screening approach of initially identifying the gene of interest and then establishing a linkage with its phenotype (Carroll, 2016). Thus, the progress of plant biotechnology and crop improvement necessitated the discovery of new tools and techniques for targeted genetic engineering (Bilichak and Eudes, 2016).

The ability of controlled alteration of a complex genome transformed biological research in the late 1980s and early 1990s (Cathoman and Joung, 2008). Genome editing involves the introduction of genetic changes at predefined locations in the genome. Genetic changes comprise gene excision, gene disruption, sequence substitution and integration of external sequences induced at specific sites (Carroll, 2014; Kim and Kim, 2014; Sander and Joung, 2014; Weeks et al., 2016). The ability to incorporate specific changes in the genome is important from the point of view of the range of applications it can influence, including elucidation of gene function; treatment of genetic diseases; and improvement of economic traits in plants, animals or microbes for industry or agriculture (Bonawitz et al., 2019).

The underlying technology involved in the precise modification of the genome is known as gene targeting. Historically, cellular homologous recombination (HR) pathway was the chosen method to achieve the objective of gene targeting. Researchers had established the methods for gene targeting in yeast (Rothstein, 1983; Scherer and Davis, 1979) and mice (Capecchi, 2005; Thomas et al., 1986), based on the HR process between an endogenous target and an externally introduced DNA molecule (Carroll, 2016). Technical as well as biological impediments made it challenging to utilize these methods for other organisms. Efficient HR in eukaryotes has been limited to chicken DT40 cells, yeast, mouse embryonic stem (ES) cells and the moss *Physcomitrella patens* (Bilichak and Eudes, 2016; Puchta 2002; Sakuma and Woltjen 2012).

In the early 1990s, when the whole-genome sequencing was gaining popularity, a desire for specific manipulation of genes was also escalating. The search for reliable tools for targeted genetic alterations led to the understanding that the phenomenon of cellular DNA repair could be suitably exploited to tailor the need of biologists. Usually, the presence of double-stranded breaks (DSBs) in chromosomal DNA has been found to be damaging and hence undergo repair processes in the cell (Chapman et al., 2012). It was found that in addition to HR, DSBs also stimulate the non-homologous end-joining (NHEJ) process, an error-prone repair mechanism. Thus, expanding the efficiency and range of gene targeting involves the creation of breaks or damages in the target DNA, making it more susceptible to repair processes (Carroll, 2016) (Figure 2.1). Pioneering studies in plants and animals with the yeast enzyme I-Sce-I, a rare cutter, revealed that the rate of HR increased by 1000-folds when subjected to DNA cleavage at endogenously created positions (Choulika et al., 1995; Cohen-Tannoudji et al., 1998).

Generation of DSBs targeted at particular regions entails the involvement of a nucleoprotein complex designed in such a way that it can bind to the sequence of interest (Nakatsukasa et al., 2005). A nonspecific endonuclease domain may be fused with a programmable DNA-binding motif for the precise targeting and introduction of DSBs (Mahfouz et al., 2014). Development of such novel enzymes with specific cleavage activity known as designer or engineered endonucleases was carried out robustly in the past decades, and currently, four types of engineered nucleases are primarily utilized for genome editing purposes: (1) engineered homing endonucleases/meganucleases (EMNs), (2) zinc finger nucleases (ZFNs), (3) transcription activator-like effector nucleases (TALENs) and (4) clustered regularly interspaced short palindromic repeats (CRISPR)/CRISPR-associated protein 9 (Cas9). All of these technologies have been used in the plant system to introduce precise modifications at the predefined positions in the genome successfully over several years.

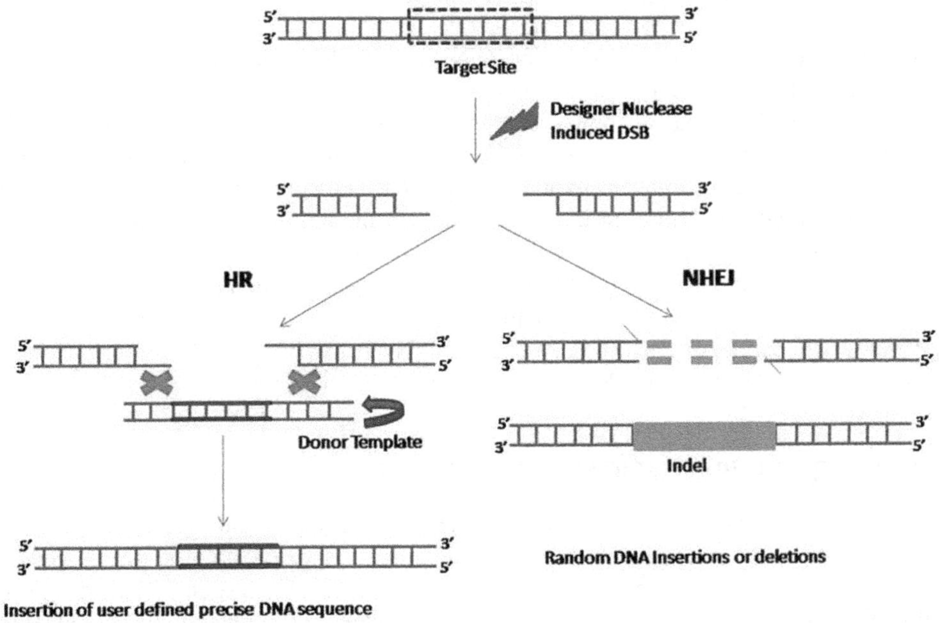

FIGURE 2.1 Repair outcomes of genomic double-stranded breaks induced by programmable nucleases. Double-stranded break is induced at a precise target site (shown in red dotted rectangle). In the presence of a homologous donor DNA sequence (maroon), repair can proceed by HR using the donor sequence as template. Alternatively, the break can be repaired by NHEJ, sometimes leading to insertions, deletions and base substitutions at the site. (Adapted from Chandrasegaran and Caroll (2016).)

2.2 ORIGIN OF ZFNs

ZFNs do not occur naturally but are derived from natural components. Srinivas Chandrasegaran and his colleagues played a major role in designing the ZFNs. In three articles published in the Proceedings of the National Academy of Sciences of the United States between 1992 and 1996, he conceived the idea and finally developed a radically new type of endonuclease (Li et al., 1992; Kim and Chandrasegaran 1994; Kim et al., 1996). In the early 1990s, they found that FokI, a type II restriction enzyme, has two separate domains: one DNA recognition domain and the other DNA cleavage domain (Li et al., 1992). This finding led to the conceptualization that linking the nonspecific cleavage domain to alternative DNA-binding module could result in conferring novel specificity traits as was demonstrated by fusion with the homeo-box from the ultrabithorax transcription factor of *Drosophila* (Kim and Chandrasegaran, 1994).

The advent of new genetic engineering tools enabled the isolation and characterization of DNA-binding proteins in prokaryotes and eukaryotes in the 1980s, while their structure was determined at the beginning of the 1990s. Studies described different families of DNA-binding protein harbouring diverse types of DNA-binding motifs like helix-turn-helix, helix-loop-helix, zinc fingers and others. The term 'zinc finger' was coined based on the existence of an amino acid sequence in these proteins, which binds a zinc atom and is viewed as fingers sensing a DNA molecule (Morange, 2017); it was found to be a part of a large family of DNA-binding proteins. NMR and X-ray diffraction studies carried out by Pavletich and Pabo (1991) revealed the structures of some of these members; it was found that these domains contained multiple zinc fingers, each recognizing three nucleotides on the DNA molecule independently. In an attempt to design a chimeric enzyme, Chandrasegaran S. fused two diverse sets of zinc fingers to the FokI cleavage domain, and demonstrated a successful redirected cutting (Kim et al., 1996). These fusions were the first ZFNs and paved the way for future research.

2.3 ADVANTAGES OF ZFNs

Target sequence restrictions are minimal for ZFNs, can be designed for a majority of genes and genomic regions of interest and are hence considered to be an attractive tool of genome editing. ZFNs show certain advantages in comparison with CRISPR technology. The capacity of ZFNs to attach themselves and initiate cleavage at most locations with a high degree of fidelity is in contrast to the CRISPR/Cas system, which necessitates the presence of a protospacer adjacent motif (PAM) sequence at the target, thus limiting the targets for genome editing (Jinek et al., 2012; Ledford, 2017). Additionally, studies have depicted the off-target effects of CRISPR, which might be due to the presence of the short 20-bp gRNA/PAM sequence that drives the target specificity (Fu et al., 2013; Hsu et al., 2013) – a feature that may limit the extensive use of this technology in polyploid plant genomes with paralogous genes. High-quality ZFN designs have also been found to minimize the off-target effects (Davies et al., 2017).

2.4 ZFN: STRUCTURE, DESIGN AND ASSEMBLY

ZFNs are formed by combinations of the nonspecific DNA cleavage domain (derived from the FokI restriction endonuclease) with zinc finger proteins. Dimers of ZFNs promote the targeted DNA DSBs, which further induce the pathways involved in DNA damage response. This designed zinc finger domain has a specific binding specificity, which guides ZFN towards the specific genomic site (Gaj et al., 2013). Gene modification by ZFNs has been successfully applied in animals such as zebrafish (Doyon et al., 2008; Meng et al., 2008), human cells (Kim et al., 1996) and plants like tobacco (Townsend et al., 2009), maize (Shukla, et al., 2009) and *Arabidopsis* (Osakabe et al., 2010). Researchers developed new sequence specificities by combining Fn domain (non-sequence-specific cleavage domain) of FokI type II restriction endonuclease to another DNA-binding domain (Porteus and Carroll, 2005).

2.4.1 ZFN Structure

ZFN is comprised of three domains: (1) non-sequence-specific cleavage domain at the C-terminal mediated by FokI nuclease, (2) DNA-binding zinc finger domain Cys_2-His_2 (C_2H_2) present at the N-terminal essential for its specificity and (3) peptide linker that joins zinc finger domain and nuclease. The restriction enzyme FokI breaks DNA double strand as a catalytic dimer. Thus, two zinc finger subunits must bind to the gene target sequence in the opposite orientation, leading to FokI dimerization for DNA cleavage (Chou et al., 2012).

2.4.2 Zinc Finger Motif

Klug and coworkers were first to discover the zinc finger motif in 1986 in transcription factor IIIA (Miller et al., 1985; Diakun et al., 1986). In eukaryotic genome, Cys_2-His_2 (C_2H_2) zinc finger domain is DNA-binding domain (composed of 30 amino acids) that binds to a Zn atom (Maeder and Gersbach 2016). A compact ββα structure is formed by zinc finger domain and α-helical portion of each finger by contacting 3 or 4 bp in the major groove of the DNA in the presence of Zn molecule. To understand the DNA recognition, the crystal structure of Zif268 is crucial, and it provides a framework for constructing DNA-binding motif (Gersbach et al., 2014).

2.4.3 Structure of FokI

The FokI (catalytic domain of ZFN) has physically separable activities like cleavage and binding (Lie et al., 1992). FokI endonuclease enzyme (65.4 kDa) is found in *Flavobacterium okeanokoites*. FokI exists in the form of monomer, which recognizes a specific asymmetric DNA sequence,

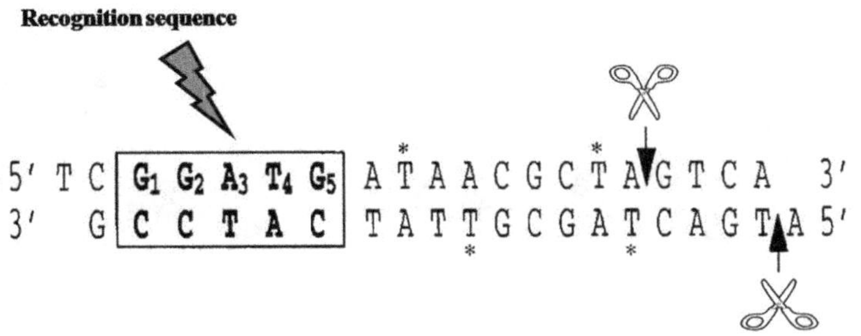

FIGURE 2.2 FokI recognition sequence and cleavage site. (Adapted from Wah et al. (1997).)

i.e. 5′-GGATG-3′, and transforms into genomic scissors which cleave phosphodiester groups present in DNA at 9 and 13 bps away from the enzyme recognition site. The FokI endonuclease consists of 579 amino acids, and its recognition domain contains three smaller D_1, D_2 and D_3 subdomains. Subdomain D_1 includes major groove of DNA, which recognizes base pairs at the 3′ end of the enzyme recognition sequence. D_1 consists of eight alpha-helices, two loops, three β-sheets and N-terminal arm. Three helices α_4, α_5, α_6 and L_1 and L_2 loops form a modified helix-turn-helix motif. D_1 connects to D_2, which interacts with the base pairs present at the 5′-end of the recognition sequence that consists of six helices, three β-sheets (β_1, β_2 and β_5), two β hairpins (β_3 and β_4) and a short loop L_1. The recognition domain sequesters the cleavage domain through a set of protein–protein interactions, which are electrostatic and governed by subdomains D_2 and D_3 of the recognition domain (Figure 2.2) (Wah et al., 1997).

2.4.4 ZFN-Mediated DNA Cleavage

In the previous section, we have already discussed the structure and catalytic domain of ZFN. ZFN target site is made up of two target half-sites, which are separated by spacer DNA made up of 5 or 6 bp between binding sites. Binding of ZFN within DNA is directed by a set of three to six zinc finger tandem repeats, and each of them recognizes the bases of target DNA site (Pavletich and Pabo, 1991). When two ZFN subunits dimerize, an activation of nucleases occurs, which cut down DNA within the spacer sequence producing DSB (Figure 2.3) (Morton et al., 2006).

2.5 ZFN: DESIGN AND ASSEMBLY

Zinc finger domain of ZFN is crucial for the site-specific cleavage. Advancement in gene editing tools developed a number of strategies for generating ZFN. These strategies include ‘modular assembly’ (Pavletich and Pabo, 1991), ‘sequential context-sensitive selection’ (Meng et al., 2008), ‘bipartite library strategy’ (Islam et al., 2001), ‘oligomerized pool engineering (OPEN)’ (Foley et al., 2009), ‘context-dependent assembly (CoDA)’ (Sander et al., 2011) and the ‘2 + 2’ approach (Davis and Stokoe, 2010).

i. **Modular assembly**: It is the first type of ZFN approach that is based on the parallel selection method and uses the phage display system (Pavletich and Pabo, 1991). In modular assembly, it was revealed that zinc finger domain binds the sequences 5′GNN3′, 5′ANN3′ and 5′CNN3′. This is simpler and shows an efficient binding affinity in lower eukaryotes (Dreier et al., 2001).
ii. **Oligomerized pool engineering:** This is a robust and efficient ZFN design. This approach involves screening of ZFN for each target DNA triplet followed by the random PCR assembly and selection based on the bacterial two-hybrid system. This strategy is costly and requires expertise (Maeder et al., 2008).

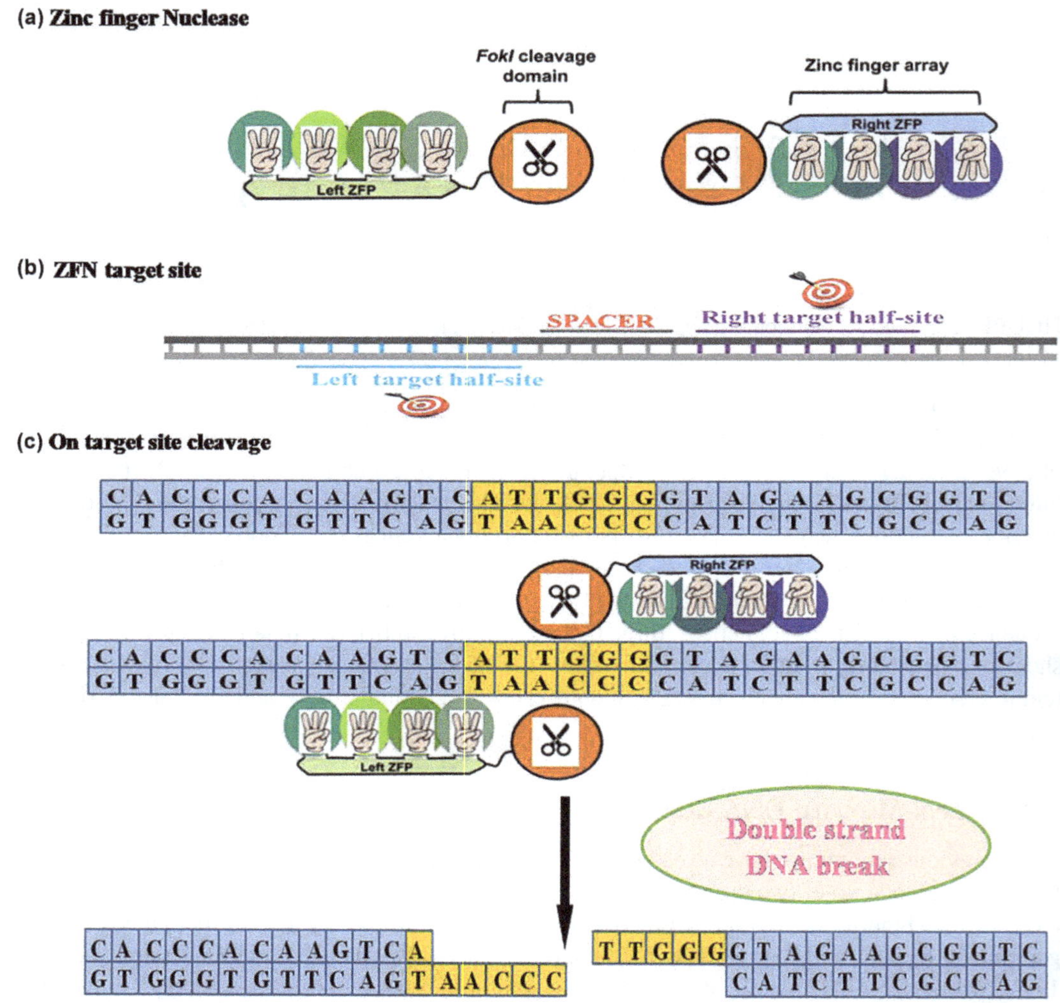

FIGURE 2.3 ZFN-mediated cleavage. Domains D1, D2, D3. (Adapted from Rahman et al. (2011) and Jo et al. (2015).)

iii. **Context-dependent assembly**: 'CoDA' is less specific than 'OPEN' approach but it is easier to develop at a very low cost (Sander et al., 2011).
iv. **2 + 2 approach:** As the name suggests, it involves two steps in which four ZFN subunits are associated with the pre-existing two finger units of known DNA (Davis and Stokoe, 2010).

The design and assembly of safer nucleases improve the binding specificities of DNA with zinc finger domain by optimizing the linker sequence, which connects the zinc finger domain with chimeric endonuclease FokI.

2.6 ZFN-MEDIATED TARGETED MUTAGENESIS

Targeted mutagenesis is a molecular biology technique that in recent times has found a huge applicability in the modification of crop plants for the improvement of agricultural traits by changing the specific DNA sequence of a particular gene. It deals with the structural biology of informational biomolecules like nucleic acid (DNA and RNA) and protein and remarkably plays a crucial role in genome editing as well as protein engineering. This approach facilitates an efficient and targeted genetic modification by inducing the DSB in a precise location in the genomic sequence.

Induction of DSB at the specific site by ZFN activates the repair process either by NHEJ or by homology-directed repair pathway (HDR) (Lieber, 2010). Exploitation of both the pathways can lead to a precise and targeted genome modification process via gene addition and deletion.

NHEJ has a robust repair mechanism that enables the joining of the DNA ends obtained from a DSB triggered by ZFN (Lieber, 2010). The complete set of proteins involved in the NHEJ pathway keeps the two DNA ends nearby and cuts the DNA to make them amenable for the subsequent enzyme-based ligation to complete the repair process. The innate nature of NHEJ includes the absence of an intact copy (a homologous string of nucleotide) to re-establish the hereditary information at DSBs (Figure 2.4). However, it does result in the intermittent formation of unreliable repair products of small-nucleotide deletion or insertions (indels) (Kim and Kim, 2011). Thus, in the event that the protein-coding region of the genome harbours the target site of the ZFN, small indels may emerge due to the unreliable repair mechanism of DSBs induced by ZFNs, inducing a frameshift mutation in the target gene that ultimately will result in a truncated version of the target protein (Porteus and Carroll 2005).

However, the homology-directed DSB repair uses the template DNA homologous to DSB locus to re-establish genetic material in an error-less approach (San Filippo et al., 2008). Nowadays, foreign

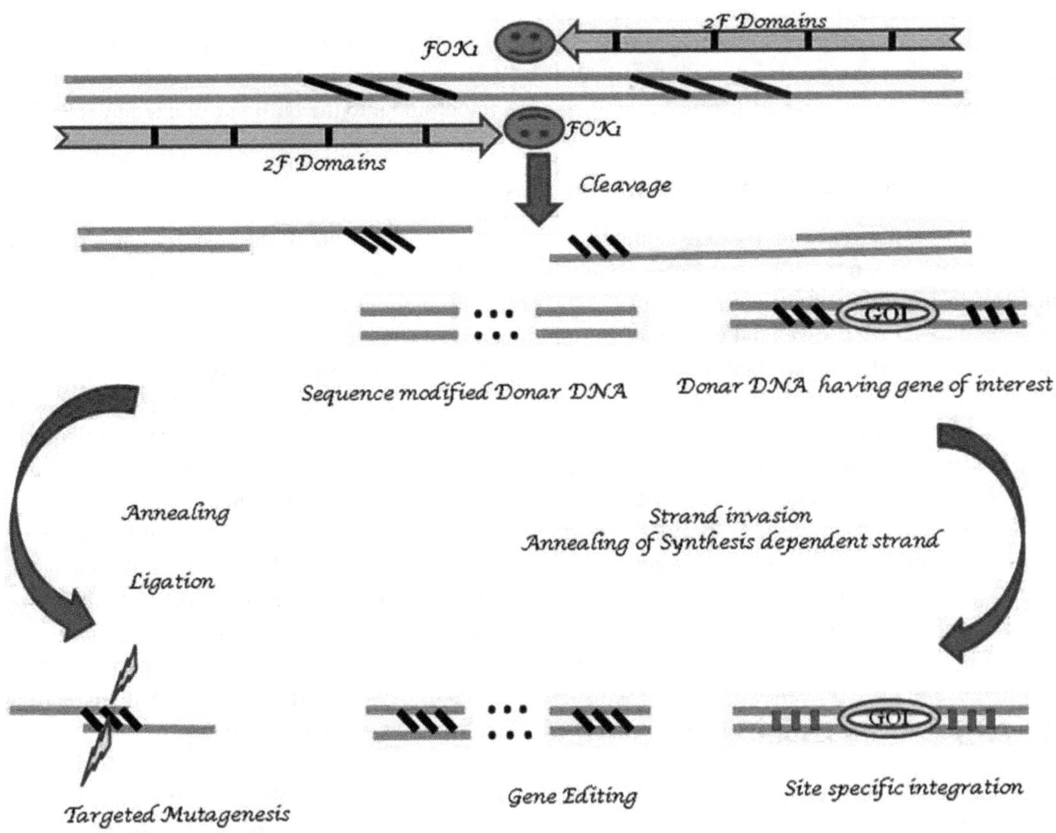

FIGURE 2.4 ZFN-mediated targeted mutagenesis by the repair of double-stranded break (DSB). Zinc finger nucleases (ZFNs) bind to a target sequence, so that dimerization of FokI nuclease takes place. DNA repair processes induced after the ZFN cleavage produced DSB. Error-prone non-homologous end-joining (NHEJ) is a preferable choice in the absence of donor template DNA that can result in 'targeted mutagenesis' (left). The presence of homologous sequence in donor DNA triggers, homology-directed repair (HDR), can result in 'gene editing' (middle) or 'site-specific integration' (right). (Adapted from Petolino (2015).)

DNA or gene of interest is delivered into cells that can elicit the HR approach between the target and the donor DNA to accomplish the process of gene targeting even without inducing site-specific DSBs. The DNA sequences in the donor template facilitate HDR of DSBs comprising targeted mutations at precise genomic locations. The donor template DNA involved in HDR response consists of ~750- to 1000-kb stretches homologous to the region flanking the genomic cleavage site and the sequence of interest. However, in somatic cells of plants, recurrence of HDR appears to be exceptionally low; hence, recognition and segregation of such alteration can be a very big objection to achieve the target. Hence, HDR process is remarkably limited in most of the eukaryotic species, with mouse ES cells and yeast being the only exceptions (Johnson and Jasin 2001).

Thus, knockout of a target gene by ZFNs following NHEJ is an advantageous approach over HDR. The higher frequency of NHEJ over HDR makes it the ideal choice for repair mechanisms in higher eukaryotes (Porteus and Carroll 2005). Thus, in a ZFN-expressing cell system, the target site of the ZFN undergoes numerous cycles of restriction and digestion followed by repair; as a result, an enormous increase in the small indels mediated through NHEJ may be observed. The increased frequency of indels may create the target-specific mutations repetitively.

NHEJ-mediated repair mechanisms operational in ZFN-induced DSBs are used to create simple gene knockouts, and may also be exploited to accomplish complex trait modifications and elaborate genome engineering. The two DSBs are induced in a site-specific manner by ZFN repairs through NHEJ-created chromosomal alterations within a subset of cells, which may result in the damage or even loss in genetic instruction between two DSB sites if both are located on the same chromosome. However, a translocation event between the two chromosomes may occur if the sites for DSB are situated on different chromosomes (Sollu et al., 2010). ZFNs can be utilized for the co-delivery of small fragments of DNA into the cell, which may further be integrated into the DSB site created by the ZFN.

2.6.1 Sequence Modification of Reporter and Genes

Reporter genes provide an aid to the quantitative expression of a gene of interest, which is difficult to assay. They are not found in the endogenous system but are able to produce a protein that has less noticeable or prompt reaction in cell culture and organism. Sequence modification of reporter genes governed by ZFN involves the editing of pre-engineered selectable marker genes (Cai et al., 2009). Wright et al., (2005) described the sequence modification of a reporter gene in tobacco protoplast facilitated by the co-delivery of a gene construct coding the ZFN sequence for cleavage between the genes encoding the GUS reporter gene and neomycin phosphotransferase (NPTII)-selectable marker proteins and a donor template DNA strand comprising sequences homologous to the flanking region of the ZFN cleavage site and the missing GUS/NPTII gene sequence.

Site-specific formation of DSB at the reporter sequence followed by the subsequent linearization by donor template DNA via NHEJ is achieved through co-delivery of ZFN construct and donor template DNA. The work by Tovkach et al. (2009) may be taken as an example where the gene encoding β-glucuronidase (GUS) has a stop codon within a cleavage site of ZFN; an integration of *Agrobacterium* strain into *Arabidopsis* shows GUS straining region having the corresponding ZFN cassette. As a result, the target site sequence analysis reveals the presence of the stop codon in the GUS open reading frame, which undergoes a modification by the single-nucleotide substitution and deletion.

A prerequisite for sequence modification of endogenous genes using ZFN is the capacity to deliver macromolecules like DNA, RNA and protein effectively into cells to promote the specific DNA sequence modification. Various plant transformation techniques such as polyethylene glycol method (Wright et al., 2005), microparticle bombardment (Ainley et al., 2013) and *Agrobacterium*-mediated transformation (Cai et al., 2009) can be used to deliver the ZFN for the modification of the genes. An ideal delivery system comprises an active ZFN, a donor template DNA to facilitate the targeted DSB and successive DNA repair. ZFNs should be designed in a way to facilitate the cleavage of endogenous genomic sequences for the promotion of site-specific mutations; e.g., cleavage of the second exon in the maize inositol-1,3,4,5,6-pentakisphosphate kinase-1 (IPK-1) gene is accomplished through the delivery of a gene sequence encoding the ZFN specific to the target along with a donor template DNA having the PAT gene (lacking the promoter sequence) along a 'stutter' sequence flanked 815 bp away from the homologous sequence of IPK1 (Shukla et al., 2009). As a result, the target-specific cleavage at IPK1 sequence and specific binding of the promoter provide the site-specific integration of the PAT gene and resistance against the herbicide.

Even genes involved in DNA repair can be mutated by ZFNs in such a way to enhance the resolution as well as mutation and targeting frequencies. Successful delivery of ZFNs into diverse genetic backgrounds for the mutation of specific genes involved in DNA repair process has been achieved; e.g., the gene encoding the protein SMC6B associated with DSB repair at ADH1 (alcohol dehydrogenase) locus in *Arabidopsis* can be mutated and integrated into the genetic background using sister chromatids to increase the ZFN-mediated mutation frequency (Qi et al., 2013a). The example so cited can be taken as a proof to conclude that in spite of the intricacy of the comprehensive genome editing system, engineering of DNA repair pathway tends to regulate ZFN-mediated cleavage sequence and is one of the most potential paths to target and generate mutagenesis.

2.6.2 Gene Deletions

ZFN-driven gene deletion creates the DSBs in DNA, following the NHEJ process leading to heritable gene silencing in the target gene. ZFN-assisted cleavage results in the deletion of larger sequence of DNA along with small NHEJ-induced mutations. Deletion of the particular gene of interest using ZFN involves the reporter construct with ZFN cleavage site. The formation of DSB at the target site indicates the consequent expression of the concerned ZFN gene; as a result, fragments of the reporter undergo a recombination via either NHEJ or HDR, and the deletion of the desired sequence can be achieved in the reporter fragment (Cai et al., 2009). To achieve the targeted deletion of the specific gene, the target locus must have to be engineered in such a way to express selectable marker, reporter gene and a unique recognition site for ZFN controlled by the specific promoter. Further, a constitutive expression of ZFN in a particular tissue leads to a single to large deletion and even larger deletion (Figure 2.5).

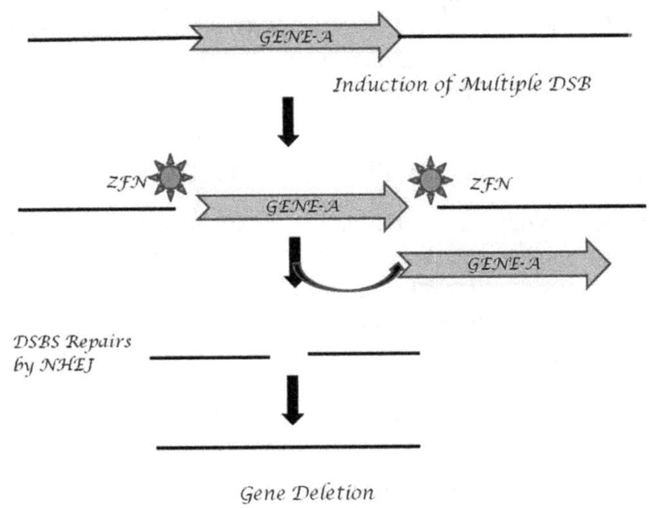

FIGURE 2.5 ZFN-mediated gene deletion. Two concomitant DSBs created by ZFN pairs, flanked a target site of the gene, further repaired via NHEJ, ultimately created targeted deletions of genes between the both sites. (Adapted from Kim and Kim (2011).)

2.6.3 Gene Disruption

Gene disruption is the process of knockout of specific genes to enable the editing of that gene using the designed nucleases. ZFN is such a programmable nuclease that can disrupt the gene in a simple way to carry out the process of genome editing. DSB is generated at the specific site within the genome by ZFN, followed by the repair process predominantly via the NHEJ pathway. Activation of the NHEJ pathway generates small insertions and deletions (indels) at the cleavage site or adjacent to the cleavage site in the genome. Formation of indels accounts for frameshift mutations, which leads to the disruptions of the gene. Incorporation of error during the DNA repair influences the disruption or abolishment of the function of a specific gene or genomic region. Such type of gene disruption can be used to scrutinize and explore the various biochemical pathways involved in nuclear factor signalling (Kweon et al., 2013), glycosylation pathways (Steentoft et al., 2011) and methylation of protein (Kernstock et al., 2012). Hence, these biochemical pathways can be utilized to knock out the targeting gene bypassing the selection procedure for the desirable event. Therefore, gene disruption can be a helpful approach to elucidate a particular genetic factor in the plant cells by analysing the knockout gene phenotype and the isogenic control.

2.7 APPLICATIONS OF ZFN FOR PLANT GENOME ENGINEERING

Successful application of engineered nucleases for plant genome editing by the stimulation of either NHEJ- or HDR-mediated repair has been well documented (Bonawitz et al., 2019). Apart from the major role it plays in strengthening the basic research on plant sciences, this technology also has widespread implications for crop improvement and breeding by helping to create novel traits through the deletion of deleterious genes or those that repress agronomically beneficial traits (Davies et al., 2017). The primary usage of NHEJ in plants involves the disruption of target genes, whereas the targeted gene replacement, integration or modification has traditionally utilized donors designed for HDR (Kim and Kim, 2014; Maruyama et al., 2015; Sander and Joung, 2014). The procedures for donor DNA and nuclease delivery to target cells, identification and/or selection for targeted cells, and to regenerate mature plants from targeted cells, though, have differed extensively.

Genome editing in crop plants poses a challenge to researchers because of the dependency on an effective and well-structured transformation methodology and the capability for designing an efficient and unique engineered endonuclease to target a distinct locus in the polyploid genome. The first successful report of genome editing by ZFN technology in crops has been documented in 2009 wherein the ZFN-mediated disruption of an *IPK1* gene that encodes an inositol-1, 3, 4, 5, 6-penta*kis*phosphate 2-kinase, the flux-controlling enzyme of phytate biosynthesis in maize seeds, was attempted. Table 2.1 enlists some of the major genome editing successes through ZFN in crop plants.

2.8 POTENTIAL LIMITATIONS OF ZFN TECHNOLOGY

Even though the ZFN technology has been found to be effective for genome editing in a number of model plants and crop species, the design of the zinc finger modules poses a challenge to the researchers due to the various complex interactions that occur between amino acid residues and the base pairs of the sequence that is targeted (Kumar and Jain, 2015). Additionally, the assembly and testing of ZFNs is a costly affair. Certain drawbacks of the technology include the following:

i. **Cost involved and time taken to engineer**: ZFNs have been used for the purpose of genome editing for a long time (Miller et al., 1985). Their widespread application, though, has been limited by the requirement to computationally predict and then iteratively engineer and test for sequence specificity (Costa et al., 2017). Such elaborate assembly and testing slows the process of development of ZFNs in comparison with TALEN or CRISPR/Cas system, and is also usually more expensive to pursue for a given target. This constitutes the major reason why this technique has fallen out of favour in recent years.

TABLE 2.1
List of Some Successful Gene Modification Events through ZFN Technology

Species	Genome Modification Technique	Targeted Gene/Locus	Trait	References
Tobacco	Gene editing using ZFN	SurA, SurB (sulphonylurea)	ALS inhibiting herbicide resistance	Townsend et al. (2009)
	ZFN-two finger modules	Chitinase	PAT herbicide tolerance	Cai et al. (2009)
	ZFN	Reporter gene	Restoring function to a defective GUS:NPTII reporter gene	Wright et al. (2005)
	ZFN	SurA, SurB (sulphonylurea)	Herbicide resistance	Maeder et al. (2008)
	ZFN	GUS	Deletion of gene sequence	Petolino et al. (2010)
	ZFN			Baltes et al. (2014)
Maize	ZFN-two finger modules	IPK1	Reduced phytate production	Shukla et al. (2009)
	ZFN-two finger modules	Zein protein 15		Shukla et al. (2009)
	ZFN-trait landing pads	Maize genome	Transgene stacking	Ainley et al. (2013)
Petunia	ZFN-induced targeted mutagenesis			Marton et al. (2010)
Arabidopsis	ZFN-mediated modification of endogenous gene driven by HSP promoter	ABI4 (ABA insensitive-4)	Abscisic acid and glucose insensitivity	Osakabe et al. (2010)
	ZFN-mediated modification of endogenous gene under the control of oestrogen-inducible promoter	Alcohol dehydrogenase (ADH1) and transparent Testa 4	Allyl alcohol resistance and lack of anthocyanins in the seed coat	Zhang et al. (2010)
	ZFN	EcoRI restriction site	Loss of restriction site	Lloyd et al. (2005)
Tobacco	ZFN	GUS	Deletion of gene sequence	Petolino et al. (2010)
Arabidopsis	ZFN	Egg cell	RAD 54 protein involved in DNA repair	Even-Faitelson et al. (2011)
	ZFN	ADH1	DNA repair protein	Qi et al. (2013b)
	ZFN	Protoporphyrinogen oxidase (PPOX)	Resistance to the herbicide butafenacil	de Pater et al. (2013)
	ZFN			Chen et al. (2014)
Soybean	ZFN-mediated modification of endogenous gene under the control of oestrogen-inducible promoter	DICER like DCL4α and DCL4β	Aborted seed	Curtin et al. (2011, 2013)
	ZFN			Curtin et al. (2013)
Canola	ZFN	β-Ketoacyl-ACP synthase II	Less palmitic acid content	Gupta et al. (2012)

ii. **Targeting specificity:** Reports of the off-target binding have also been a major deterrent in the extensive use of ZFN in genome editing. ZFNs may have high affinity for a unique sequence of interest within the genome that has been targeted for digestion, yet a low probability for the off-target binding and cleavage of other DNA sequences resulting in undesirable modifications in the genome still exists (Costa et al., 2017). In the event of such an occurrence, the whole-genome sequencing needs to be carried out to determine the points of disruption, which is both time-consuming and expensive. In addition, during the knockout experiments, in order to demonstrate that a particular phenotype is a result of a desired genome modification by ZFN, the phenotype needs to be rescued with the reintroduction of the functional gene (Costa et al., 2017).

2.9 CONCLUSION

Use of ZFNs and other designer nucleases for the precise targeting of DSBs has helped in opening up a myriad amount of possibilities of the efficient genome editing in prokaryotes as well as eukaryotes. Utilization of genome editing processes is likely to have a profound impact on both functional genomics and applied crop improvement (Petolino, 2015). Advancement of the scientific knowledge behind the genome structure and function as a result of precise DNA sequence modifications will eventually help in better understanding of the association between a particular gene and its associated phenotype, especially in cases where the mutant is unknown or does not exist (Bilichak and Eudes, 2016).

Applied crop improvement or commercial product development will also be improved by the adaption of genome editing technologies. Reverse genetic approach will be boosted by the ability to make the predefined modifications instead of creating random changes. Non-transgenic events will give rise to products, which will be cost-effective and also eliminate the need to screen multiple unproductive events that might otherwise be generated due to random insertions. The cost and time required for the product development may also be substantially reduced by the targeted transgene integration to the predetermined genomic loci through the trait stacking approach (Bortesi and Fischer, 2015).

Innumerable possibilities for the applications of ZFN technology to the future plant research exist. As we recognize and comprehend the inherent complexities of the genome, the opportunity for the application of biotechnological tools based upon the understanding of the processes expands as well. Zinc finger proteins constitute a set of potential tools tailor-made to realize the advantages offered by the growing body of information and exploit it for the discovery and creation of new traits for the purpose of improvement of plants.

Thus, genome editing exhibits a great potentiality for both basic and applied plant biological sciences and will continue to play a key role in the next-generation research activities.

2.10 FUTURE PROSPECTS

Advancement in the fields of genome editing over the last decades has provided valuable insights into various aspects of biological research, including gene and protein functionality studies. The true potential of the technology, though, can only be realized if certain challenges can be addressed in future. Development of novel tools and techniques that may be capable of introducing genomic modifications in the absence of DNA breaks constitutes the primary concern and needs to be taken care of. In addition, further improvements to delivery and efficiency need to be made in future for this technology to have an impact in the applied research. It can be said with certainty that the future holds countless prospects for the applications of ZFN technology to plant biotechnology. As we undertake futuristic research and gain in-depth knowledge about the complexities of the genome, the opportunity for zinc fingers as a versatile tool for plant biotechnology expands.

REFERENCES

Ainley, W. M., L. S. Dent, M. E. Welter, et al. "Trait stacking via targeted genome editing." *Plant Biotechnology Journal* 11, no. 9 (2013): 1126–1134.

Baltes, N. J., J. G. Humanes, T. Cermak, P. A. Atkins, and D. F. Voytas. "DNA replicons for plant genome engineering." *The Plant Cell* 26, no. 1 (2014): 151–163.

Bilichak, A., and F. Eudes. "Genetic engineering of plants using Zn fingers, TALENs, and CRISPRs." In *Genome Stability*, edited by I. Kovalchuk and O. Kovalchuk, pp. 187–201. New York, Cambridge: Academic Press (2016).

Bonawitz, N. D., W. M. Ainley, A. Itaya, et al. "Zinc finger nuclease-mediated targeting of multiple transgenes to an endogenous soybean genomic locus via non-homologous end joining." *Plant Biotechnology Journal* 17, no. 4 (2019): 750–761.

Bortesi, L., and R. Fischer. "The CRISPR/Cas9 system for plant genome editing and beyond." *Biotechnology Advances* 33, no. 1 (2015): 41–52.

Cai, C. Q., Y. Doyon, W. M. Ainley, et al. "Targeted transgene integration in plant cells using designed zinc finger nucleases." *Plant Molecular Biology* 69, no. 6 (2009): 699–709.

Capecchi, M. R. "Gene targeting in mice: Functional analysis of the mammalian genome for the twenty-first century." *Nature Reviews Genetics* 6, no. 6 (2005): 507–512.

Carroll, D. "The development and use of zinc-finger nucleases." In *Genome Editing: Advances in Experimental Medicine and Biology*, edited by T. Cathomen, M. Hirsch, and M. Porteus, pp. 15–28. New York: Springer (2016).

Cathomen, T., and J. K. Joung. "Zinc-finger nucleases: The next generation emerges." *Molecular Therapy* 16, no. 7 (2008): 1200–1207.

Chandrasegaran, S., and D. Carroll. "Origins of programmable nucleases for genome engineering." *Journal of Molecular Biology* 428, no. 5 (2016): 963–989.

Chapman, J. R., M. R. G Taylor, and S. J. Boulton. "Playing the end game: DNA double-strand break repair pathway choice." *Molecular Cell* 47, no. 4 (2012): 497–510.

Chen, W., Y. Qian, X. Wu, Y. Sun, X. Wu, and X. Cheng. "Inhibiting replication of begomoviruses using artificial zinc finger nucleases that target viral-conserved nucleotide motif." *Virus Genes* 48, no. 3 (2014): 494–501.

Chou, S. T., Q. Leng, and A. J. Mixson. "Zinc finger nucleases: Tailor-made for gene therapy." *Drugs of the Future* 37, no. 3 (2012): 183.

Choulika, A., A. Perrin, B. Dujon, and J. F. Nicolas. "Induction of homologous recombination in mammalian chromosomes by using the I-SceI system of Saccharomyces cerevisiae." *Molecular and Cellular Biology* 15, no. 4 (1995): 1968–1973.

Cohen-Tannoudji, M., S. Robine, A. Choulika, et al. "I-SceI-induced gene replacement at a natural locus in embryonic stem cells." *Molecular and Cellular Biology* 18, no. 3 (1998): 1444–1448.

Costa, J. R., B. E. Bejcek, J. E. McGee, A. Fogel, K. R. Brimacombe, and R. Ketteler. "Genome editing using engineered nucleases and their use in genomic screening." In *Assay Guidance Manual [Internet]*, edited by G. S. Sittampalam, A. Grossman, and K. Brimacombe. Bethesda, MD: Eli Lilly & Company and the National Center for Advancing Translational Sciences (2017).

Curtin, S. J., F. Zhang, J. D. Sander, et al. "Targeted mutagenesis of duplicated genes in soybean with zinc-finger nucleases." *Plant Physiology* 156, no. 2 (2011): 466–473.

Curtin, S. J., J. E. Anderson, C. G. Starker, et al. "Targeted mutagenesis for functional analysis of gene duplication in legumes." In *Legume Genomics*, edited by R. J. Rose, pp. 25–42. Totowa, NJ: Humana Press (2013).

Davis, D., and D. Stokoe. "Zinc finger nucleases as tools to understand and treat human diseases." *BMC Medicine* 8, no. 1 (2010): 42.

Davies, J. P., S. Kumar, and L. S. Dent. "Use of zinc-finger nucleases for crop improvement." *Progress in Molecular Biology and Translational Science* 149 (2017): 47–63.

de Pater, S., J. E. Pinas, P. J. J. Hooykaas, and B. J. van der Zaal. "ZFN-mediated gene targeting of the Arabidopsis protoporphyrinogen oxidase gene through Agrobacterium-mediated floral dip transformation." *Plant Biotechnology Journal* 11, no. 4 (2013): 510–515.

Diakun, G. P., L. Fairall, and A. Klug. "EXAFS study of the zinc-binding sites in the protein transcription factor IIIA." *Nature* 324, no. 6098 (1986): 698–699.

Doyon, Y., J. M. McCammon, J. C. Miller, et al. "Heritable targeted gene disruption in zebrafish using designed zinc-finger nucleases." *Nature Biotechnology* 26, no. 6 (2008): 702–708.

Dreier, B., R. R. Beerli, D. J. Segal, J. D. Flippin, and C. F. Barbas. "Development of zinc finger domains for recognition of the 5′-ANN-3′ family of DNA sequences and their use in the construction of artificial transcription factors." *Journal of Biological Chemistry* 276, no. 31 (2001): 29466–29478.

Even-Faitelson, L., A. Samach, C. Melamed-Bessudo, N. Avivi-Ragolsky, and A. A. Levy. "Localized egg-cell expression of effector proteins for targeted modification of the Arabidopsis genome." *The Plant Journal* 68, no. 5 (2011): 929–937.

Foley, J. E., J. R. J. Yeh, M. L. Maeder, et al. "Rapid mutation of endogenous zebrafish genes using zinc finger nucleases made by Oligomerized Pool ENgineering (OPEN)." *PLoS One* 4, no. 2 (2009): e4348.

Fu, Y., J. A. Foden, C. Khayter, et al. "High-frequency off-target mutagenesis induced by CRISPR-Cas nucleases in human cells." *Nature Biotechnology* 31, no. 9 (2013): 822–826.

Gaj, T., C. A. Gersbach, and C. F. Barbas III. "ZFN, TALEN, and CRISPR/Cas-based methods for genome engineering." *Trends in Biotechnology* 31, no. 7 (2013): 397–405.

Gersbach, C. A., T. Gaj, and C. F. Barbas III. "Synthetic zinc finger proteins: The advent of targeted gene regulation and genome modification technologies." *Accounts of Chemical Research* 47, no. 8 (2014): 2309–2318.

Gupta, M., R. C. DeKelver, A. Palta, et al. "Transcriptional activation of Brassica napus β-ketoacyl-ACP synthase II with an engineered zinc finger protein transcription factor." *Plant Biotechnology Journal* 10, no. 7 (2012): 783–791.

Hsu, P. D., D. A. Scott, J. A. Weinstein, et al. "DNA targeting specificity of RNA-guided Cas9 nucleases." *Nature Biotechnology* 31, no. 9 (2013): 827.

Isalan, M., A. Klug, and Y. Choo. "A rapid, generally applicable method to engineer zinc fingers illustrated by targeting the HIV-1 promoter." *Nature Biotechnology* 19, no. 7 (2001): 656–660.

Jinek, M., K. Chylinski, I. Fonfara, M. Hauer, J. A. Doudna, and E. Charpentier. "A programmable dual-RNA–guided DNA endonuclease in adaptive bacterial immunity." *Science* 337, no. 6096 (2012): 816–821.

Jo, Y., H. Kim, and S. Ramakrishna. "Recent developments and clinical studies utilizing engineered zinc finger nuclease technology." *Cellular and Molecular Life Sciences* 72, no. 20 (2015): 3819–3830.

Johnson, R. D., and M. Jasin. "Double-strand-break-induced homologous recombination in mammalian cells." *Biochemical Society Transactions* 29, no. 2 (2001): 196–201.

Kernstock, S., E. Davydova, M. Jakobsson, et al. "Lysine methylation of VCP by a member of a novel human protein methyltransferase family." *Nature Communications* 3, no. 1 (2012): 1–11.

Kim, Y. G., and S. Chandrasegaran. "Chimeric restriction endonuclease." *Proceedings of the National Academy of Sciences of the United States of America* 91, no. 3 (1994): 883–887.

Kim, S., and J. S. Kim. "Targeted genome engineering via zinc finger nucleases." *Plant Biotechnology Reports* 5, no. 1 (2011): 9–17.

Kim, Y. G., J. Cha, and S. Chandrasegaran. "Hybrid restriction enzymes: Zinc finger fusions to Fok I cleavage domain." *Proceedings of the National Academy of Sciences of the United States of America* 93, no. 3 (1996): 1156–1160.

Kim, Y., J. Kweon, A. Kim, et al. "A library of TAL effector nucleases spanning the human genome." *Nature Biotechnology* 31, no. 3 (2013): 251.

Kumar, V., and M. Jain. "The CRISPR–Cas system for plant genome editing: Advances and opportunities." *Journal of Experimental Botany* 66, no. 1 (2015): 47–57.

Ledford, H. "Why the CRISPR patent verdict isn't the end of the story." *Nature News* (2017). doi:10.1038/nature.2017.21510.

Li, L., L. P. Wu, and S. Chandrasegaran. "Functional domains in Fok I restriction endonuclease." *Proceedings of the National Academy of Sciences of the United States of America* 89, no. 10 (1992): 4275–4279.

Lieber, M. R. "The mechanism of double-strand DNA break repair by the nonhomologous DNA end-joining pathway." *Annual Review of Biochemistry* 79(2010): 181–211.

Lloyd, A., C. L. Plaisier, D. Carroll, and G. N. Drews. "Targeted mutagenesis using zinc-finger nucleases in Arabidopsis." *Proceedings of the National Academy of Sciences of the United States of America* 102, no. 6 (2005): 2232–2237.

Maeder, M. L., and C. A. Gersbach. "Genome-editing technologies for gene and cell therapy." *Molecular Therapy* 24, no. 3 (2016): 430–446.

Maeder, M. L., S. Thibodeau-Beganny, A. Osiak, et al. "Rapid "open-source" engineering of customized zinc-finger nucleases for highly efficient gene modification." *Molecular Cell* 31, no. 2 (2008): 294–301.

Mahfouz, M. M., A. Piatek, and C. N. Stewart Jr. "Genome engineering via TALENs and CRISPR/Cas9 systems: Challenges and perspectives." *Plant Biotechnology Journal* 12, no. 8 (2014): 1006–1014.

Marton, I., A. Zuker, E. Shklarman, et al. "Nontransgenic genome modification in plant cells." *Plant Physiology* 154, no. 3 (2010): 1079–1087.

Meng, X., M. B. Noyes, L. J. Zhu, N. D. Lawson, and S. A. Wolfe. "Targeted gene inactivation in zebrafish using engineered zinc-finger nucleases." *Nature Biotechnology* 26, no. 6 (2008): 695–701.

Miller, J., A. D. McLachlan, and A. Klug. "Repetitive zinc-binding domains in the protein transcription factor IIIA from Xenopus oocytes." *The EMBO Journal* 4, no. 6 (1985): 1609–1614.

Morange, M. "What history tells us XLIV: The construction of the zinc finger nucleases." *Journal of Biosciences* 42, no. 4 (2017): 527–530.

Morton, J, M. W. Davis, E. M. Jorgensen, and D. Carroll. "Induction and repair of zinc-finger nuclease-targeted double-strand breaks in Caenorhabditis elegans somatic cells." *Proceedings of the National Academy of Sciences of the United States of America* 103, no. 44 (2006): 16370–16375.

Nakatsukasa, T., Y. Shiraishi, S. Negi, M. Imanishi, S. Futaki, and Y. Sugiura. "Site-specific DNA cleavage by artificial zinc finger-type nuclease with cerium-binding peptide." *Biochemical and Biophysical Research Communications* 330, no. 1 (2005): 247–252.

Osakabe, K., Y. Osakabe, and S. Toki. "Site-directed mutagenesis in Arabidopsis using custom-designed zinc finger nucleases." *Proceedings of the National Academy of Sciences of the United States of America* 107, no. 26 (2010): 12034–12039.

Pavletich, N. P., and C. O. Pabo. "Zinc finger-DNA recognition: Crystal structure of a Zif268-DNA complex at 2.1 A." *Science* 252, no. 5007 (1991): 809–817.

Petolino, J. F. "Genome editing in plants via designed zinc finger nucleases." *In Vitro Cellular & Developmental Biology-Plant* 51, no. 1 (2015): 1–8.

Petolino, J. F., A. Worden, K. Curlee, et al. "Zinc finger nuclease-mediated transgene deletion." *Plant Molecular Biology* 73, no. 6 (2010): 617–628.

Porteus, M. H., and D. Carroll. "Gene targeting using zinc finger nucleases." *Nature Biotechnology* 23, no. 8 (2005): 967–973.

Puchta, H. "Gene replacement by homologous recombination in plants". *Plant Molecular Biology* 48 (2002): 173–182.

Qi, Y., X. Li, Y. Zhang, et al. "Targeted deletion and inversion of tandemly arrayed genes in Arabidopsis thaliana using zinc finger nucleases." *G3: Genes, Genomes, Genetics* 3, no. 10 (2013a): 1707–1715.

Qi, Y., Y. Zhang, F. Zhang, et al. "Increasing frequencies of site-specific mutagenesis and gene targeting in Arabidopsis by manipulating DNA repair pathways." *Genome Research* 23, no. 3 (2013b): 547–554.

Rahman, S. H., M. L. Maeder, J. K. Joung, and T. Cathomen. "Zinc-finger nucleases for somatic gene therapy: The next frontier." *Human Gene Therapy* 22, no. 8 (2011): 925–933.

Rothstein, R. J. "One-step gene disruption in yeast." In *Recombinant DNA*, edited by R. Wu, L. Gro-Sman, and K. Moldaver, Methods in Enzymology, vol. 101, pp. 202–211. Cambridge, MA: Academic Press (1983).

Sakuma, T., and K. Woltjen. "Nuclease-mediated genome editing: At the front-line of functional genomics technology." *Development, Growth & Differentiation* 56, no. 1 (2014): 2–13.

San Filippo, J., P. Sung, and H. Klein. "Mechanism of eukaryotic homologous recombination." *Annual Review in Biochemistry* 77 (2008): 229–257.

Sander, J. D., E. J. Dahlborg, M. J. Goodwin, et al. "Selection-free zinc-finger-nuclease engineering by context-dependent assembly (CoDA)." *Nature Methods* 8, no. 1 (2011): 67–69.

Scherer, S., and R. W. Davis. "Replacement of chromosome segments with altered DNA sequences constructed in vitro." *Proceedings of the National Academy of Sciences of the United States of America* 76, no. 10 (1979): 4951–4955.

Shukla, V. K., Y. Doyon, J. C. Miller, et al. "Precise genome modification in the crop species Zea mays using zinc-finger nucleases." *Nature* 459, no. 7245 (2009): 437–441.

Şöllü, C., K. Pars, T. I. Cornu, et al. "Autonomous zinc-finger nuclease pairs for targeted chromosomal deletion." *Nucleic Acids Research* 38, no. 22 (2010): 8269–8276.

Stadler, L. J. "Genetic effects of X-rays in maize." *Proceedings of the National Academy of Sciences of the United States of America* 14, no. 1 (1928a): 69.

Stadler, L. J. "Mutations in barley induced by X-rays and radium." *Science* 68, no. 1756 (1928b): 186–187.

Steentoft, C., S. Y. Vakhrushev, M. B. Vester-Christensen, et al. "Mining the O-glycoproteome using zinc-finger nuclease–glycoengineered SimpleCell lines." *Nature Methods* 8, no. 11 (2011): 977.

Thomas, K. R., K. R. Folger, and M. R. Capecchi. "High frequency targeting of genes to specific sites in the mammalian genome." *Cell* 44, no. 3 (1986): 419–428.

Tovkach, A., V. Zeevi, and T. Tzfira. "A toolbox and procedural notes for characterizing novel zinc finger nucleases for genome editing in plant cells." *The Plant Journal* 57, no. 4 (2009): 747–757.

Townsend, J. A., D. A. Wright, R. J. Winfrey, et al. "High-frequency modification of plant genes using engineered zinc-finger nucleases." *Nature* 459, no. 7245 (2009): 442–445.

Wah, D. A., J. A. Hirsch, L. F. Dorner, I. Schildkraut, and A. K. Aggarwal. "Structure of the multimodular endonuclease *Fok I* bound to DNA." *Nature* 388, no. 6637 (1997): 97–100.

Wright, D. A., J. A. Townsend, R. J. Winfrey Jr, et al. "High-frequency homologous recombination in plants mediated by zinc-finger nucleases." *The Plant Journal* 44, no. 4 (2005): 693–705.

Zhang, F., M. L. Maeder, E. Unger-Wallace, et al. "High frequency targeted mutagenesis in Arabidopsis thaliana using zinc finger nucleases." *Proceedings of the National Academy of Sciences of the United States of America* 107, no. 26 (2010): 12028–12033.

3 TALENs: Customizable Molecular Scissors for Tailoring Newer Types of Genomes in Plants

Ravi Prakash Saini and N. Manjunatha
ICAR-Indian Grassland and Fodder Research Institute

Manoj Kumar
ICAR-Central Institute for Research on Cotton Technology

Pawar Deepak Viswanath
ICAR-Directorate of Weed Research

CONTENTS

3.1 INTRODUCTION

Genome editing (GE) is currently one of the most popular technologies of molecular biology. It has revolutionized the art of genome engineering through its target site-specific approach. GE allows researchers to perform site-specific alterations in the genome or gene of an organism in a more predictable and accurate way. These alterations range from one or few nucleotides change (insertions, deletions or base substitutions) to the entire gene insertion or deletion. These modifications are executed by programmed sequence-specific nucleases (SSNs), which cause DNA double-stranded breaks (DSBs), at the target sites in the genome, which are then repaired by the inherent DNA repair mechanisms of the cell.

DNA repair at the site of DSBs takes place either through homologous recombination (HR) or through non-homologous end-joining (NHEJ) as mentioned in Figure 3.1. HR facilitates large alterations, i.e. gene replacement or insertion, whereas NHEJ leads to insertions or deletions (indels)

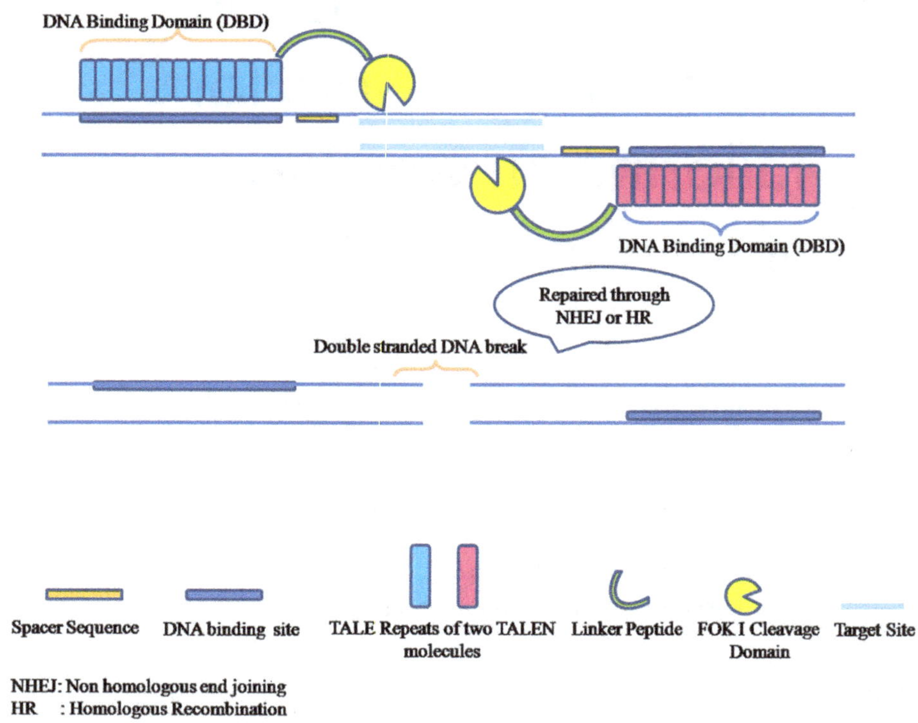

FIGURE 3.1 A pair of TALEN molecules inducing a double-stranded DNA break at the target site in the genome.

or base substitutions of few nucleotides at one or few places (Yamamoto et al., 2018). Probability of either repair mechanism depends upon the presence of other endogenous (within the cell) or exogenous factors (from outside) during the time of repair. Despite being a decade-old technology, GE is still popular due to the continuous refinements and modifications, which are improving its accessibility and applicability among researchers. Moreover, it has a wider range of applications too particularly in the field of molecular gene regulation.

Among various GE technologies (GETs), zinc finger nucleases (ZFNs) were the first to develop followed by meganucleases (MGNs), transcription activator-like effector nucleases (TALENs) and the latest clustered regularly interspaced palindromic repeats (CRISPR)-associated 9 (Cas9) (Mushtaq et al., 2020; Hua et al., 2019). These SSNs are regarded as GE technologies (GETs).

Among several commercially available GETs, opting for a particular and more convenient GET depends upon a number of factors like target trait or character that needs to be edited, number of genes governing it and their sequences (Hua et al., 2019). For instance, oil quality in soya bean, cottonseed and rapeseed, and amylose content in rice or maize can be improved by targeting a single gene, whereas the agronomic traits attributing to yield and biotic/abiotic stress resistance are controlled by numerous quantitative trait loci (QTLs), which are potential targets for GETs (Zhang et al., 2018; Zuo and Li, 2014). GETs can be employed for regulating the expression of gene at transcriptional, post-transcriptional or translational level to attain a particular character in the targeted crop (Hua et al., 2019).

In this chapter, we are mainly focusing on the application of TALENs for altering the genome of plants for achieving the desired trait. TALENs have been widely used for improving the qualitative and quantitative traits in plants (Christian et al., 2013; Zhou et al., 2015; Itokawa et al., 2016). In crop improvement, the first successful example of TALEN-mediated GE was against biotic stress in rice where bacterial blight-resistant mutant lines of rice were developed by disrupting the bacterial blight susceptibility gene *OsSWEET14* (Li et al., 2012; Nowicka et al., 2018). Similarly, fragrance is another important economic character of rice and has been improved using TALENs, without any adverse effect on the yield and other quantitative traits (Shan et al., 2015). In potato, TALEN-mediated targeted disruption of the *VACUOLAR INVERTASE* gene *(VInv)* resulted in an increased cold storability and processing of the potato tubers (Clasen et al., 2015).

Commercially available GETs differ from each other with respect to the recognition mechanism, target site length, DNA binding factor, endonuclease involved in cleavage and functional applicability. Furthermore, these GETs also vary with respect to ease of design, specificity levels of nucleases, ease of gene delivery, cost and targeting efficiency. For example, ZFNs and TALENs involve DNA/protein interaction, whereas CRISPR-Cas system involves DNA/RNA interaction for recognizing their target site. CRISPR-Cas leads to gene knockout, gene knock-in and transcriptional regulations, whereas ZFNs and TALENs result in gene knockout and transcriptional regulation. The target efficiencies of ZFNs are considered low compared to those of TALENs and CRISPR-Cas due to complex interactions between amino acid residues and base pairs of the target sequence (Bilichak and Eudes, 2016). The comparison of GETs has been reported by various researchers (Shanmugam et al., 2019) with respect to various parameters, and is presented in the Venn diagram in Figure 3.2.

3.2 TALEN-MEDIATED GENOME EDITING

Like ZFN, TALEN is a fusion protein molecule synthesized by combining two functionally independent domains, i.e. DNA binding domain – comprising transcriptional activator-like effector (TALE) repeats – and the DNA cleavage domain of the FokI restriction enzyme joined through a linker peptide as described in Figure 3.3.

TALE proteins commonly known as effector molecules are naturally found in plant pathogenic bacteria of the genus *Xanthomonas* (Bogdanove et al., 2010; Scholze and Boch, 2011). These effector molecules are moved into the diseased plant cells via type III secretion system and contain eukaryotic nuclear localizing signals (NLS), which facilitate their movement from cell's cytoplasm to the nucleus and once inside the nucleus, induce the transcription of genes that are essential for pathogen growth and spread by binding to their cis-acting elements (within 60 bp of the start codon) (Boch and Bonas, 2010). Figure 3.4 shows a typical structure of TALE consisting of a translocation domain at the N-terminal end and an NLS at the C-terminal end with an acidic transcriptional activation domain and a central DNA binding domain.

DNA binding domain of each TAL effector consists of few to 33.5 repeats also known as monomer, arranged in tandem. Each monomer is typically made up of nearly identical 34 amino acids but the amino acids differ primarily at 12th and 13th positions; amino acids at these positions are known as repeat variable diresidues (RVDs), which determine the binding specificity of the respective monomer to a particular nucleotide pair (Moscou and Bogdanove, 2009; Boch and Bonas, 2010). Thus, each monomer or TALE repeat recognizes a single base pair in a contiguous DNA sequence through its corresponding RVDs and establishes a simple 1:1 code for the protein-to-DNA interaction. The binding behaviour and the binding affinity of various RVDs with their corresponding base pair are clearly mentioned in Table 3.1 (Boch et al., 2009; Moscou and Bogdanove, 2009; Miller et al., 2011; Streubel et al., 2012; Meckler et al., 2013).

Combination of different monomers within a TALE molecule decides its binding specificity to a particular DNA sequence (Boch and Bonas, 2010). In order to decipher whether the entire structure of TALE molecule (from the N-terminal end to the C-terminal end) is needed for DNA binding, several deletion variants with truncated N- and C-terminal ends were generated (Miller et al., 2011;

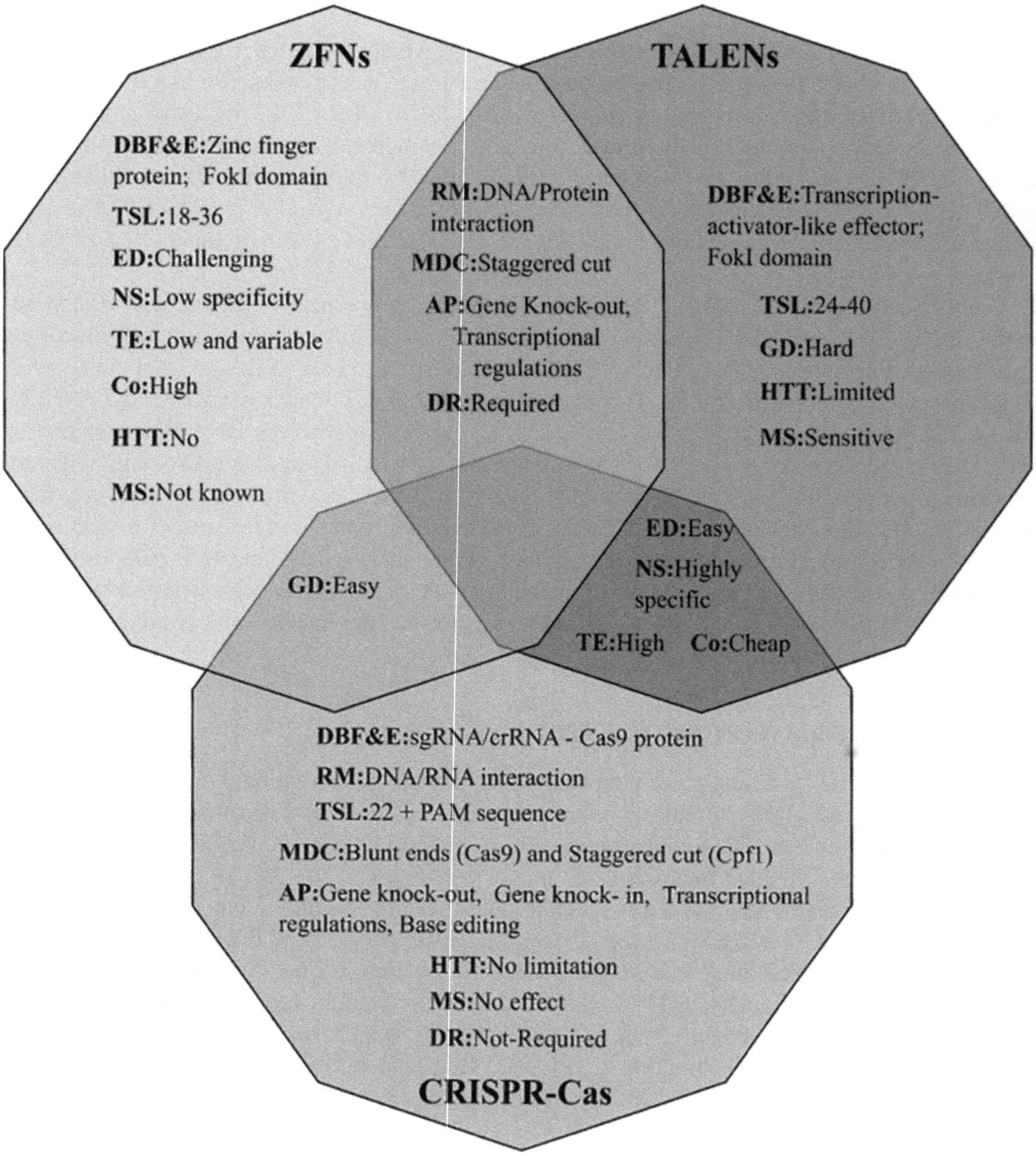

FIGURE 3.2 Comparison of the three popular GETs (ZFN, TALEN and CRISPR-Cas) with respect to recognition mechanism, target site length, DNA binding factor, endonuclease involved in cleavage, functional applicability, ease of design, specificity levels of nucleases, ease of gene delivery, cost and targeting efficiency. Key to the short form: GETs, genome editing tools/technologies; ZFNs, zinc finger nucleases; TALENs, transcription activator-like effector nucleases; CRISPR, clustered regularly interspaced palindromic repeats; DBF&E, DNA binding factor and endonuclease; RM, recognition mechanism; TSL, target site length (bp); MDC, mechanism of DNA cleavage pattern; AP, functional applicability; ED, ease of design; NS, nuclease specificity; GD, gene delivery; TE, targeting efficiency; Co, cost; HTT, high-throughput targeting; MS, methylation (CpG)-sensitive; and DR, dimerization requirement.

Mussolino et al., 2011). In both the studies, it was found that the truncated form of natural TALE molecule (repeat units having at least 150 amino acid residues at the N-terminal end and nearly 20 amino acid residues at the C-terminal end) is enough for editing. This information helped researchers to reduce the size of TALEN molecules, thus easing their delivery inside the cell.

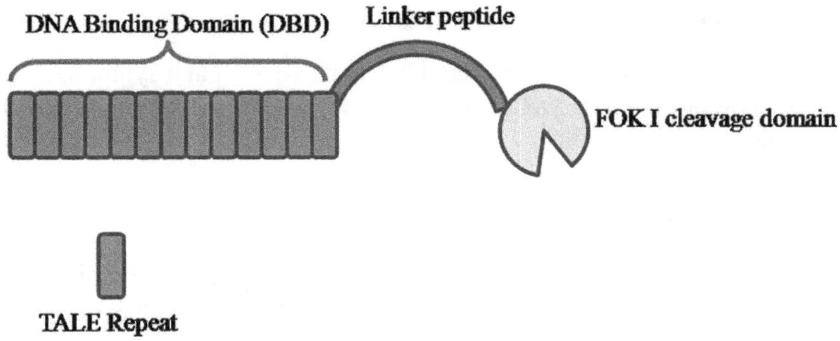

FIGURE 3.3 Structure of TALEN molecule comprising DNA binding domain connected to FokI cleavage domain through a linker peptide.

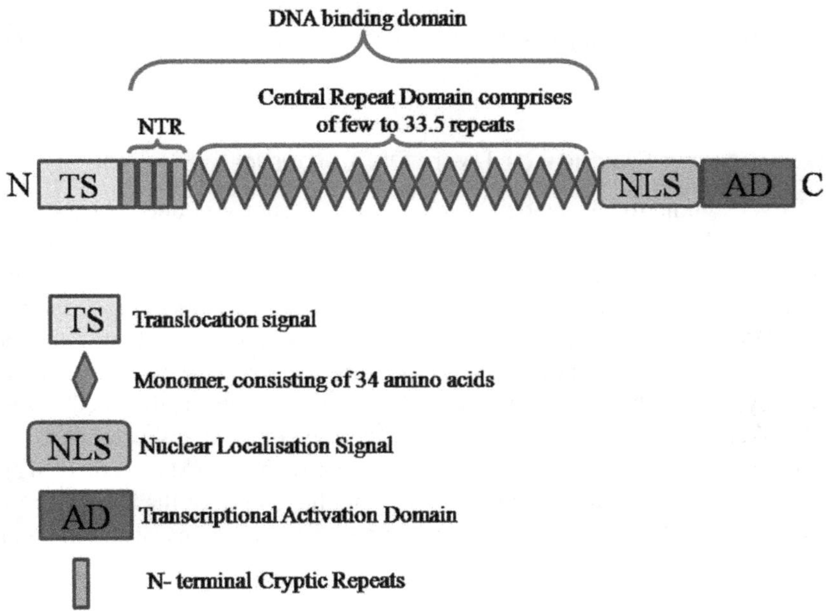

FIGURE 3.4 Detailed structure of TALE molecule, where N and C represents the N-terminal and C-terminal domains, respectively.

TABLE 3.1
Cipher Code between RVDs of TALE Monomers and Corresponding Nucleotide at Target Site

RVDs at 12th and 13th Positions in a Monomer	Interacting Nucleotide at the Target Site	Binding Affinity	Interaction
NI	A	Weak	Specific
HD	C	Strong	Specific
NH/NK	G	Intermediate/weak	Specific
NG	T	Strong	Specific
NN	G/A	Strong/weak	Degeneracy
NS	A, C and G	-	Degeneracy

Source: Data from Boch et al. (2009) and Moscou and Bogdanove (2009).

N, Asparagine; I, Isoleucine; H, Histidine; D, Aspartic acid; K, Lysine; G, Glycine; S, Serine.

3.3 LIMITATIONS OF TALEN TECHNOLOGY

For the commercial utility of any GET, it should be easy to design and synthesize, has flexibility regarding the selection of the target DNA sequence and is highly specific in its action with minimum off-target effects.

3.3.1 Ease of Synthesis

The highly repetitive nature of TALE repeats complicates the construction of new TALEs vectors. Even a skilled researcher requires a time of about 1 week for its construction. The assembly of TALE DBD from the nearly identical TALE monomers is difficult by employing regular cloning techniques (PCR and restriction ligation) and leads to the development of several other platforms for the same purpose. Most of these platforms use the hierarchical ligation strategy, in which several DNA fragments (monomers) are ligated in a sequential way to get the multimer arrays. Joining of monomers at the junction point utilizes various combinations of degenerate codons and type II restriction enzymes (Cermak et al., 2011; Li et al., 2011; Morbitzer et al., 2011; Weber et al., 2011).

3.3.2 Flexibility in the Selection of Target DNA Sequence

Whether TALEN can bind to any genomic sequence of our choice or does it work in a sequence-context manner? TALEN molecules are more flexible in their genomic site selection as compared to firstly discovered GE tools, ZFNs. In ZFNs, decoding of individual zinc finger against the corresponding DNA nucleotides is not very accurate (one ZF recognizes the three nitrogenous bases in DNA). Moreover, it changes when individual zinc fingers are combined to form the customized ZFN molecule (individual zinc fingers interact with each other and alter their binding efficiency). The only limitation with TALEN nuclease sites is the need of thymine residue before the 5′-end of the target sequence. The tryptophan residue located at 232nd position in the N-terminal end was responsible for the interaction with 5′-T (Mak et al., 2012). However, this limitation can be easily managed either by varying the length of the spacer or by searching for the mutant forms of the TALEN that are capable of binding to any of the three nucleotides, A, G or C (Lamb et al., 2013).

3.3.3 Minimum Off-Target Effects

There are several reasons that can lead to the off-target effects, which are given in the following subsections.

3.3.3.1 Target Site Occurrence in the Genome

An important point that needs to be confirmed before the selection of the target site is to determine its occurrence in the genome through various bioinformatics tools. Repetitive sequences as well as sequences having high homology with other genomic regions are always better to avoid as the target site (Nemudryi et al., 2014).

3.3.3.2 Variable Binding Strength of DNA Binding Domain with Its Target Site

Binding affinities of RVDs of each TALE repeat determine the overall binding strength of TALEN molecule to its respective target site. Although codes of various RVDs are very well known, their binding strength may vary; for example, HD and NN RVDs form strong hydrogen bonds, while NK and NI form weak hydrogen bonds with their respective nucleotide, which may lead to their binding with other noncanonical target sites containing few mismatches from the original target site.

3.3.3.3 Degeneracy of Code

Similar to the genetic code where one amino acid can be coded by more than one triplet code, degeneracy exists between the respective RVDs and their nucleotides where one RVD can able to recognize more than one nucleotide although with altered affinities. As mentioned above, NN can bind to both A and G with more preference towards G. This degeneracy may lead to an incorrect binding of TALEN.

3.3.3.4 Homodimerization of FokI Domain

Dimerization of a pair of TALEN molecules, having different DNA binding domain and which bind on either side of the double-stranded cleavage site, is essential for editing the target site in the genome. This kind of dimerization is known as heterodimerization but sometimes similar TALEN molecules dimerize and form homodimers, which may lead to the off-target cleavage. This limitation can be resolved by inducing point mutations at selective sites in the FokI domain, which could prevent homodimerization without disturbing its nuclease activity as heterodimers (Gabriel et al., 2011).

3.3.3.5 Promiscuity of Short Linker Peptide

A short linker peptide that connects DNA binding domain to the cleavage domain sometimes spans different spacers instead of the one chosen in the target site (Gabriel et al., 2011; Pattanayak et al., 2011).

3.4 CONCLUSION AND FUTURE PROSPECTS OF TALEN TECHNOLOGY

GE with designer nucleases is an excellent way to modify the complex genomes and determine the gene function. TALEN-mediated gene editing has distinct advantages over ZFN and MGNs. It is a simple design with high target specificity, low toxic effects and unlimited options for the target site selection, and has no context-dependent effects. With the advent of RNA-guided clustered regularly interspaced short palindromic repeat (CRISPR) approach, TALEN technology may become the second choice for gene editing but its outstanding potential for research and therapy remains undisputed. TALEN technology has been successfully utilized in the field of agriculture, medicine and energy to achieve the desired results (Satheesh et al., 2019).

REFERENCES

Bilichak, A., F. Eudes. "Genetic engineering of plants using Zn fingers, TALENs, and CRISPRs." In *Genome Stability*, edited by I. Kovalchuk, O. Kovalchuk, pp. 187–201. Cambridge, MA: Academic Press (2016).

Boch, J., U. Bonas. "Xanthomonas Avrbs3 family-type III effectors: Discovery and function." *Annual Review of Phytopathology* 48 (2010): 419–436.

Boch, J., H. Scholze, S. Schornack, et al. "Breaking the code of DNA binding specificity of TAL-type III effectors." *Science* 326 (2009): 1509–1512.

Bogdanove, A. J., S. Schornack, T. Lahaye. "TAL effectors: Finding plant genes for disease and defense." *Current Opinion in Plant Biology* 13 (2010): 394–401.

Cermak, T., E. L. Doyle, M. Christian, et al. "Efficient design and assembly of custom TALEN and other TAL effector-based constructs for DNA targeting." *Nucleic Acids Research* 39, no. 12 (2011): e82.

Christian, M., Y. Qi, Y. Zhang, D. F. Voytas. "Targeted mutagenesis of *Arabidopsis thaliana* using engineered TAL effector nucleases." *G3: Genes, Genomes, Genetics* 3, no. 10 (2013): 1697–1705.

Clasen, B. M., T. J. Stoddard, S. Luo, et al. "Improving cold storage and processing traits in potato through targeted gene knockout." *Plant Biotechnology Journal* 14, no. 1 (2015): 169–176.

Gabriel, R., A. Lombardo, A. Arens, et al. "An unbiased genome-wide analysis of zinc-finger nuclease specificity." *Nature Biotechnology* 29, no. 9 (2011): 816–823.

Hua, K., J. Zhang, J. R. Botella, et al. "Perspectives on the application of genome editing technologies in crop breeding." *Molecular Plant* 12, no. 8 (2019): 1047–1059.

Itokawa, K., O. Komagata, S. Kasai, K. Ogawa, T. Tomita. "Testing the causality between CYP9M10 and pyrethroid resistance using the TALEN and CRISPR/Cas9 technologies." *Scientific Reports* 6 (2016): 24652.

Lamb, B. M., A. C. Mercer, C. F. Barbas, 3rd. "Directed evolution of the TALE N-terminal domain for recognition of all 5′ bases." *Nucleic Acids Research* 41, no. 21 (2013): 9779–9785.

Li, T., S. Huang, X. Zhao, et al. "Modularly assembled designer TAL effector nucleases for targeted gene knockout and gene replacement in eukaryotes." *Nucleic Acids Research* 39, no. 14 (2011): 6315–6325.

Li, T., B. Liu, M. H. Spalding, D. P. Weeks, B. Yang. "High-efficiency TALEN-based gene editing produces disease-resistant rice." *Nature Biotechnology* 30, no. 5 (2012): 390–392.

Mak, A. N., P. Bradley, R. A. Cernadas, A. J. Bogdanove, B. L. Stoddard. "The crystal structure of TAL effector PthXo1 bound to its DNA target." *Science* 335, no. 6069 (2012): 716–719.

Meckler, J. F., M. S. Bhakta, M. S. Kim, et al. "Quantitative analysis of TALE-DNA interactions suggests polarity effects." *Nucleic Acids Research* 41, no. 7 (2013): 4118–4128.

Miller, J. C., S. Tan, G. Qiao, et al. A TALE nuclease architecture for efficient genome editing." *Nature Biotechnology* 29, no. 2 (2011): 143–148.

Morbitzer, R., J. Elsaesser, J. Hausner, T. Lahaye. "Assembly of custom TALE-type DNA binding domains by modular cloning." *Nucleic Acids Research* 39, no. 13 (2011): 5790–5799.

Moscou, M. J., A. J. Bogdanove. "A simple cipher governs DNA recognition by TAL effectors." *Science* 326 (2009): 1501.

Mushtaq, M., S. Mukhtar, A. Sakina, et al. "Tweaking genome-editing approaches for virus interference in crop plants." *Plant Physiology and Biochemistry* 147 (2020): 242–250.

Mussolino, C., R. Morbitzer, F. Lutge, N. Dannemann, T. Lahaye, T. Cathomen. "A novel TALE nuclease scaffold enables high genome editing activity in combination with low toxicity." *Nucleic Acids Research* 39, no. 21 (2011): 9283–9293.

Nemudryi, A. A., K. R. Valetdinova, S. P. Medvedev, S. M. Zakian. "TALEN and CRISPR/Cas genome editing systems: Tools of discovery." *Actanaturae* 6, no. 3 (2014): 19–40.

Nowicka, B., J. Ciura, R. Szymańska, J. Kruk. "Improving photosynthesis, plant productivity and abiotic stress tolerance: Current trends and future perspectives." *Journal of Plant Physiology* 231 (2018): 415–433.

Pattanayak, V., C. L. Ramirez, J. K. Joung, D. R. Liu. "Revealing off-target cleavage specificities of zinc-finger nucleases by in vitro selection." *Nature Methods* 8, no. 9 (2011): 765–770.

Satheesh, V., H. Zhang, X. Wang, M. Lei. Precise editing of plant genomes: Prospects and challenges. *Seminars in Cell & Developmental Biology* 96 (2019): 115–123.

Scholze, H., J. Boch. "TAL effectors are remote controls for gene activation." *Current Opinion in Microbiology* 14 (2011): 47–53.

Shan, Q., Y. Zhang, K. Chen, K. K. Zhang, C. Gao. Creation of fragrant rice by targeted knockout of the OsBADH2gene using TALEN technology. *Plant Biotechnology Journal* 13, no. 6 (2015): 791–800.

Shanmugam, S., H. H. Ngo, Y. R. Wu. "Advanced CRISPR/Cas-based genome editing tools for microbial biofuels production: A review." *Renewable Energy* 149 (2019): 1107–1119.

Streubel, J., C. Blücher, A. Landgraf, J. Boch. "TAL effector RVD specificities and efficiencies." *Nature Biotechnology* 30, no. 7 (2012): 593–595.

Weber, E., R. Gruetzner, S. Werner, C. Engler, S. Marillonnet. "Assembly of designer TAL effectors by golden gate cloning." *PLoS One* 6, no. 5 (2011): e19722.

Yamamoto, T., S. Kashojiya, S. Kamimura, et al. "Application and development of genome editing technologies to the solanaceae plants." *Plant Physiology and Biochemistry* 131 (2018): 37–46.

Zhang, J., H. Zhang, J. R. Botella, J. K. Zhu. "Generation of new glutinous rice by CRISPR/Cas9-targeted mutagenesis of the Waxy gene in elite rice varieties." *Journal of Integrative Plant Biology* 60, no. 5 (2018): 369–375.

Zhou, J., Z. Peng, J. Long, et al. "Gene targeting by the TAL effector PthXo2 reveals cryptic resistance gene for bacterial blight of rice." *The Plant Journal* 82, no. 4 (2015): 632–643.

Zuo, J., J. Li. "Molecular genetic dissection of quantitative trait loci regulating rice grain size." *Annual Review of Genetics* 48, no. 1 (2014): 99–118.

4 Mechanism of CRISPER/Cas9-Mediated Genome Editing: Scope and Opportunities

Antim
ICAR-Indian Institute of Wheat and Barley Research
Chaudhary Charan Singh Haryana Agricultural University

Ashish
ICAR-Central Soil Salinity Research Institute
Chaudhary Charan Singh Haryana Agricultural University

Bhudev Singh Tyagi, H. M. Mamrutha,
Sonia Sheoran, and Om Prakash Gupta
ICAR-Indian Institute of Wheat and Barley Research

CONTENTS

4.1 INTRODUCTION

Variations in DNA sequences and epigenetic variations are the inherited variations present in all organisms except few viruses where RNA is the carrier of the genetic information. Though most of us are aware of the DNA and RNA being hereditary units to carry information from parents to progenies, epigenetic variations are referred to as heritable changes of the hereditary information without the alteration in the DNA/RNA sequence of an individual. In the last few years, these variations have been extensively studied by scientists due to the advent of cheaper sequencing methods and efficient laboratory techniques. The next step is obviously to harness these heritable variations for humankind. Though we have studied previously gene editing systems like ZFNs (zinc finger nucleases) (see Chapter 2) and TALENs (transcription activator-like effector nucleases) (see Chapter 3), which advanced as the molecular biology techniques evolved, the arrival of CRISPR/Cas9 in 2012 made gene editing faster, cheaper and precise. This most recent technique targets DNA as well as epigenetic variations, which is a landmark for future genome editing. Discovery of clustered regularly interspaced short palindromic repeats (CRISPR) and DNA repeat arrays was made as an adaptive defence mechanism system in archaea and bacteria against viruses. These arrays are functional when associated with CRISPR-associated proteins (Cas9 proteins). Basically, in vivo CRISPR/Cas9 incorporates two domains: for instance, a protein (Cas9 nuclease) and a programmable single guide RNA (sgRNA). sgRNA is a unit of two RNAs: a developed CRISPR RNA (crRNA) and a partially corresponding trans-activating tracrRNA. A sequence of DNA can be targeted by Cas unit only when NGG protospacer-adjacent motifs, i.e. PAM sequences, are present. These essential components of this system have been discussed one by one in this chapter.

We have also elaborately discussed the evolution of CRISPR/Cas as an adaptive defence mechanism in archaea and bacteria; the mechanism of CRISPR/Cas9-mediated genome editing in plants; and the comparison of CRISPR/Cas with other techniques of genome editing like meganucleases, ZFNs and TALENs, and also about the modified form of CRISPR, i.e. prime editing.

4.2 CRISPR: HISTORY AND ADAPTIVE DEFENCE MECHANISM IN BACTERIA AND ARCHAEA

Immune system of animals and prokaryotes is supposed to be the most advanced system as a lot of pathways are triggered after the attack of any pathogen. Prokaryotes, much simpler than a eukaryotic system, also have developed an immune system to keep away bacteriophage or plasmid (Horvath and Barrangou, 2010). One of the systems is called CRISPR/Cas, which consists of non-repetitive spacer sequences unique in nature and adjacent to spacer are present 6–20 genes, which encode for Cas proteins (Lillestøl et al., 2006). The repeats found in each CRISPR locus have 23–47 numbers of base pairs and are highly conserved in nature. Horvath and Barrangou revealed that spacers are of 21–72 base pairs in length and composed of extrachromosomal elements. In 1987, some scientists cloned and sequenced iap genes in *Escherichia coli*. Later on, studies on iap genes reported repeats having 29 nucleotide base pairs, which were separated by non-repetitive spacer sequences (Nakata et al., 1989). This was the first report of CRISPR in history followed by many reports for the presence of the locus in many other organisms (Mojica et al., 2000).

Before 2002, various terms like direct variable repeats (DVR), short regularly spaced repeats (SPSR), long tandemly repeated repetitive sequences (LTRR), large clusters of tandem repeats (LCTR), spacer interspersed direct repeats (SPIDR) and tandem repeats (TREP) were frequently used to address the structures like CRISPR. Later, Jensen and his colleagues used the term 'CRISPR' for this particular structure of loci.

Cas or CRISPR-associated genes present on the one end of the loci code for proteins. Spacer sequences present in CRISPR loci are similar to the sequences of the intruding bacteriophage or plasmid (Bolotin et al., 2005; Pourcel et al., 2005). For the first time, the biological function of CRISPR was exhumed in 2005 by three different researcher's groups who observed random sequences between identical CRISPR repeats showing homology to the DNA sequences of invading virus or plasmids (Mojica et al., 2005). It was suggested from the studies that CRISPR may play a role in the prokaryotic defence system, which might be analogous to the RNA interference (RNAi) system in the eukaryotes despite no similarity between RNAi system components and Cas proteins (Ishino et al., 1987). CRISPR locus for the first time was elaborately illustrated in 1993 by Francisco Mojica (1993–2005) at University of Alicante, Spain. He also coined the term 'CRISPR' through correspondence with Ruud Jansen, who first used the term in print in 2002. He observed that the contrasting repeat sequences described by various groups of scientists have a common set of features which are now known to be hallmark sequences of CRISPR/Cas system. Later in 2005, he described in his paper these sequences as snippets from invading bacteriophage genome (Mojica et al., 2005) and assumed CRISPR to be the adaptive immune system in bacteria. Another group of researchers (Pourcel et al., 2005) published similar findings about CRISPR system. Bolotin et al. (2005) were studying the bacteria *Streptococcus thermophilus* genome sequence, when they found an unusual CRISPR locus. The locus was similar to the previously reported systems but contained novel Cas genes and lacked some of the previously known genes. One of the newly predicted genes was showing a nuclease action and later termed as Cas9. Spacer sequences, homologous to the virus genome sequences, were found to be sharing a common sequence at one end by Bolotin and his colleagues, and termed as 'protospacer-adjacent motif' (PAM). PAM sequences are involved essentially in the target sequence recognition.

A group of scientists, John van der Oost and his colleagues, studied and elaborated the interference of CRISPR/Cas9 system while invasion of any bacteriophage. They disclosed how in *Escherichia coli* at the time of bacteriophage infection, phage-derived spacer sequences are transcribed into small RNAs, i.e. CRISPR RNAs (crRNAs) that guide Cas9 proteins to the target DNA.

When people were thinking CRISPR/Cas9 system equivalent to RNAi silencing mechanism, which can only target RNA, at that time studies of Marraffini and Sontheimer (2008) confirmed that this system targets DNA molecule not RNA in their study. Later in 2009, Hale and his colleagues presented this system targeting RNA also. If we talk about the discovery and development of CRISPR/Cas9 system, it will be a very lengthy description to include in a single chapter. So to give you an idea of the journey of its origin and advancement, we have compiled the work of few eminent researchers in Table 4.1.

4.3 PROCESS OF CRISPR-Cas ACQUIRED IMMUNE SYSTEM IN BACTERIA

In bacteria, CRSIPR/Cas is a cascade of genes containing CRISPRs, unique sequences non-repetitive in nature, and 6–20 genes adjacent to it, which code for Cas proteins. There are also presently repetitive sequences of 23–47 bp, highly conserved along with the spacer sequences (21–72 bp) composed of extra chromosomal elements (Horvath and Barrangou, 2010). The spacer units in CRISPR loci of most of the species have <50 repeats except in the species like *Chloroflexus* where there may be 375 repeats in a spacer unit. In bacteria and archaea, the CRISPR system has a role in adaptive immunity, and the mechanism of action happens in three stages, which are explained in Table 4.2. But before studying the mechanism of action of this system in bacteria, we must be acquainted with the term used to describe its various components. The following section describes the components of CRISPR/Cas9 systems in brief.

TABLE 4.1
A Brief History and Milestones in CRISPR Discovery

1.	Ishino et al. (1987)	Discovery of the presence of clustered repeats in *E. coli*
2.	Nakata et al. (1989)	Specific study of the clustered repeats in enterobacteria
3.	Groenen et al. (1993)	Use of CRISPR for *M. tuberculosis* typing
4.	Mojica et al. (1995)	Identification of CRISPRs in *Haloferax*
5.	Masepohl et al. (1996)	Description of CRISPRs in cyanobacteria
6.	Hoe et al. (1999)	Use of CRISPR for *S. pyogenes* rapid genetic subtyping
7.	Mojica et al. (2000)	Widespread occurrence of CRISPRs in archaea and bacteria
8.	Makarova et al. (2002)	DNA repair hypothesis for Cas genes
9.	Jansen et al. (2002a,b)	Term 'CRISPR' coined; description of Cas genes
10.	Pourcel et al. (2005), Mojica et al. (2005), and Bolotin et al. (2005)	Spacer sequences are probably snippets from foreign genome
11.	Makarova et al. (2006)	Putative RNAi-based mechanism proposed
12.	Barrangou et al. (2007)	Establishment of CRISPR/Cas9 as adaptive immune system of *S. thermophilus* (a bacteria) against viruses
13.	Deveau et al. (2008) and Horvath et al. (2008)	Discovery of the protospacer-adjacent motif in *S. thermophilus*
14.	Brouns et al. (2008)	Immunity is mediated by small CRISPR RNAs in *E. coli*
15.	Marraffini and Sontheimer (2008)	Plasmid dsDNA is the target in *S. epidermidis*
16.	Hale et al. (2009)	ssRNA is the target in *P. furiosus*
17.	Garneau et al. (2010)	Precise cleavage of target DNA by Cas9 in *S. thermophilus*
18.	Deltcheva et al. (2011)	Discovery of the tracrRNA in *S. pyogenes*
19.	Sapranauskas et al. (2011)	*S. thermophilus* CRISPR-Cas system transmission successfully to provide an immunity in *E. coli* (heterologous transfer)
20.	Gasiunas et al. (2012) and Jinek et al. (2012)	Reprogrammable, Cas9-based DNA cleavage
21.	Cong et al. (2013), Jiang et al. (2013a), and Mali et al. (2013)	Cas9-based editing of human and bacterial genomes
22.	Nishimasu et al. (2014) and Anders et al. (2014)	Cas9 structure
23.	Hu et al. (2014)	Cas9-based eradication of latent HIV infection
24.	Zetsche et al. (2015)	Nuclease single RNA-guided nuclease activity of Cpf1
25.	Pawluk et al. (2016)	Discovery of anti-CRISPR proteins against Cas9 in *N. meningitides*
26.	2016, USDA	Edited crops are different from GMOs
27.	2017, US patent office	Broad institute gets CRSPR/Cas9 patent
28.	Jian-kui (2019)	Created the first gene-edited babies having resistance to HIV

TABLE 4.2
Mechanism of CRISPR/Cas Immunity in Bacteria

First stage	It is referred as adaptation (Garneau et al., 2010; Marraffini and Sontheimer, 2010), immunization or spacer acquisition (Karginov and Hannon, 2010; Van der Oost et al., 2009) because in this stage fragments of viral or plasmid double-stranded DNA (spacers) are inserted on host DNA at the leader end of the CRISPR cascade. The spacer acquirement by host DNA is due to Cas1 and Cas2 proteins, which are transcribed by the genes present near the vicinity of the CRISPR cascade
Second stage	This stage is described by the biosynthesis of crRNA (small CRISPR RNAs) because of its synthesis from pre-CRISPR RNA (pre-crRNA) maturation. Some researchers name it as biogenesis of crRNA. Pre-CRISPR RNA (pre-crRNA) is transcribed by RNA polymerase (RNAP) from the CRISPR locus and cleaved by endoribonucleases into crRNAs. The crRNAs act as guide RNAs as described by Brouns et al. (2008) and Carte et al. (2008); or prokaryotic silencing RNA (psiRNAs) (Carte et al., 2008; Makarova et al., 2006)

(*Continued*)

TABLE 4.2 (*Continued*)
Mechanism of CRISPR/Cas Immunity in Bacteria

Third stage	The final stage has been called as interference by Deveau et al. (2010) or immunity acquirement stage by Garneau et al. (2010). The invading organism has the target sequences which specifically have to match to crRNA for interference to occur. As described by Brouns et al. (2008), crRNA present in a multiprotein complex recognizes and forms base pairing with the invader's genome, which must have a perfect or almost-perfect reciprocity with it. This base pairing initiates the cleavage of crRNA-invader complex, which ultimately hinders the process of infection by foreign invading particles. Though, any mismatches in base pairing of target DNA and crRNA may break the immunity

4.3.1 Main Components of CRISPR/Cas9 System

a. **Cas9 endonuclease**: It is the enzyme/protein that is able to modify DNA in its active form by recognizing its sequence and making a cut at that site. It has various forms that differ in their functions like DNA binding and making cut in the single strand or double strand. It identifies DNA sequences having NGG protospacer motifs only.
b. **CRISPR RNA (crRNA)**: Its major role is in locating a specific site on the target DNA, which is accomplished by guide RNA present in it. It also has a tracrRNA binding domain (hairpin loop form). Binding of tracrRNA to crRNA forms an active complex.
c. **Trans-activating crRNA (tracrRNA)**: It was discovered for the very first time in *Streptococcus pyogenes*. It binds to crRNA forming an active complex, which targets the specific site in DNA, and this process is called the maturation of crRNA. crRNA/tracrRNA hybrid function is to guide endonuclease Cas9 to the target DNA site.
d. **sgRNA**: It is tracrRNA/crRNA hybrid where tracrRNA is responsible for Cas9 endonuclease activity and crRNA binds to the target sequences. The two units of this complex are joined by tetraloop.
e. **Repair template:** It is the sequence in the DNA molecule, which acts as a guide for the cellular repair machinery allowing the insertion of a specific DNA sequence.
f. **sgRNA scaffold vectors**: sgRNA is having two components, i.e. a ~20-nt target sequence to the target genomic site and a scaffold, which act as carriers of the targeting sequences. After the selection of a target site, target-specific sgRNA oligos (forward and reverse oligos) can be synthesized and inserted into the sgRNA scaffold vector. The oligos have 20-nucleotide target sites with 5′ and 3′ overhangs complementary to the digested scaffold vector.

4.4 TYPES OF CRISPR/Cas SYSTEM

As we have read in previous sections that CRISPR/Cas was discovered as an adaptive immunity mechanism in bacterial and archaeal genomes, this system has been divided into two classes based on the characteristics of nuclease effector proteins and further categorized into six types (Types I, II, III, IV, V and VI) based on specific Cas endonuclease. Nuclease effectors play a role at the interference stage and are divided into Class 1 type and Class 2 type of effectors. Class 1 type of effectors are multi-subunit Cas protein complexes, which are found in Type I, III and IV CRISPR systems, while Class 2 includes Types II, V and VI and have a single protein effector module as a common feature. Class 1 system is present in both archaea and bacteria but Class 2 system is predominantly existing in archaea (Chylinski et al., 2014). Both types of the CRISPR systems can target DNA and RNA. A detailed classification of types and subtypes of CRISPR-Cas organization has been provided by Makarova et al. (2015), Koonin et al. (2017) and Shmakov et al. (2017). Here, we will briefly describe the classification of CRISPR/Cas systems.

4.4.1 Class 1 CRISPR Systems

This class of the CRISPR system is characterized by multiple effector modules, which are responsible for RNA recognition and crRNA binding. Types I, III and IV have multi-subunit Cas protein complex effectors and dominantly exist in 90% of CRISPR loci of bacteria and archaea. Among these three types, Types I and III are more common and are responsible for crRNA biogenesis, target identification and cleavage by binding to it, and ultimately insertion of the spacer sequences to the CRISPR cascade.

Type I: It has been further divided into seven different subtypes (I-A, I-B, I-C, I-U, I-D, I-E and I-F). Subtypes I-A, I-B and I-C contain Cas5, Cas7 and Cas8 loci that have the ability to unwind DNA and RNA–DNA complexes in order to facilitate double-stranded breaks (DSBs) at the target site. The different subtypes have effector molecules made up of different components. Further, Type I systems can only target DNA.
Type III: It is further divided into four subtypes, and all of these subtypes are having Cas10 as common in their effector module. Cas10 encodes something similar to an RNA recognition motif called Palm and a cyclase domain responsible for cutting. Type III systems can recognize both DNA and RNA for cleavage. Although there are far fewer subtypes of Type III than of Type I, each Type III subtype is more diverse in terms of the Cas operon structure.
Type IV: It has a small effector molecule but multiprotein having Cas5, Cas7 and Cas8. It lacks domains capable of target cleavage or spacer insertion.

4.4.2 Class 2 CRISPR Systems

Characterized by the occurrence of a single effector module and studied under three types and nine subtypes, they only represent 10% of the CRISPR loci. Types II, V and VI are the types of CRISPR systems in Class 2 (Cas9 is a Class 2 system). Unlike Class 1, they are only found in bacteria. These CRISPR/Cas systems can target both DNA and RNA, depending on the type.

Type II: The most frequently utilized genome engineering endonuclease is a Type II system. It is characterized by the presence of Cas9, as well as ancillary proteins Cas1 and Cas2. Cas9, the commonly used genome engineering endonuclease, is a Type II system. This type of endonuclease requires a tracrRNA to function. For the first time, Type II system was discovered in *Streptococcus pyogenes* where it was shown to specifically cut the DNA *in vitro* and in eukaryotic cells (Gasiunas et al. 2012; Cong et al., 2013; Jinek et al., 2012; Mali et al., 2013).
Type V: This type of CRISPR/Cas has Cas9 endonuclease as an effector module. Like Cas9, Cas12 targets DNA for editing. Similar to Type II systems, Type V also requires tracrRNA for function. There is the difference in the type of cut made by these two endonucleases as Cas9 makes blunt cuts in the vicinity of PAM (Jinek et al., 2012), while Cas12a and Cas12b make staggered cuts away from PAM at the end of the target DNA (Zetsche et al., 2015).
Type VI: Type VI is the only Class 2 system that targets RNA for editing. Cas13 is a Type VI endonuclease, and it only enables the editing of RNA. The ribo-endonuclease effector is very different structurally than other Class 2 effectors and also lacks RuvC domain, which is present in both Cas9 and Cas12 effector molecules.

4.5 COMPARISON AMONG SEQUENCE-SPECIFIC NUCLEASES

The discovery of DSBs, induced by enzymes, was made by Rudin et al. (1989), followed by its elaboration by Donoho et al. (1998). It changed the field of genome editing forever and led to the development of more precise and efficient HR-based targeting system. DSBs in the DNA of the nucleus naturally activate DNA repair machinery present in the cell and repair the breaks through two mechanisms, i.e. NHEJ (non-homologous end-joining) and HDR (homology-directed repair). Symington and Gautier studied the DSB end resection and repair pathway choice in detail in 2011.

Since the discovery of DNA repair system, many efforts have been made to develop sequence-specific nucleases (SSNs) that can catalyse genome editing in organisms by making DSBs. Mainly four types of programmable SSNs have been developed to date, which are described below:

a. Meganucleases
b. ZFNs
c. TALENs
d. CRISPR-associated (Cas) nucleases

DSBs in the genome are utilized as a site where any of the foreign DNA sequence can be inserted and made to express. Though in nature, DSBs are cytotoxic lesions, which cause mutation in the cell or sometime death of the cell due to an improper repairment. The cell naturally uses the two pathways to repair these lesions, i.e. NHEJ and HDR. The different types of CRISPR/Cas systems use either NHEJ or HDR as DSB repair pathway. The use of these two DNA repair pathways in the cell depends on the phase of the cell cycle and the nature of the ends made by SSNs, i.e. blunt or staggered. In Types I, II and III of CRISPR/Cas systems, it is the NHEJ pathway which is predominant, while in Types IV and V, HDR pathway is used to repair cuts in the DNA. The DSBs made by four types of SSNs can be repaired through the inherent NHEJ pathway in the cell, which results in an altered DNA sequence with the deletion of the DNA sequence or insertion of it. But if a donor DNA template is provided, the HDR repair pathway of the cell inserts a donor sequence into the locus and leads to a repair of the cut, thus leaving edited gene or target site.

4.5.1 Genome Editing by Non-Homologous End-Joining (NHEJ)

NHEJ method of DNA repair naturally occurs in both prokaryotic and eukaryotic cells; however, they differ in functional components (Lieber, 2010). 'Non-homologous end-joining' term was for the first time used by Moore and Harber (1996). This pathway requires nuclease to delete damaged DNA, polymerase to fill the gap with new DNA and a ligase to ligate DNA strands together. Ku is the first protein to bind the DNA after a break arising in it, and Ku bound to DNA is referred as Ku:DNA complex, which serves as a node at which the nuclease, polymerase and ligase of NHEJ can bind.

In a typical cell cycle, the NHEJ repair pathway is preferred because it repairs the breaks without a homologous repair template. It has made the method popular than HDR for gene disruption or new sequence insertion at the target sites. This method of DNA repair can make donor DNA sequence insertion in a homology-independent manner so one of its potential applications may be gene stacking in plant breeding to improve a plant genotype for better phenotypic expression.

4.5.2 Precision Genome Editing by Using HDR Pathway

Despite the wide application of NHEJ to knock out genes, it has been criticized due to low precision for genome engineering. HDR method is more precise and can be used for specific point mutations, insertions or replacement of target DNA sequences into the genome. HDR repair pathway is activated in the S and G2 phases of the cell cycle. For this pathway to function, there is a requirement of a DNA template with homology to the repair site. The template sequence could be from a sister chromatid or an exogenous DNA sequence. The exogenous DNA sequence can be a single-stranded DNA with a desired sequence modification to be incorporated at the break repair site (Salsman and Dellaire, 2017). Though it is a very precise method of gene targeting in plants, it is difficult to perform due to its low efficiency and the need of a donor template delivery in plant cells. But there are reports which say if the donor template concentration is enhanced in the cell, then HDR efficiency improves. Gemini virus replicons based on wheat dwarf virus (WDV) have been engineered to increase the number of copies of the donor template per cell, which improved the frequency of HDR-mediated gene insertion in many plant species (Gil-Humanes et al., 2017). There are few other reports of its application in plant breeding: Butt et al. (2017) have used chimeric sgRNA molecules (sgRNA along with repair template sequences) to increase the efficiency of HDR in rice.

For the first time, meganucleases were examined for genome editing. These SSNs are present in bacteria, archaea and eukaryotes naturally. Meganucleases consisting of a single protein identify a 12-nucleotide-long DNA sequence and make a cut leaving a DSB, which is repaired through NHEJ or HDR pathway (Silva et al., 2011). The meganucleases have been applied for genome editing of *Zea mays* and *Nicotiana* spp (Baltes and Voytas, 2015). The major limitation in their application as genome editing tools is the difficulty in changing their specificity. A brief comparison between the four major SSNs is described in Table 4.3.

Among these four SSNs, CRISPR systems are the most recent developments which have wide applications like gene knock-ins and knockouts, translocations and chromosomal deletions (Salsman and Dellaire, 2017). The CRISPR/Cas systems have the capability to target multiple sites simultaneously using multiple sgRNAs and single Cas9 or Cpf1 protein. This has been referred as multiplexing or multiplex editing by various researchers and is supposed to change the future of genome engineering. Various approaches have been suggested to achieve multiplexing most remarkable due to Cpf1 capability to process its own crRNA in plants (Wang et al., 2017b). Another application of CRISPR systems, which can change plant breeding methods, is their use for pyramid breeding owing to their ability to target multiple traits simultaneously.

TABLE 4.3
Brief Comparison among Four SSNs

	Meganucleases	ZFN	TALEN	CRISPR
Identified from	Microbial mobile genetic elements	Gene expression regulators in eukaryotes	Plant pathogenic bacterial species *Xanthomonas*	Adaptive immunity system of bacteria and archaea
Length of target site	14–40 bp (mostly 12 nt)	18–36 nucleotide (including 5–7 spacer bp)	50–60 nucleotide (including 14–18 bp)	19–22 nucleotide (mostly 20 nt)
Spacer nucleotide	No spacer	5–7 bp	2–24 bp (generally 14–16 bp)	No spacer
Binding principle	Protein–DNA	Protein–DNA	Protein–DNA	RNA–DNA
Accuracy (off-target rate)	Specific, with no off-target effects. Unable to target sequences located on separate homologous or heterologous chromosomes	Target any genomic sequence with very few off-target effects	Target any genomic sequence with limited off-target effects	Target multiple sites altogether but target selection needs PAM sequence
Recognition rule	Complex	1 module per 3 bp	1 module per 1 bp	1 base per 1bp

(*Continued*)

TABLE 4.3 (*Continued*)
Brief Comparison among Four SSNs

	Meganucleases	ZFN	TALEN	CRISPR
Size of mRNA transcript	Short	Short	Long	Long
Efficiency	Low due to the need to first introduce a known cleavage site into the region of interest	Low	Relatively low	Highly efficient
Multiplexing	Very difficult	Very difficult	Limited	Highly feasible
Design flexibility	No flexibility	Flexible design and specificity improved by adding three to six fingers to the left and right of zinc finger arrays	Flexible design and specificity improved by the addition of 12–21 repeat variable diresidue (RVD) units to the right and left of TALE units	It has no such flexibility in design
Assembly	Difficult to modify, binding and restriction sites in the same unit	Expensive assembly and inefficient selection-based methods	Easier assembly due to kits and cloning reagents at moderate price	Inexpensive and easy
Specificity	Due to target recognition domain/nuclease domain	Improved by heterodimerization of FokI nuclease domains	Same as ZFNs	CrRNA/Cas9 protein single unit used so high chances of its off-target binding
Mutagenic throughput	Decrease as the distance from initial DSB increases	Low	Moderate	High
Core component	Nuclease protein	ZFA-FokI fusion protein molecules	TALE-FokI fusion protein molecules	sgRNA and Cas9
Work mode/ delivery	One meganuclease for one target site	A pair of ZFN for one target site	One pair of TALEN for one target site	Single or multiple sgRNA and only one Cas9 nuclease
Design	Difficult to design	Moderate	Easy	Very easy
Construction	Difficult and complex construction	Difficult, protein engineering for every single target	Relatively easy	Very easy
Targeting seq	Difficult to change the target sequence specificity	Specificity to bind a DNA sequence is easily manipulated	Easily manipulated to bind the specific sequence of DNA	Very easily moulded to bind to a particular sequence
Affordability	Costly and laborious	Costly and laborious	Affordable but time-consuming	Affordable

4.6 ADVANCES IN GENOME EDITING

4.6.1 Anti-CRISPR

Till now we have studied that CRISPR/Cas9 is a powerful adaptive immunity mechanism in bacteria against bacteriophages. But the highly mutating genome of the bacteriophage viruses does not let cleave their genome so easily. After 2013, many reports spurted the claiming presence of genes encoding anti-CRISPR proteins (Arcs) to silence the CRISPR/Cas systems (Hynes et al., 2017; Harrington et al., 2017). Alexander Hynes and Genevieve detailed about the phage genes responsible for blocking the CRISPR/Cas9 system. While cloning phage genome (>80% covered), they came across AcrIIA5 gene encoding 140-amino acid protein and having a coiled motif. This gene product blocks the nuclease action of Cas9 from *S. pyogenes* and thus protects the phage DNA or RNA. Lucas Harrington and Kevin Doxzen tested the specificities of the proteins encoded by AcrIIC1 and AcrIIC3 through biochemical assays, and the structure of the protein of AcrIIC1 gene revealed that proteins differ in host selectivity and mechanism of action. The proteins from AcrIIC3 genes prevent Cas9 binding to the target DNA sequence; this gene is specific to *N. meningitidis*; on the contrary, AcrIIC1 gene products inhibit the cleavage of the DNA bound to Cas9. Thus, Acrs use as 'off-switch' will eliminate the off-target effects of Cas9, which is a major limitation of the CRISPR/Cas9 system presently.

4.6.2 Prime Editing

Prime editing is a strategy to introduce deletions and insertions (indels) and developed newly by Anzalone et al. (2019). In most of the human diseases, it remains difficult to introduce desired mutations correctly due to INDEL formation upon the repairment of Cas-generated DSBs. Off-targeting has been also a problem, which has been reported in many recent studies performed on mice and zebrafish (Auer et al., 2014). CRISPR-Cas9-based 'prime editing' solves this problem by counting on the process of reverse transcription to edit the genes. In general, Cas9 ability to target a sequence is due to guide RNAs (sgRNAs). Anzalone and his colleagues disabled nickase form of Cas9 such that it could not form DSB by a reverse transcriptase. They edited sgRNAs (known as pegRNAs or prime editing guide RNAs here) to serve as a template for precise editing through reverse transcription. Upon the induction of a nick by Cas9 domain on target sequence, the reverse transcriptase starts the polymerization of the edited DNA segment from the pegRNA template into the template DNA strand. Thus, a heteroduplex with the non-edited strand is formed, which is further converted into the edited sequence by the DNA repair pathways. The prime editing system is testified to accommodate ≥30-bp-long RT templates, thus eliminating the constraints of specificity to target a site, by PAM sequences' availability in the vicinity of the CRISPR array.

4.6.3 Epigenome Editing

Hereditary information carrier is not only DNA and RNA, but there are other mechanisms like methylation, demethylation and hydroxyl-methylation of DNA, histone modification, non-coding RNA and chromatin remodelling, which inherit information from parents to the progeny. Broadly, these are referred as epigenetic markers and play an important role in the formation of Barr bodies, neurodegeneration, tumour formation, genome imprinting and gene regulation ultimately.

ZFN and TALEN, fused with trans-activators, are the feasible methods for epigenome editing but as we know, the binding specificity of these two is predominantly affected by amino acid sequences of their repeat domains, which ultimately implies to the fact that to change a genomic target every time the amino acid sequence has to be changed. This protein synthesis step meant to change amino acid sequence is very costly and laborious. CRISPR/Cas9 system is not only cost-effective but can also be manipulated easily because simply we have to redesign the sgRNA sequence and synthesize the new expression cassette. Here, the main focus stays on the Cas9 protein fusion with the transcription activator or repressor domain (epieffector or epigenetic effector).

Epigenome editing using CRISPR/Cas9 is based on the DNA-binding ability of the Cas9 protein. Cas9 nuclease is deactivated (dCas9) and can only bind with the DNA. The whole complex is referred as dCas9–epieffector fusion complex and proving itself an efficient epigenome editing tool. Xie et al. (2018) have reviewed the epigenetic techniques provided by CRISPR/Cas9.

4.7 CRISPR/Cas EDITING IN PLANTS

Humans always look for the plants with the best traits in and carry forward their population as a variety. Selection of the best plants for further generations has been an old age practice. With the publication of remarkable studies of Gregor J. Mendel, people started looking for genes responsible for specific traits and their genetics in plants. A whole new branch of science, *i.e.* plant breeding, is now based on the controlled hybridization of the best plants and selection among the progenies. Earlier plant breeding methods like cross-breeding (8–10 years), transgene breeding (8–12 years) and mutation breeding (8–10 years) take sufficiently large amount of time with the drawbacks like undesirable traits/genes transfer. The efficacy of plant breeding to provide higher-yielding varieties with specific quality traits, disease and pest resistance in a short span of time has been seriously questioned. Thus, there was a dire need for target-specific techniques, which can transform plants in a short period of time so that the time to develop a variety reduced significantly. Scientists tried and developed genome editing techniques (4–6 years) for traits modification and discovered many tools like meganucleases, ZFNs, TALENs and the most recent CRISPR/Cas9. Since the discovery of CRISPR, it has been studied and applied vastly by scientists in plants and animals' modification. In this section, we will deal with the approaches used to transform plants to produce both Cas9 and a target-specific gRNA. There are two major CRISPR/Cas systems used for plant genome engineering: i.e. Cas9 and Cpf1. In CRISPR/Cas9, the protein targeted is Cas9, and in CRISPR/Cpf1, it is the Cpf1. A Cas9-sgRNA can be engineered such that it can target any virus causing disease in plants. This engineered complex is responsible for producing a targeted immune response leading to host resistance against the DNA virus.

4.7.1 Applications of CRISPR/Cas9 for Improvement of Crop Plants

This laboratory-based method is producing significant results in the field in many crops by providing them immunity against many deadly pathogens like Gemini viruses. This technique has subjugated the crop improvement field in the past decade. Some of the applications are mentioned in Table 4.4.

Hybrid of crRNA with tracrRNA forms a complex with the Cas9 protein and cleaves only those target DNA sequences which are adjacent to short sequences known as protospacer-adjacent motifs (PAMs). Cas9 has two conserved nuclease domains, i.e. an HNH nuclease domain (cleaves the strand complementary to the crRNA) and a RuvC-like nuclease domain (which makes cut at the non-complementary strand).

4.7.2 Procedure of CRISPR/Cas9-Mediated Genome Editing in Crop Plants

In this section, we have briefly described the procedure for the target site selection and designing of sgRNA for CRISPR/Cas9-mediated genome editing in plants. CRISPR/Cas9 system makes a possible regeneration of genome-edited plants within 13–20 weeks. A brief workflow for genome editing using CRISPR/Cas9 is explained in Figure 4.1.

4.7.2.1 Cas9 Target Site Selection

RNA-guided Cas9 targeting specificity has been studied by Hsu et al. (2013). There are three major components that need to be observed while selecting for the target site.

TABLE 4.4
Recent Applications of CRISPR/Cas9 in Plants

Scientists	Crops	Genes targeted	Remarks
Jiang et al. (2013b)	Rice	OsSWEET14 OsSWEET11	Bacterial blight susceptibility addressed
Wang et al. (2014)	Wheat	TaMLO-A1, TaMLO-B1 TaMLO-D1	Powdery mildew resistance
Li et al. (2018)	Sorghum	k1C	Kafirin levels reduced in seed, leading to the improvement in its protein quality and digestibility
Chandrasekaran et al. (2016)	Cucumber	eIF4E	Resistance to broad virus
Jia et al. (2016)	*Arabidopsis*	CBF1and CBF1/CBF2	Cold tolerance
Li et al. (2016)	Rice	Gn1a and DEP1	Grain number per spike increased
		GS3	Larger grain size
		IPA1	Dense erect panicles
Gomez et al. (2019)	Cassava	elF4E isoforms nCBP-1 and nCBP-2	Improved resistance in cassava against brown streak disease
Wang et al. (2017a)	Tomato	SlMAPK3	Enhanced drought tolerance
Lou et al. (2017)	Rice	SAPK2	ABA-sensitive & drought stress-tolerant phenotypes
Shi et al. (2017)	Maize	ARGOS8	Improvement in grain yield under drought
Jia et al. (2017)	Citrus	CsLOB1	Canker resistance in citrus
Soyk et al. (2017)	Tomato	SP5G	Reduced days to flowering and improvement in determinate growth habit of tomatoes in the field
Andersson et al. (2017)	Potato	GBSS	Reduced amylose content and increased amylopectin/amylose ratio
Kim et al. (2018)	Wheat	TaDREB2 and TaERF3	Provide an insight about the functioning of genes in abiotic stress response
Cai et al. (2018)	Soya bean	GmFT2a	Delayed flowering time
Wang et al. (2018)	Grape	VvWRKY52	Enhanced resistance to *Botrytis cinerea*
Zhang et al. (2018)	Rice	Waxy	Decreased amylose content
Tripathi et al. (2019)	Plantain cv. Gonja Manjaya (cooking banana)	eBSV	Inactivation of eBSV (integrated endogenous banana streak virus in B genome)

a. 5′-NGG-PAM sequence should be always present at the end of 20-bp target sequence. It is very essential for sgRNA to identify the PAM sequences, then only it will be able to guide Cas9 to the target site. sgRNA identify the PAM sequences by making Watson–Crick base pairing in 20-bp target sequence.
b. If gene function disruption is the aim, then avoid the target sequences present at 3′ end of the introns. Because targeting these sequences would not interrupt the gene function.
c. To minimize the off-target effects, the BLAST search of genome for 22 nucleotide sequences (20 nt sgRNA and 2 nt NGG-PAM) is found helpful in many studies. It ensures that we have a unique target sequence to start with the procedure. Over the time, many other web-based tools also have been developed like 'CRISPRPLANT' (http://www.genome.arizona.edu/crispr/) and 'CasOT' (http://eendb.zfgenetics.org/casot/) to select unique target sites in major crop plants (Xiao et al., 2014; Xie et al., 2014).

Selection of desired target sequence on genomic DNA

⇩

Construction of sgRNA (Bioinformatics tools used for construction and promoter such as U3p or U6p used for its expression)

⇩

Insertion and cloning of sgRNA-oligos into sgRNA-scaffold vectors for transient expression in protoplasts (4days)

⇩

Protoplast isolation (1day)

⇩

In-vitro protoplast transformation using PEG and genomic DNA extraction (2–3 h hands-on time; 2 days total time)

⇩

Mutation detection in protoplast (3 days)

⇩

Stable transformation and regeneration of plants (13- 17 weeks)

⇩

Detection and Sequencing of indels in edited plants (3 days)

FIGURE 4.1 Procedure of CRISPR/Cas9-facilitated genome editing by transforming protoplast in plants.

4.7.2.2 sgRNA Scaffold Vectors' Designing and Cloning

As one selects the target site as per their experiment, forward and reverse oligos can be designed. We have discussed earlier that the oligos are 20-bp target sites with 5′ and 3′ overhangs complementary to the digested scaffold vector. Shan et al. (2014) have explained the stepwise procedure for the cloning of target-specific sgRNA oligos into sgRNA scaffold vectors for a transient expression in protoplasts. Here, we provide you a brief account of the procedure for the sgRNA cloning.

a. sgRNA oligos (target-specific reverse and forward) are designed.
b. The dilution of each reverse and forward oligo is prepared by keeping in mind the final concentration.
c. A mixture of annealing buffer and forward and reverse oligos is used for annealing of oligos in a thermocycler at a particular temperature for a fixed time.

d. Digest the scaffold plasmid like pTaU6-sgRNA for wheat and pOsU3-sgRNA for rice with buffer (like AaRI) in a reaction mixture.
e. The reaction mixture digestion is carried out at 37°C for 3–16 hours. As the time for digestion is enhanced, the better results are achieved.
f. Agarose gel electrophoresis of the digested sample on a 1.2% (wt/vol) agarose gel in TAE buffer describes the results of effective digestion. The bands are cut out under UV light to purify the vector DNA from them.
g. Ligation of the annealed sgRNA oligos and digested vector is carried out in the presence of ligation buffer, ligase and ddH_2O. The reaction mixture incubation for an hour at 22°C completes the step.
h. Competent *E. coli* cells are used to transform the ligation reactions as per the manufacturer's instructions. The cells are plated onto ampicillin-containing LB plates and left overnight for incubation at 37°C.
i. Two to three colonies are enough, to get DNA for plasmid sequencing, which is important to verify that the clones have the correct sgRNA.
j. After verifying the correct sgRNA present in clones, the correct clones are used to isolate plasmids. It takes 1 day for protoplast isolation alone. The protoplast quantity and quality are quite crucial for a transient protoplast transformation. Shan and his colleagues have provided a clear description of plasmid extraction in rice and wheat in 2014.

4.7.2.3 PEG-Mediated Transformation of Protoplasts and Extraction of Genomic DNA

It takes 2–3 hours hands-on time and 2 days of the total time. Generally, a transformation mix, having sgRNAs, is prepared and protoplasts are added to it. One more important constituent of this mix is PEG, which is freshly prepared. After the process is over, DNA is extracted from protoplasts and used for the detection of mutations.

4.7.2.4 Mutation Detection in Protoplasts

Control samples are first used to test each pair of primers. A reaction for PCR is set up using high-fidelity polymerase like Pfu polymerase, and the control samples are used as negative controls here. Agarose gel electrophoresis (1.2%) prepared in TAE buffer is used to check the mutation in the protoplast samples.

4.7.2.5 Stable Transformation in Rice with the Help of Biolistic Bombardment Method Takes 13–17 Weeks

It uses 6- to 9-week-old embryogenic calli to transform. After the completion of transformation step, the calli are transferred to recovery media, which is followed by incubation of these calli in a selection medium. These calli are regenerated in rooting medium and shooting medium. Finally, the hardening of the regenerated plantlets is carried out before exposing them to field conditions.

4.7.2.6 In the Final Step, DNA from the Transformed Plants Is Extracted and Subjected to PCR Amplification

Check the PCR products on agarose gel to ensure amplification, and sequence the products to determine the precise nature of mutation.

Protocols for CRISPR-mediated genome editing in plants by protoplast transformation in rice and wheat have been studied by Shan et al. (2014). Liang et al. (2018) and Wang et al. (2014) have also described the application of this method of genome editing.

4.8 FUTURE PROSPECTS

CRISPR/Cas9 system is an RNA-guided genome editing system and has risen as a versatile and robust tool for genome editing via point mutations, knockout, knock-in, epigenome editing and fine tuning of gene regulation. It is proving fast method of developing stress-tolerant, superior varieties in plants and is being utilized in a wide range of species owing to its cost-effectiveness and easy-to-use technology. It is widely being used as a tool for addressing diseases, abiotic stresses like heat, cold, drought and salinity; and in the improvement of quality traits of crop plants.

To improve the efficiency of this genome editing system, researchers need to work upon few aspects. Off-target minimization by monitoring the genome-wide presence of such target sequences and avoiding the selection of target sequences with homology to many other sites will improve the performance of the CRISPR/Cas9 method of genome editing. Some of the issues, like rapid discovery of the genetic basis of important traits, improving gene replacement and insertion rate and developing effective delivery methods for CRISPR/Cas9 delivery to the plants, can be solved with the help of synthetic biology approaches combined with genomics and phenomics of plants. Problems arising due to mindset of people, like genome-edited crops that have DNA sequences of bacteria or viral origin, hence called transgenic, may hamper their public acceptance. As this technique is capable of editing plants without inserting a foreign DNA sequence, it can solve the acceptance problem arising with transgenic crops. Other ethical and scientific problems arising with transgenic crops can be resolved by scientific community and governments, by coming together and developing a regulatory system for the use of the technique in human welfare.

REFERENCES

Anders, C., O. Niewoehner, A. Duerst, and M. Jinek, "Structural basis of PAM-dependent target DNA recognition by the Cas9 endonuclease." *Nature* 513, no. 7519 (2014): 569–573.

Andersson, M., H. Turesson, A. Nicolia, et al. "Efficient targeted multiallelic mutagenesis in tetraploid potato (Solanum tuberosum) by transient CRISPR-Cas9 expression in protoplasts." *Plant Cell Reports* 36, no. 1 (2017): 117–128.

Anzalone, A. V., P. B. Randolph, J. R. Davis, et al. "Search-and-replace genome editing without double-strand breaks or donor DNA." *Nature* 576 (2019): 149–157.

Auer, T. O., K. Duroure, A. De Cian, J. P. Concordet, and F. Del Bene. "Highly efficient CRISPR/Cas9-mediated knock-in in zebrafish by homology-independent DNA repair." *Genome Research* 24, no. 1 (2014): 142–153.

Baltes, N. J., and D. F. Voytas. "Enabling plant synthetic biology through genome engineering." *Trends in Biotechnology* 33, no. 2 (2015): 120–131.

Barrangou, R., C. Fremaux, H. Deveau, et al. "CRISPR provides acquired resistance against viruses in prokaryotes" *Science* 315, no. 5819 (2007): 1709–1712.

Bolotin, A., B. Quinquis, A. Sorokin, and S. D. Ehrlich. "Clustered regularly interspaced short palindrome repeats (CRISPRs) have spacers of extrachromosomal origin." *Microbiology* 151, no. 8 (2005): 2551–2561.

Brouns, S. J. J., M. M. Jore, M. Lundgren, et al. "Small CRISPR RNAs guide antiviral defense in prokaryotes." *Science* 321, no. 5891 (2008): 960–964.

Butt, H., A. Eid, Z. Ali, et al. "Efficient CRISPR/Cas9-mediated genome editing using a chimeric single-guide RNA molecule." *Frontiers in Plant Science* 8 (2017): 1441.

Cai, Y., L. Chen, X. Liu, et al. "CRISPR/Cas9-mediated targeted mutagenesis of GmFT2a delays flowering time in soya bean." *Plant Biotechnology Journal* 16, no. 1 (2018): 176–185.

Carte, J., R. Wang, H. Li, R. M. Terns, and M. P. Terns. "Cas6 is an endoribonuclease that generates guide RNAs for invader defense in prokaryotes." *Genes and Development* 22, no. 24 (2008): 3489–3496.

Chandrasekaran, J., M. Brumin, D. Wolf, et al. "Development of broad virus resistance in non-transgenic cucumber using CRISPR/Cas9 technology." *Molecular Plant Pathology* 17, no. 7 (2016): 1140–1153.

Chylinski, K., K. S. Makarova, E. Charpentier, and E. V. Koonin. "Classification and evolution of type II CRISPR-Cas systems." *Nucleic Acids Research* 42, no. 10 (2014): 6091–6105.

Cong, L., F. A. Ran, D. Cox, et al. "Multiplex genome engineering using CRISPR/Cas systems." *Science* 339, no. 6121 (2013): 819–823.

Deltcheva, E., K. Chylinski, C. M. Sharma, et al. "CRISPR RNA maturation by trans-encoded small RNA and host factor RNase III." *Nature* 471, no. 7340 (2011): 602–607.

Deveau, H., R. Barrangou, J. E. Garneau, et al. "Phage response to CRISPR-encoded resistance in Streptococcus thermophilus." *Journal of Bacteriology* 190, no. 4 (2008): 1390–1400.

Deveau, H., J. E. Garneau, and S. Moineau. "CRISPR/Cas system and its role in phage-bacteria interactions." *Annual Review of Microbiology* 64 (2010): 475–493.

Donoho, G., M. Jasin, and P. Berg, "Analysis of gene targeting and intrachromosomal homologous recombination stimulated by genomic double-strand breaks in mouse embryonic stem cells." *Molecular and Cellular Biology* 18, no. 7 (1998): 4070–4078.

Garneau, J. E., M. È. Dupuis, M. Villion, et al. "The CRISPR/Cas bacterial immune system cleaves bacteriophage and plasmid DNA." *Nature* 468, no. 7320 (2010): 67–71.

Gasiunas, G., R. Barrangou, P. Horvath, and V. Siksnys. "Cas9–crRNA ribonucleoprotein complex mediates specific DNA cleavage for adaptive immunity in bacteria." *Proceedings of the National Academy of Sciences* 109, no. 39 (2012): E2579–E2586.

Gil-Humanes, J., Y. Wang, Z. Liang, et al. "High-efficiency gene targeting in hexaploid wheat using DNA replicons and CRISPR/Cas9." *The Plant Journal* 89, no. 6 (2017): 1251–1262.

Gomez, M. A., Z. D. Lin, T. Moll, et al. "Simultaneous CRISPR/Cas9-mediated editing of cassava eIF 4E isoforms nCBP-1 and nCBP-2 reduces cassava brown streak disease symptom severity and incidence." *Plant Biotechnology Journal* 17, no. 2 (2019): 421–434.

Groenen, P. M., A. E. Bunschoten, D. V. Soolingen, and, J. D. V Errtbden. "Nature of DNA polymorphism in the direct repeat cluster of Mycobacterium tuberculosis; application for strain differentiation by a novel typing method." *Molecular Microbiology* 10, no. 5 (1993): 1057–1065.

Hale, C. R., P. Zhao, S. Olson, et al. "RNA-guided RNA cleavage by a CRISPR RNA-Cas protein complex." *Cell* 139, no. 5 (2009): 945–956.

Harrington, L. B., K. W. Doxzen, E. Ma. "A broad-spectrum inhibitor of CRISPR-Cas9." *Cell* 170, no. 6 (2017): 1224–1233.

Hoe, N., K. Nakashima, D. Grigsby, et al. "Rapid molecular genetic subtyping of serotype M1 group A Streptococcus strains." *Emerging Infectious Diseases* 5, no. 2 (1999): 254.

Horvath, P. and R. Barrangou. "CRISPR/Cas, the immune system of bacteria and archaea." *Science* 327, no. 5962 (2010): 167–170.

Horvath, P., D. A. Romero, A. C. Coûté-Monvoisin, et al. "Diversity, activity, and evolution of CRISPR loci in Streptococcus thermophilus." *Journal of Bacteriology* 190, no. 4 (2008): 1401–1412.

Hsu, P. D., D. A. Scott, J. A. Weinstein, et al. "DNA targeting specificity of RNA-guided Cas9 nucleases." *Nature Biotechnology* 31, no. 9 (2013): 827–832.

Hu, W., R. Kaminski, F. Yang, et al. "RNA-directed gene editing specifically eradicates latent and prevents new HIV-1 infection." *Proceedings of the National Academy of Sciences* 111, no. 31 (2014): 11461–11466.

Hynes, A. P., G. M. Rousseau, M. L. Lemay, et al. "An anti-CRISPR from a virulent streptococcal phage inhibits Streptococcus pyogenes Cas9." *Nature Microbiology* 2, no. 10 (2017): 1374–1380.

Ishino, Y., H. Shinagawa, K. Makino, M. Amemura, and A. Nakata. "Nucleotide sequence of the iap gene, responsible for alkaline phosphatase isozyme conversion in *Escherichia coli*, and identification of the gene product." *Journal of Bacteriology* 169, no. 12 (1987): 5429–5433.

Jansen, R., J. D. V. Embden, W. Gaastra, and L. M. Schouls. "Identification of genes that are associated with DNA repeats in prokaryotes." *Molecular Microbiology* 43, no. 6 (2002a): 1565–1575.

Jansen, R., J. D. V. Embden, W. Gaastra, and L. M. Schouls. "Identification of a novel family of sequence repeats among prokaryotes." *OMICS: A Journal of Integrative Biology* 6, no. 1 (2002b): 23–33.

Jia, Y., Y. Ding, Y. Shi, X. Zhang, Z. Gong, and S. Yang. "The CBFs triple mutants reveal the essential functions of CBFs in cold acclimation and allow the definition of CBF regulons in Arabidopsis." *New Phytologist* 212, no. 2 (2016): 345–353.

Jia, H., Y. Zhang, V. Orbović, et al. "Genome editing of the disease susceptibility gene Cs LOB 1 in citrus confers resistance to citrus canker." *Plant Biotechnology Journal* 15, no. 7 (2017): 817–823.

Jiang, W., D. Bikard, D. Cox, F. Zhang, and L. A. Marraffini. "RNA-guided editing of bacterial genomes using CRISPR-Cas systems." *Nature Biotechnology* 31, no. 3 (2013a): 233–239.

Jiang, W., H. Zhou, H. Bi, M. Fromm, B. Yang, and D. P. Weeks. "Demonstration of CRISPR/Cas9/sgRNA-mediated targeted gene modification in Arabidopsis, tobacco, sorghum and rice." *Nucleic Acids Research* 41, no. 20 (2013b): e188.

Jinek, M., K. Chylinski, I. Fonfara, M. Hauer, J. A. Doudna, and E. Charpentier. "A programmable dual-RNA–guided DNA endonuclease in adaptive bacterial immunity." *Science* 337, no. 6096 (2012): 816–821.

Karginov, F. V. and G. J. Hannon. "The CRISPR system: Small RNA-guided defense in bacteria and archaea." *Molecular Cell* 37, no. 1 (2010): 7–19.

Kim, D., B. Alptekin, and H. Budak. CRISPR/Cas9 genome editing in wheat. *Functional and Integrative Genomics* 18, no. 1 (2018): 31–41.

Koonin, E. V., K. S. Makarova, and F. Zhang. "Diversity, classification and evolution of CRISPR-Cas systems." *Current Opinion in Microbiology* 37 (2017): 67–78.

Li, M., X. Li, Z. Zhou, et al. "Reassessment of the four yield-related genes Gn1a, DEP1, GS3, and IPA1 in rice using a CRISPR/Cas9 system." *Frontiers in Plant Science* 7 (2016): 377.

Li, A., S. Jia, A. Yobi, et al. "Editing of an alpha-kafirin gene family increases, digestibility and protein quality in sorghum." *Plant Physiology* 177, no. 4 (2018): 1425–1438.

Liang, Z., K. Chen, Y. Zhang, et al. "Genome editing of bread wheat using biolistic delivery of CRISPR/Cas9 in vitro transcripts or ribonucleoproteins." *Nature Protocols* 13, no. 3 (2018): 413.

Lieber, M. R. "The mechanism of double-strand DNA break repair by the nonhomologous DNA end-joining pathway." *Annual Review of Biochemistry* 79 (2010): 181–211.

Lillestøl, R., P. Redder, R. A. Garrett, and K. I. M. Brügger. "A putative viral defence mechanism in archaeal cells." *Archaea* 2, no. 1 (2006): 59–72.

Lou, D., H. Wang, G. Liang, and D. Yu. "OsSAPK2 confers abscisic acid sensitivity and tolerance to drought stress in rice." *Frontiers in Plant Science* 8 (2017): 993.

Makarova, K. S. and E. V. Koonin. "Evolution and classification of CRISPR-Cas systems and cas protein families." In: Barrangou, R. and van der Oost, J. (eds), *CRISPR-Cas Systems*, pp. 61–91. Springer, Berlin, Heidelberg (2013).

Makarova, K. S., L. Aravind, N. V. Grishin, I. B. Rogozin, and E. V. Koonin. "A DNA repair system specific for thermophilic Archaea and bacteria predicted by genomic context analysis." *Nucleic Acids Research* 30, no. 2 (2002): 482–496.

Makarova, K. S., N. V. Grishin, S. A. Shabalina, Y. I. Wolf, and E. V. Koonin. "A putative RNA-interference-based immune system in prokaryotes: Computational analysis of the predicted enzymatic machinery, functional analogies with eukaryotic RNAi, and hypothetical mechanisms of action." *Biology Direct* 1, no. 1 (2006): 7.

Makarova, K. S., Y. I. Wolf, O. S. Alkhnbashi, et al. "An updated evolutionary classification of CRISPR–Cas systems." *Nature Reviews Microbiology* 13, no. 11 (2015): 722–736.

Mali, P., Yang, L., K. M. Esvelt, et al. "RNA-guided human genome engineering via Cas9." *Science* 339, no. 6121 (2013): 823–826.

Marraffini, L. A. and E. J. Sontheimer. "CRISPR interference limits horizontal gene transfer in staphylococci by targeting DNA." *Science* 322, no. 5909 (2008): 1843–1845.

Marraffini, L. A. and E. J. Sontheimer. "CRISPR interference: RNA-directed adaptive immunity in bacteria and archaea." *Nature Reviews Genetics* 11, no. 3 (2010): 181–190.

Masepohl, B., K. Görlitz, and H. Böhme. "Long tandemly repeated repetitive (LTRR) sequences in the filamentous cyanobacterium Anabaena sp. PCC 7120." *Biochimica et Biophysica Acta (BBA)-Gene Structure and Expression* 1307, no. 1 (1996): 26–30.

Mojica, F. J. M., C. Ferrer, G. Juez, and F. Rodriguez-Valera. "Long stretches of short tandem repeats are present in the largest replicons of the Archaea Haloferax mediterranei and Haloferax volcanii and could be involved in replicon partitioning." *Molecular Microbiology* 17, no. 1 (1995): 85–93.

Mojica, F. J., C. Díez-Villaseñor, E. Soria, and G. Juez. "Biological significance of a family of regularly spaced repeats in the genomes of archaea, bacteria and mitochondria." *Molecular Microbiology* 36, no. 1 (2000): 244–246.

Mojica, F. J., J. García-Martínez, and E. Soria. "Intervening sequences of regularly spaced prokaryotic repeats derive from foreign genetic elements." *Journal of Molecular Evolution* 60, no. 2 (2005): 174–182.

Moore, J. K. and J. E. Haber. "Cell cycle and genetic requirements of two pathways of nonhomologous end-joining repair of double-strand breaks in Saccharomyces cerevisiae." *Molecular and Cellular Biology* 16, no. 5 (1996): 2164–2173.

Nakata, A. T. S. U. O., M. I. T. S. U. K. O. Amemura, and K. O. Z. O. Makino. "Unusual nucleotide arrangement with repeated sequences in the *Escherichia coli* K-12 chromosome." *Journal of Bacteriology* 171, no. 6 (1989): 3553–3556.

Nishimasu, H., F. Ann Ran, P. D. Hsu, et al. "Crystal structure of Cas9 in complex with guide RNA and target DNA." *Cell* 156 (2014): 935–949.

Pawluk, A., N. Amrani, Y. Zhang, et al. "Naturally occurring off-switches for CRISPR-Cas9." *Cell* 167, no. 7 (2016): 1829–1838.

Pourcel, C., G. Salvignol, and G. Vergnaud. "CRISPR elements in Yersinia pestis acquire new repeats by preferential uptake of bacteriophage DNA, and provide additional tools for evolutionary studies." *Microbiology* 151, no. 3 (2005): 653–663.

Rudin, N., E. Sugarman, and J. E. Haber. "Genetic and physical analysis of double-strand break repair and recombination in Saccharomyces cerevisiae." *Genetics* 122, no. 3 (1989): 519–534.

Salsman, J. and G. Dellaire. "Precision genome editing in the CRISPR era." *Biochemistry and Cell Biology* 95, no. 2 (2017): 187–201.

Sapranauskas, R., G. Gasiunas, C. Fremaux, R. Barrangou, P. Horvath, and V. Siksnys. "The Streptococcus thermophilus CRISPR/Cas system provides immunity in *Escherichia coli*." *Nucleic Acids Research* 39, no. 21 (2011): 9275–9282.

Shan, Q., Y. Wang, J. Li, and C. Gao. "Genome editing in rice and wheat using the CRISPR/Cas system." *Nature Protocols* 9, no. 10 (2014): 2395–2410.

Shi, J., H. Gao, H. Wang, et al. "ARGOS 8 variants generated by CRISPR-Cas9 improve maize grain yield under field drought stress conditions." *Plant Biotechnology Journal* 15, no. 2 (2017): 207–216.

Shmakov, S., A. Smargon, D. Scott, et al. "Diversity and evolution of class 2 CRISPR–Cas systems." *Nature Reviews Microbiology* 15, no. 3 (2017): 169–182.

Silva, G., L. Poirot, R. Galetto, et al. "Meganucleases and other tools for targeted genome engineering: perspectives and challenges for gene therapy." *Current Gene Therapy* 11, no. 1 (2011): 11–27.

Soyk, S., N. A. Müller, S. J. Park, et al. "Variation in the flowering gene SELF PRUNING 5G promotes day-neutrality and early yield in tomato." *Nature Genetics* 49, no. 1 (2017): 162–168.

Symington, L. S. and J. Gautier. "Double-strand break end resection and repair pathway choice." *Annual Review of Genetics* 45 (2011): 247–271.

Tripathi, J. N., V. O. Ntui, M. Ron, S. K. Muiruri, A. Britt, and L. Tripathi. "CRISPR/Cas9 editing of endogenous banana streak virus in the B genome of Musa spp. overcomes a major challenge in banana breeding." *Communications Biology* 2, no. 1 (2019): 1–11.

Van der Oost, J., M. M. Jore, E. R. Westra, M. Lundgren, and S. J. Brouns. "CRISPR-based adaptive and heritable immunity in prokaryotes." *Trends in Biochemical Sciences* 34, no. 8 (2009): 401–407.

Wang, L., L. Chen, R. Li, et al. "Reduced drought tolerance by CRISPR/Cas9-mediated SlMAPK3 mutagenesis in tomato plants." *Journal of Agricultural and Food Chemistry* 65, no. 39 (2017a): 8674–8682.

Wang, M., Y. Mao, Y. Lu, X. Tao, and J. K. Zhu. "Multiplex gene editing in rice using the CRISPR-Cpf1 system." *Molecular Plant* 10, no. 7 (2017b): 1011–1013.

Wang, Y., X. Cheng, Q. Shan, et al. "Simultaneous editing of three homoeoalleles in hexaploid bread wheat confers heritable resistance to powdery mildew." *Nature Biotechnology* 32, no. 9 (2014): 947–951.

Wang, X., M. Tu, D. Wang, et al. "CRISPR/Cas9-mediated efficient targeted mutagenesis in grape in the first generation." *Plant Biotechnology Journal* 16, no. 4 (2018): 844–855.

Xiao, A., Z. Cheng, L. Kong, et al. "CasOT: A genome-wide Cas9/gRNA off-target searching tool." *Bioinformatics* 30, no. 8 (2014): 1180–1182.

Xie, K., J. Zhang, and Y. Yang. "Genome-wide prediction of highly specific guide RNA spacers for CRISPR–Cas9-mediated genome editing in model plants and major crops." *Molecular Plant* 7, no. 5 (2014): 923–926.

Xie, N., Y. Zhou, Q. Sun, and B. Tang. "Novel epigenetic techniques provided by the CRISPR/Cas9 system." *Stem Cells International* 2018, no. 113 (2018): 1–12.

Zetsche, B., J. S. Gootenberg, O. O. Abudayyeh, et al. "Cpf1 is a single RNA-guided endonuclease of a class 2 CRISPR-Cas system." *Cell* 163, no. 3 (2015): 759–771.

Zhang, J., H. Zhang, J. R. Botella, and J. K. Zhu. "Generation of new glutinous rice by CRISPR/Cas9-targeted mutagenesis of the Waxy gene in elite rice varieties." *Journal of Integrative Plant Biology* 60, no. 5 (2018): 369–375.

5 Genome Editing for Improvement of Wheat and Millets

Ram Sewak Singh Tomar
Rani Lakshmi Bai Central Agricultural University

Sushma Tiwari
Rajmata Vijayaraje Scindia Krishi Vishwa Vidyalaya

Prabha Singh
ICAR-Indian Grassland and Fodder Research Institute

K. Bhojaraja Naik
ICAR-Indian Institute of Seed Research

Anil Kumar
Rani Lakshmi Bai Central Agricultural University

CONTENTS

5.1 INTRODUCTION

Wheat (*Triticum aestivum L.*) is a largely consumed cereal crop by most of the world population. It is the prime source of energy and staple food crop. Wheat production in India has increased to a record of 101.20 million tonne (MT) for the year 2018–2019. It is up by 1.3% from the previous year 2017–2018. It is the time where the overall food grain production marginally decreased from 285.01 MT in 2018–2019 to 283.37 MT in 2019–2020 (www.financialexpress.com). The wheat is a cereal grass that belongs to the genus *Triticum* and family Poaceae. It is one of the primordial and most supreme of the cereal crops. The uses of wheat are immense: bread is made by common wheat (*Triticum aestivum*), pasta and macaroni by durum wheat (*T. durum*) and cake, pastries and flours by softer-type club wheat (*T. compactum*). Besides, wheat is used in the manufacturing of products like malt, dextrose, gluten, alcohol and other products of industry. There are two major types of wheat depending on the season of cultivation that are winter and spring, the nutritional constitution of wheat grain varies with differences in climate and soil. However, on an average, the wheat kernel contains water (12%), carbohydrates (70%), protein (12%), fat (2%), minerals (1.8%) and crude fibres (2.2%) with small quantities of hormones like thiamine, riboflavin, niacin and vitamin A. The common wheat or chapatti-making wheat is mostly used for breadmaking. The wheat cultivated in dry climates generally has a protein content of 11%–15% and strong gluten (elastic protein), whereas wheat grown in humid conditions has a protein content of about 8%–10% and weak gluten. The wheat from these regions is suitable for making cakes, crackers, cookies, pastries and household flours. Durum wheat semolina (from the endosperm) is used for making pastas, or alimentary pastes.

The wheat crop exhibits an extensive adaptability to a wide range of agro-climatic conditions. But an ever-rising population and increase in demand with diminishing land is the major concern nowadays. The change in the climatic conditions in addition to increased biotic and abiotic factors is leading to aggravate its production and productivity. Due to the large and complex genome size of wheat, it was difficult and time-consuming process to completely sequence (IWGSC, 2018). The wheat genome sequence will facilitate in deciphering the genome function and loci involved in different plants related to various activities like developmental pathways, immune system, quality aspects etc. Wheat plants deal with several challenges in coping with biotic as well as abiotic stresses. The most prone is its susceptibility to fungal, bacterial, viral and other pathogen encounters in addition to drought, salt, cold, submergence stress as well as lodging problems. These factors lead to heavy yield losses. To characterize and study these traits is very difficult because of their complex nature and multi-genic inheritance. However, researchers have found out several traits whose loci and mechanisms are now known in other model crops. These mechanisms and pathways are used in wheat improvement programmes. It involves both conventional and modern molecular tools and techniques. Traditional crop improvement methods depend mainly upon environmental conditions in addition to labour-intensive and slow process. The use of modern tools and techniques, which are independent of environmental factors in association with conventional methods, will be feasible to develop a new and improved plant.

The modern plant improvement technology is an interdisciplinary approach, which uses both molecular genetic tools and methodologies with conventional approaches for crop improvement. Modern plant breeding strategies include marker-assisted selection (MAS), marker-assisted backcrossing (MABC), marker-assisted recurrent selection (MARS), genome-wide selection (GWS) or genome selection (GS). These methods of molecular plant breeding continue to evolve from time to time and are of great interest in crop improvement projects. The use of genetic markers is based on allelic forms of genes (loci), which are a hereditary unit that can be transferred from one generation to other. At times, the use of mutagenic agents to generate a variability in crop also plays a significant role. Nevertheless, it has also limitations like non-specificity and random nature, concurrent effect on large numbers of genes, changes in chromosomal structure etc. However, to restrict on these limitations, several genome editing systems comprising sequence-specific nucleases (SSNs) like zinc finger nucleases (ZNFs), transcription activator-like effector nucleases (TALENs), oligonucleotide-directed mutagenesis (ODM) and clustered regularly interspersed short palindromic repeats (CRISPER) systems have been evolved. These modern molecular techniques are easy, accurate and efficient. This is the reason various researchers and workers are endorsing these techniques and achieving success (Mao et al., 2018). Among the several genome editing technologies, CRISPR/Cas9 is an adaptable mechanism good for use in genome editing, widely in use because of its characteristic feature of high success rate. Multiple genes are targeted using CRISPR/Cas9 system based on the information of small RNAs (Doudna and Charpentier 2014). Nowadays, these techniques are commonly used in a variety of cereal crops (Zhu et al., 2017; Ansari et al., 2020), horticultural crops (Karkute et al., 2017), oilseed crops (Wayne et al., 2019) and several other crops for improving the number of traits. The genome editing has been done for various biotic or abiotic stress tolerance in cereals like *Triticum aestivum*, *Nicotiana benthamiana*, *Arabidopsis thaliana*, *Oryza sativa* and *Sorghum bicolor* (Kim et al., 2018; Kumar et al., 2019). The common wheat (*Triticum aestivum* L.) is comprised of a complex genome using A, B and D. Due to its hexaploid nature, it is an important model for genome editing system. It will serve as a model plant for studying and optimizing conditions in such crop plants. This chapter will focus on the applications of genome editing techniques in wheat having a polyploid genome, for its improvement.

5.2 STRATEGIES TO IMPROVE RESISTANCE TO BIOTIC STRESSES

Wheat crop is at risk to various biotic stresses like bacteria, fungi, viruses, insects, pests and herbicide toxicity. The impact of these stresses is so severe, and these stresses have consequences on the plant growth, development, seed quality and yield of the crop. To solve the problem of stresses perception and generate new solutions in exploiting DNA allowed the successful development of genetically modified crop plants. In the beginning, the biggest challenge was complexity of large genome, which was not possible through conventional strategies. Also, the introduction of foreign genes and development of genetically modified plants is another important achievement that was not possible through traditional methods of crop improvement. These approaches led to the inception of methods like genomics and proteomics, which helped in understanding the regulatory networks and development of genetically modified plants against biotic stresses (Agarwal and Rakwal, 2011).

Several biotechnological tools are available nowadays that can be used to develop resistance against these stresses by reducing loss and improving yield. This has led to the identification of genes and QTLs, cloning and transfer into wheat conferring resistance to various insects and pests. Recent emergence of genome editing technologies, specifically CRISPR/Cas9, has revolutionized the speed of functional characterization of genes that will ultimately lead to the generation of improved varieties having an enhanced tolerance against various pests and stresses (Liang et al., 2016). With this new technology, varietal improvement is possible that was limited due to the restricted availability of natural and mutant strains. This section mainly describes how genetically biotic stress tolerance in wheat and CRISPR/Cas9 technology can be efficiently applied in establishing resistance to these pests and pathogens by highlighting a significant progress achieved during the past years (Table 5.1).

TABLE 5.1
Application of Genome Editing Techniques in Wheat to Improve Biotic and Abiotic Stress Tolerance

S.N.	Target Gene	Target Trait	References
1	GW2	Increased grain weight and protein content	Zhang et al. (2016)
2	TaGW2	Decrease in grain size and weight	Bednarek et al. (2012)
3	TaGW2	Increases in the grain width and weight	Hong et al. (2014)
4	α-Gliadins	GM crop 'Butte 86'	Altenbach and Allen (2011)
5	γ-Gliadin	Technological properties of the dough	Gil-Humanes et al. (2012)
6	α-, γ- and ω-Gliadins	Low molecular weight (LMW) glutenins	Barro et al. (2016)
7	GTF1 and GTF2	Resistance to the powdery mildew fungus *Blumeria graminis*	Nowara et al. (2010)
8	FcGls1, FcFmk1, FcChsV	Resistance to *Fusarium graminearum*	Chen et al. (2015)
9	FaMyo5	*F. asiaticum* provided disease resistance	Song et al. (2018)
10	PsFUZ7	Resistance to stripe rust	Zhu et al. (2017)
11	PsCPK1	Resistance to stripe rust	Qi et al. (2018)
12	TaMLO	Powdery mildew resistance	Upadhyay et al. (2013)

(*Continued*)

TABLE 5.1 (*Continued*)
Application of Genome Editing Techniques in Wheat to Improve Biotic and Abiotic Stress Tolerance

S.N.	Target Gene	Target Trait	References
13	TaMLO	Powdery mildew resistance	Wang et al. (2014)
14	TaLpx-1	Resistance to *Fusarium graminearum*	Nalam et al. (2015)
15	TaLpx 1 and TaLox 2	Genome editing in wheat using the CRISPR/Cas system	Shan et al. (2014)
16	TaALS	Herbicide resistant	Zong et al. (2018)
17	EDR1	Powdery mildew resistance	Zhang et al. (2017)
18	TaGASR7, TaGW2, TaDEP1	Enhancing grain size and yield	Wang et al. (2018)

5.3 RECENT DEVELOPMENTS AT MOLECULAR LEVEL AGAINST BIOTIC STRESS TOLERANCE

The information of genes, expression and position of trait loci was explored using conventional techniques in hybridization, selection and hybrid breeding for selected traits. These breeding approaches completely rely on the available natural differences in the predominating gene pool. But with the development and introduction of new and advanced techniques at the molecular level by genome editing, the role of novel nucleases that targeted DNA double-stranded break (DSB) at the specific loci was disclosed. In view of the fast growth and potential connotations, numbers of research experiments were conducted in various plants to develop improved cereal crops with intense stress tolerance using genome editing. In order to manipulate the stress response pathways in wheat, it is a preliminary requirement to know the components playing an important role in such pathways. Several studies have contributed in understanding genes and regulatory network responsible for plant's defence against pathogens or insects. The crop counteracts the stress by down-regulating and up-regulating various genes and transcription factors as well as by managing the plant regulatory pathways. Jouanin et al. (2020) carried out a study on targeted gene editing using CRISPR/Cas9 on wheat gliadins resulting in producing progeny with features like silenced, deleted and edited gliadins. These edited gliadins will finally minimize the vulnerability of patients to coeliac disease (CD) epitopes. The CD is the most common disease associated with wheat in which an autoimmune reaction is endemic in nearly 1%–2% of the human population. Genome editing technology in wheat lines results in fewer gluten genes, or gluten genes with deactivated CD epitopes in plants result in a variability in edited and deleted genes. Based on the basic understanding of stress-responsive pathways and genes involved in them, suitable strategies can be developed for improving the stress tolerance in rice. CRISPR/Cas tools have been effectively used lately for this purpose, and the developments in this field are described in the section below.

5.3.1 Development of Resistance Against Fungal Diseases

In polyploidy genome, SSNs were used to modify and engineer the genetic composition resistance for various biotic and abiotic stresses. Wang et al. (2016) used TALEN and CRISPR-Cas9 to introduce specific mutations in mildew resistance locus (MLO), a homoeoallele in wheat. Due to genetic redundancy (i.e. hexaploid nature) in wheat for powdery mildew, it was difficult to search for allelic variation in all three MLO alleles but TALEN-induced mutation homoeologs in the single plant conferred heritable resistance. Then, CRISPR-Cas9 technology was used to develop genetically modified wheat with TaMLO-A1 allele. Crop improvement in wheat, a polyploid crop, is possible using the structural methodology.

5.4 IMPROVING RESISTANCE TO ABIOTIC STRESSES

Plant growth and development is severely affected by natural calamities like unpredicted rain, increase in pollution level at air, soil, and water quality with decreased soil fertility. Decrease in fertile land with an increase in population is also one factor responsible for low production. These abiotic stresses are responsible for low crop productivity. These abiotic stresses are more effective in a condition where drought is associated with salt stress resulting in extremely low yield (Gosal et al., 2010; Mohammadi et al., 2008). Previous works have been successfully proceeded so far to identify the key metabolic pathways, candidate genes, promoter motifs and QTLs involved in plant's response to abiotic stresses using transgenic and molecular breeding approaches. This section focuses on the most pertinent studies conducted in the area of abiotic stress in wheat and how genome editing technology leads to high productivity under stressed conditions.

During abiotic stress, plant undergoes a series of morphological changes, which include reduced photosynthesis, stomatal closure, increased root length, reduced leaf growth and increased reactive oxygen scavenging activity. At the molecular level, plants regulate these changes by a series of coordinated responses producing various protein classes involving enzymes, functional proteins, molecular chaperones and transcriptional factors (Usman et al., 2017). Abiotic stresses are also regulated by hormones like ethylene, abscisic acid and jasmonic acid. Mainly, two types of components work in stress response: a signalling component that controls the expression of stress-responsive genes and a functional component that actually protects plant cells from the harmful impact of stress (Todaka et al., 2012). Various abiotic stress-responsive genes have been recognized, which are expressed in such conditions. The list of genes including *Asr1*, *catalase*, *dehydrins*, *Lea*, *LIP9*, *LIP19*, *Myb* transcription factor, *RAB16*-C, *SalT*, *WS176* and *WS1724* are consistently observed to be up-regulated during abiotic stress in several studies. In addition to these genes, other important factors involved in abiotic stress response are aldehyde dehydrogenase, lipoxygenase, O-methyltransferase, RING finger protein, sugar transporter, zinc finger proteins, etc. (Rabbani et al., 2003). The response of plants against different abiotic stresses is a coordinated approach as the interlinked signalling pathways act against various stresses (Shinozaki and Yamaguchi-Shinozaki, 2000). Exposure of plants to abiotic stresses such as drought, high temperature, cold, alkalinity and application of ABA induced 15 genes commonly, implying that they are controlled with the same and also overlying signalling pathways (Rabbani et al., 2003). Many of these genes identified in rice breeding programmes have been utilized to generate stress-resistant wheat genotypes using the latest molecular tools and techniques. Since the inception of genome modification technology, it has been commonly used for improving abiotic stress tolerance in wheat.

5.4.1 Genome Editing to Improve Abiotic Stress Tolerance in Wheat

The targeted genome editing using CRISPR/Cas9 system is a successful and promising mechanism with proven efficiency in several species of plant (Gao et al., 2016). In polyploid plants, editing competence and inaccurate mutation rate is the problem (Peng et al., 2016). Reduction in yield is immense due to abiotic stresses such as drought, salt and micronutrient deficiency every year (Araus et al., 2008; Budak et al., 2013). Plants tackle against these conditions by employing an inclusive signal mechanism involving the expression of several genes (Baldoni et al., 2015). The activation/deactivation of complete molecular pathways for stress responses is based on transcription factors and expression of genes (Singh et al., 2002). Kim et al. (2018) conducted a study based on TaDREB2 and TaERF3 abiotic stress-responsive genes designated by the AP2/ERF domain (Lucas et al., 2011) for executing CRISPR/Cas9-based genome editing approach. Under drought stress, the role of DREB genes has been reported by many experiments in the past (Lucas et al., 2011; Morran et al., 2011). In wheat and barley, DREB proteins are in plenty and highly related to abiotic stress, particularly drought conditions especially in root tissue other than other plant parts (Lucas et al., 2011). The study confirmed that DREB2 and ERF3 are important genes which play an important role under

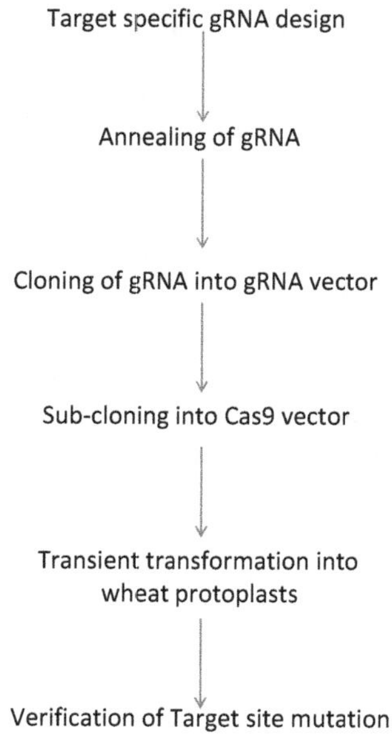

FIGURE 5.1 Procedures of genome editing using CRISPR/Cas9 in wheat.

abiotic stress conditions. The targeted genome editing showed a stable transformation and expression against stress reaction. The process involved the selection of the target sites having PAM (NGG) sequence at 3′-region and the synthesis of related oligos. These selected sgRNAs were then inserted into pJIT163-2NLSCas9 plasmid at TaU6 promoter site, resulting in sgRNA-Cas9 constructs for TaDREB2 and TaERF3. Low transformation efficiency was recorded as compared with other plant species (Shan et al., 2014). It may be due to the effect of wheat protoplast fragile nature as proven in the past (Sun et al., 2013). However, low transformation efficiency didn't affect the transformation rate, which was confirmed by mutation efficiency using the restriction enzyme digestion assay. TaDREB2 and TaERF3, two genes responsible for abiotic stress, showed the successful application of CRISPR/Cas9-based genome editing in wheat protoplast. The study established the CRISPR/Cas9 genome editing system with a transient expression of sgRNA and Cas9 protein in wheat protoplast. Its efficiency was proven with the use of restriction enzyme assay, T7 endonuclease and Sanger sequencing. This procedure confirms that CRISPR/Cas9 system is a methodical technique in manipulation in wheat genome leading to a crop improvement (Figure 5.1).

5.4.1.1 Drought Stress

Comparatively, reports of application of genome editing for improving abiotic stress are less than those of biotic stress in crops. Genome editing is primarily used for creating mutations, and therefore, it is largely used for developing biotic stress resistance by mutating susceptible genes in plants. On the other hand, genome editing is mainly being used for the functional characterization of genes involved in abiotic stress response. Nevertheless, the technique has been significantly used for understanding as well as improving abiotic stress tolerance in wheat. The main targets for genome editing to enhance abiotic stress tolerance are negative regulators playing a role in various cellular mechanisms like excessive ROS production, reduction of antioxidant activity, programmed cell death and disturbance of hormonal homeostasis (Zafar et al., 2020). The wheat ethylene-responsive factor 3 (TaERF3) and wheat dehydration-responsive element binding protein 2 (TaDREB2) were taken in a study where CRISPR/Cas9 genome editing system was applied. Restriction enzyme digestion assay, T7 endonuclease I assay and sequencing were used to confirm the targeted mutagenesis produced by CRISPR/Cas9. The *in silico* methods were used to detect the possible effect of off-targeting by developed sgRNAs, and amplicon sequencing was used to prove the targeted editing. The study proves that a combined approach of *in silico* off-target proofing and NGS can be used while editing of genes in polyploid plant genomes.

5.5 IMPROVEMENT IN YIELD USING GENOME EDITING TECHNOLOGY

The ultimate aim of agricultural research is to fulfil the demand of food by growing population through increasing crop productivity. With the abrupt change in climatic and soil conditions, production is declining day by day. The ultimate aim of all the breeding programmes is to enhance the yield of a crop by reducing effects of environmental stress as well as by increasing the water- and nutrient-use efficiency. The yield enhancement in wheat is possible only with the improvement consisting of combined approach of all the yield-associated traits like number of effective tillers in plant, number of grains in spike, spike length, grain length, width and weight, etc. Large number of genes and QTLs governing various traits directly or indirectly contributing to the yield of wheat have been identified and utilized in breeding programmes. Genome editing techniques can further strengthen the breeding programmes by validating the function and role of these identified genes/QTLs. The extremely specific and relatively cheaper technology can rightly serve as a best tool to improve the grain yield traits. The technology has been exploited to improve various aspects of grain yield, and this section briefly highlights it. For instance, one such study in wheat improvement was conducted by Wang et al. (2014). Several genes identified in rice for grain size and grain weight in rice were found to have a similar control in common wheat (Dong et al., 2014; Zhang et al., 2019). Wang et al. reported a gene in wheat TaGW7, which is ortholog to rice OsGW7 gene responsible for shorter and wider grains (Wang et al., 2015a). Gene-edited alleles of TaGW7 exhibited the similar phenotypic results to rice, producing the similar characteristics in wheat grain. The mutants of edited TaGW7 increased thousand grain weight (TGW) by almost three times in common wheat. However, it was not seen in rice line with the same gene (Wang et al., 2015b). These studies demonstrate that the important regulators of multiple traits can be altered in single cultivar using CRISPR/Cas9 technology. Thus, the genome editing technology can be exploited in dissecting the complex regulatory network of genes in the genetic background of the same cultivar and stacking the cultivated varieties with desirable traits. Thus, it opens up the new strategies to improve the yield of crops where genome editing can be efficiently used to target miRNAs in plant developmental pathways.

5.6 GENOME EDITING IN MILLETS

Current projections suggest that global population could reach 9 billion by 2050. To fulfil the growing food supply by the rising population, an estimate of 70% increase in global food production is projected by 2050. The conventional cereal crops might not be able to fulfil these requirements because the ultimate yield potential of most of these crops is currently at a plateau level. To replace the existing pressure on the conventional crops, millets are the best alternatives in near future considering food and nutritional reliability. By virtue of their exceptional nutritional profiles, they have a great potential to contribute to solve the malnourishment problems mainly in low-income, food-deficient developing countries. The use of C4 photosynthetic mechanism and potential to combat the moisture stress makes them a significant option for climate-resilient agriculture. In spite of enormous possibilities, attempts for the genetic enhancement of millets are substantially lagged behind the conventional cereal crops. Research done so far in millets has decoded their unique nutritive value, broad adjustability and phenotypic plasticity. Information on genetic mechanisms underlying the important agronomic traits of millets, including nutritional parameters, is limited. In the past, efforts were not done so efficiently to improve the crop using novel advanced molecular and genomic tools. However, with the availability of draft genome sequences in some millets, there is a wide scope for improvement. This chapter is an attempt to gather the meagre information available on the aspects of grain quality and nutraceutical development in millets by utilizing the potential of emerging genomics tools and techniques.

While the researchers and workers of the world have decided to bring 'zero hidden hunger' by 2030, around 868 million people worldwide are presently 'chronically undernourished'. This alarming condition is even increasing due to the marginal supply of nutri-enriched food to eat. People have less access to essential micronutrients, viz. iron, zinc, calcium, etc, in their diet. People from low-income countries from South-East Asia, Africa and parts of America basically rely on non-diversified staple crops for their nutritional requirements. However, a majority of these staple crops, including rice, wheat and maize, are often deficient in essential micronutrients and are unable to supply these nutrients to the people as per the recommended daily allowance (RDA) (Bouis and Welch, 2010; Saltzman et al., 2013; Tiwari et al., 2016). The nutritional security and health status of the growing population primarily depend on this plant diet. Nutrient-rich and balanced diet is prerequisite for sustainable growth, healthy living and development as the majority of the global population suffers from nutrition deficiency due to the intake of insufficient and malnutrition diet (Zhao and McGrath 2009). In under-developed and developing countries like parts of Asia and Africa where resources are limited, millets play an important role by providing ~75% of total calorie intake. These 'small-seeded grasses' include several species like finger millet (*Eleusine coracana*), pearl millet (*Pennisetum glaucum*), proso millet (*Panicum miliaceum*), foxtail millet (*Setaria italica*), barnyard millet (Echinochloa spp.), Kodo millet (*Paspalum scrobiculatum*) and little millet (*Panicum sumatrense*). These different species of millets are widely cultivated in Asian countries like India, Japan, China and some region of Korea for fodder as well as food. Being nutritionally rich similar to many crops like wheat, rice, maize and soya bean, millets include good level of nutraceuticals like dietary fibres, proteins, phosphorus, potassium, iron, calcium, zinc, vitamin B and several essential amino acids (Saleh et al., 2013).

Several procedures are available, which can enrich the crops with minerals. One of the promising approaches is selecting a crop or genotypes reactive to particular nutrients and triggering their molecular machinery for sequestrations to storage tissues. The relative responsiveness of genotypes to specific nutrients mainly depends upon their genetic potential for nutrient-use efficiency (NuUE) of crops (Liu et al., 2017). NuUE of crop plants needs to be enhanced so that these plants or cultivars would be able to produce higher yields per unit of nutrient applied under specified conditions. The vast genetic variability is found in millets compared to other cereal crops for certain minerals like zinc, iron and calcium (Muthamilarasan and Prasad 2015). These millets have characteristic of tolerance towards both biotic and abiotic stresses, which confer good insurance to the grower (Reddy et al., 2011). Minor millets have been given priority over other millets as the crop choice for quality improvement. However, the quality improvement can be achieved either by increasing the pilation of nutrients in grains or by enhancing the bioavailability of other minerals by decreasing the antinutritional property. After observing the similar work on other model crop plants, a similar work can be emphasized in millets using both traditional and conventional breeding methods in association with genomics-assisted tools and techniques.

5.7 GENOME-ENABLED APPROACHES FOR IMPROVING GRAIN QUALITY AND NUTRACEUTICAL TRAITS

Availability of polygenic variability is a prerequisite for any crop improvement programme. Phenotypic and nutritional characterization of small millets germplasm revealed an immense genetic variation for economic traits. However, further confirmation with molecular markers is required for a better understanding of allelic diversity to exploit the genetic variation more judiciously and precisely. The use of molecular markers in millets was very low in the past with little use of non-reproducible DNA marker system like random amplified polymorphic DNA (RAPD) for germplasm characterization (Patil and Kale 2013; Ramakrishnan et al., 2016; Rajendran et al., 2015). The unavailability of a significant number of molecular markers is still a limitation in millets grain quality improvement at the genomic level. Further exhaustive efforts are required for the development of new, highly polymorphic microsatellite markers that will cover the regions in the genome, which are currently poorly covered with available microsatellites (Babu et al. 2014; Rajendran et al., 2015). Exhaustive germplasm characterization for economic traits, including grain nutritional profile with next-generation sequencing (NGS)-based markers, is required, using diversity panels and core collections extracted from global germplasm collections, including diverse landraces, wild relatives for its implementation in breeding programmes. However, germplasm characterization studies utilizing NGS-based high-throughput discovery of DNA markers, especially single nucleotide polymorphisms (SNPs), are yet to be implemented in small millets on a large scale. The availability of a well-assembled reference genome has solved many problems. The physical map of finger millet genome has a total number of scaffolds of 1897 with a N50 length of 2.6 Mb and 96% single-copy orthologs. The draft genome of finger millet was released earlier to genome assembly. Likewise, in the foxtail millet genome sequence, the total length of all scaffolds is 423 Mb, with 28 Mb (6.6%) gaps (Zhang et al., 2012). The proso millet genome sequence is known to possess 55,930 protein-coding genes and 339 microRNA genes. However, genome sequence information for many nutritionally important minor types of millet such as barnyard millet, Kodo millet and little millet is still awaited.

With the advent of NGS technology, the identification of genome-wide nucleotide variations in various crops has become possible. To date, scanty information is available in small millets for uncovering the QTLs and genes governing grain quality and nutraceutical traits and their inheritance pattern. However, in some of the millet species, advances in molecular marker discovery and genotyping assays have undeciphered genomic regions governing the grain quality traits. For instance, microsatellite marker-based association mapping leads to the identification of agronomic traits, including grain quality in millets. In small millets, only the association mapping of diversity panels based on a limited set of markers was used for QTL studies. While efforts are underway in generating genomic resources in small millets, biparental and multiparental mapping populations are lacking, which can contribute in crop improvement efforts.

Considerable advances have been made in the genomic interventions in many orphan crops, through comparative genomics approaches. The genome synteny and functional similarity using comparative genomics is an approach to understand closely related species for detecting the important genes governing trait of interest. Recent availability of high-quality reference genome sequences in some of the small millets like foxtail millet (Zhang et al., 2012), finger millet and proso millet will allow a detailed structural and functional comparison with model cereal genomes such as rice for genes regulating various essential micronutrients and health-promoting bioactive compounds. Such genomic comparisons have already demonstrated their utility in identifying the genomic regions in small millets.

Recent advances in low-cost genome sequencing and array-based technologies have opened new avenues for the development of SNPs and facilitated high-throughput automated genotyping across the crop plants. Considering their immense utility in small millets genomics, the genetic make-up of global barnyard millet collection through SNP genotyping and genome editing is much more important. Use of microprojectile bombardment method for plant transformation has been reported in some of the millets.

5.8 CONCLUSION

Minor millets are nutritious crops on which population of Asia and Africa rely heavily. However, the research work to utilize highly nutritious millet crops is inadequate and insufficient to combat malnutrition. The bioavailability of nutrients, essential amino acids, minerals and vitamins needs further study. It can be done by the extensive understanding of the pathways involved in nutrient uptake, translocation and storage. The study leading to the search of molecular markers such as SNPs and InDels linked to quality traits and candidate genes will help in dissecting complex traits. The encouraging results observed in finger millet suggested that studies pertaining to transporter pathways involved in the transport and localization of macro- and micronutrients in grains of other minor millets need an exhaustive investigation through transcriptomics, proteomics and metabolomics approaches. It will ultimately excel the breeding and characterization of germplasm for new and unique nutritional traits.

With the emerging insights in genome sequences of some of the minor millets, it is crucial time that genomic regions responsible for nutritional traits should be deployed in breeding programmes. With the advancement in technologies like NGS platforms and rapid re-sequencing, thousands of small millet germplasm annotation is possible in short duration. A schematic illustration of an integrative genomics approach for the accelerated genetic improvement of grain quality and nutritional traits is proposed. On the whole, the collective approach utilizing all the omics tools like genomics, transcriptomics, proteomics and metabolomics with conventional and traditional breeding methods could promote millets as a model system for the advancement of quality traits and be used as a staple crop in world.

REFERENCES

Agrawal, G. K., and R. Rakwal. "Rice proteomics: A move toward expanded proteome coverage to comparative and functional proteomics uncovers the mysteries of rice and plant biology." *Proteomics* 11, no. 9 (2011): 1630–1649.

Altenbach, S. B., and P. V. Allen. "Transformation of the US bread wheat 'Butte 86' and silencing of omega-5 gliadin genes". *GM Crops* 2 (2011): 67–74. doi: 10.4161/gmcr.2.1.15884.

Ansari, W. A., S. U. Chandanshive, V. Bhatt, et al. "Genome editing in cereals". *International Journal of Molecular Sciences* 21 (2020): 4040. doi: 10.3390/ijms21114040.

Araus, J. L., G. A. Slafer, C. Royo, and M. D. Serret. "Breeding for yield potential and stress adaptation in cereals". *Critical Reviews in Plant Science* 27 (2008): 377–412. doi: 10.1080/07352680802467736.

Babu, B. K., P. Dinesh, P. K. Agrawal, et al. "Comparative genomics and association mapping approach for blast resistant genes in finger millet using SSRs". *PLoS One* 9, no. 6 (2014): e99182.

Baldoni, E., A. Genga, and E. Cominelli. "Plant MYB transcription factors: Their role in drought response mechanisms". *International Journal of Molecular Science* 16 (2015): 15811–15851. doi: 10.3390/ijms160715811.

Barro, F., J. C. M. Iehisa, M. J. Giménez, et al. "Targeting of prolamins by RNAi in bread wheat: Effectiveness of seven silencing-fragment combinations for obtaining lines devoid of coeliac disease epitopes from highly immunogenic gliadins". *Plant Biotechnology Journal* 14 (2016): 986–996. doi: 10.1111/pbi.12455.

Bednarek, J., A. Boulaflous, C. Girousse, et al. "Down-regulation of the TaGW2 gene by RNA interference results in decreased grain size and weight in wheat". *Journal of Experimental Botany* 63, no. 16 (2012): 5945–5955.

Bouis, H. E., and R. M. Welch. "Biofortification-A sustainable agricultural strategy for reducing micronutrient malnutrition in the global south". *Crop Science* 50 (2010): 20–32.

Budak, H., M. Kantar, and K. Yucebilgili Kurtoglu. "Drought tolerance in modern and wild wheat". *Science World Journal* (2013): 548246. doi: 10.1155/2013/548246.

Chen, Y., X. Yang, C. Y. Gu, et al. "Genotypes and phenotypic characterization of field *Fusarium asiaticum* isolates resistant to carbendazium in Anhui Province of China". *Plant Disease* 99 (2015): 342–346.

Dong, L., Wang, F., Liu, T., et al. "Natural variation of TaGASR7-A1 affects grain length in common wheat under multiple cultivation conditions". *Molecular Breeding* 34 (2014): 937–947.

Doudna, J. A., and E. Charpentier. "The new frontier of genome engineering with CRISPR-Cas9". *Science* 346, no. (6213) (2014): 1258096.

Gao, Y., L. Chaolei, L. Yuanyuan, et al. "QTL analysis for chalkiness of rice and fine mapping of a candidate gene for qACE9." *Rice* 9, no. 1 (2016): 41.

Gil-Humanes, J., F. Piston, C. M. Rosell, and F. Barro. "Significant down-regulation of γ-gliadins has minor effect on gluten and starch properties of bread wheat". *Journal of Cereal Science* 56, no. 2 (2012): 161–170.

Gosal, S. S., S. H. Wani, and M. S. Kang. "Biotechnology and crop improvement." *Journal of Crop Improvement* 24, no. 2 (2010): 153–217.

Hong, Y., L. Chen, L. P. Du, et al. "Transcript suppression of TaGW2 increased grain width and weight in bread wheat". *Functional Integrative Genomics* 14 (2014): 341–349.

International Wheat Genome Sequencing Consortium (IWGSC). "Shifting the limits in wheat research and breeding using a fully annotated reference genome". *Science* (2018). doi: 10.1126/Science.aar7191.

Jouanin Aurelie, J. W. J., L. Gilissen, J. G. Schaart, et al. "CRISPR/CAS9 gene editing of gluten in wheat to reduce gluten content and exposure reviewing methods to screen for coeliac safety". *Frontiers in Nutrition* 7 (2020): 51. doi: 10.3389/fnut.2020.00051.

Karkute, S. G., A. K. Singh, O. P. Gupta, P. M. Singh, and B. Singh. "CRISPR/Cas9 mediated genome engineering for improvement of horticultural crops." *Frontiers in Plant Science* 8 (2017): 1635.

Kim, Y.-A., H. Moon, and C.-J. Park. "CRISPR/Cas9-targeted mutagenesis of Os8N3 in rice to confer resistance to Xanthomonas oryzaepv. oryzae." *Rice* 12, no. 1 (2018): 1–13.

Kumar, R., A. Kaur, A. Pandey, H. M. Mamrutha, and G. P. Singh. "CRISPR-based genome editing in wheat: A comprehensive review and future prospects". *Molecular Biology Reports* (2019). doi: 10.1007/s11033-019-04761-3.

Liang, G., H. Zhang, D. Lou, and D. Yu. "Selection of highly efficient sgRNAs for CRISPR/Cas9-based plant genome editing." *Scientific Reports* 6, no. 1 (2016): 1–8.

Liu, D., Y. Liu, W. Zhang, X. Chen, and C. Zou. "Agronomic approach of zinc biofortification can increase zinc bioavailability in wheat flour and thereby reduce zinc deficiency in humans". *Nutrients* 9 (2017): 465.

Lucas, S., E. Durmaz, B. A. Akpnar, and H. Budak. "The drought response displayed by a DRE-binding protein from *Triticum dicoccoides*". *Plant Physiology Biochemistry* 49 (2011): 346–351. doi: 10.1016/j.plaphy.2011.01.016.

Mao, X., Y. Zheng, K. Xiao, et al. "OsPRX2 contributes to stomatal closure and improves potassium deficiency tolerance in rice." *Biochemical and Biophysical Research Communications* 495, no. 1 (2018): 461–467.

Mohammadi-Nejad, G., A. Arzani, A. M. Rezai, R. K. Singh, and G. B. Gregorio. "Assessment of rice genotypes for salt tolerance using microsatellite markers associated with the saltol QTL." *African Journal of Biotechnology* 7, no. 6 (2008): 730–736.

Morran, S., O. Eini, T. Pyvovarenko, et al. "Improvement of stress tolerance of wheat and barley by modulation of expression of DREB/CBF factors". *Plant Biotechnology Journal* 9 (2011): 230–249. doi: 10.1111/j.1467-7652.2010.00547.x.

Muthamilarasan, M., and M. Prasad. "Advances in Setaria genomics for genetic improvement of cereals and bioenergy grasses". *Theoretical Applied Genetics* 128 (2015): 1–14.

Nalam, V. J., S. Alam, J. Keereetaweep, et al. "Facilitation of *Fusarium graminearum* infection by 9-lipoxygenases in arabidopsis and wheat". *Molecular Plant-Microbe Interactions* 28, no. 10 (2015): 1142–s.

Nowara, D., A. Gay, C. Lacomme, et al. "HIGS: Host-induced gene silencing in the obligate biotrophic fungal pathogen *Blumeria graminis*." *The Plant Cell* 22, no. 9 (2010): 3130–3141.

Patil, J., and A. Kale. "Study of genetic diversity in finger millet (*Elesuine coracana L*) genotypes using RAPD markers". *International Journal of Integrative Science and Innovation Technology* 2 (2013): 31–36.

Peng, X., H. Wang, J.-C. Jang, et al. "OsWRKY80-OsWRKY4 module as a positive regulatory circuit in rice resistance against Rhizoctoniasolani." *Rice* 9, no. 1 (2016): 1–14.

Qi, T., X. Zhu, C. Tan, et al. "Host-induced gene silencing of an important pathogenicity factor P s CPK 1 in *Puccinia striiformis* f. sp tritici enhances resistance of wheat to stripe rust". *Plant Biotechnology Journal* 16 (2018): 797–807. doi: 10.1111/pbi.12829.

Rabbani, M. A., K. Maruyama, H. Abe, et al. "Monitoring expression profiles of rice genes under cold, drought, and high-salinity stresses and abscisic acid application using cDNA microarray and RNA gel-blot analyses." *Plant Physiology* 133, no. 4 (2003): 1755–1767.

Rajendran, S. R. C. K., Y. Yau, D. Pandey, and A. Kumar. "CRISPR-Cas9 based genome engineering: Opportunities in Agri-food-nutrition and healthcare". *OMICS* 19(5) (2015): 1–16.

Ramakrishnan, M., S. Antony Ceasar, V. Duraipandiyan, N. A. Al Dhabi, and S. Ignacimuthu. "Assessment of genetic diversity, population structure and relationships in Indian and non-Indian genotypes of finger millet (*Eleusine coracana (L)* Gaertn) using genomic SSR markers". *Springer Plus* 5(120) (2016):1–11.

Reddy, I. N. B. L., D. S. Reddy, M. L. Narasu, and S. Sivaramakrishnan. "Characterization of disease resistance gene homologues isolated from finger millet (*Eleusine coracana L* Gaertn)". *Molecular Breeding* 27 (2011): 315–328.

Saleh, A. S. M., Q. Zhang, J. Chen, and Q. Shen. "Millet grains: Nutritional quality, processing, and potential health benefits". *Comprehensive Reviews on Food Science and Food Safety* 12 (2013): 281–295.

Saltzman, A., E. Birol, H. E. Bouis, et al. "Biofortification: Progress toward a more nourishing future". *Global Food Security* 2 (2013): 9–17.

Shan, Q., Y. Wang, J. Li, and C. Gao. "Genome editing in rice and wheat using the CRISPR/Cas system". *Nature Protocols* 9 (2014): 2395–2410. doi: 10.1038/nprot.2014.157.

Shinozaki, K. and K. Yamaguchi-Shinozaki. "Molecular responses to dehydration and low temperature: Differences and cross-talk between two stress signaling pathways." *Current Opinion in Plant Biology* 3, no. 3 (2000): 217–223.

Singh, K. B., R. C. Foley, and L. Onate Sanchez. "Transcripton factors in plant defense and stress responses". *Current Opinion in Plant Biology* 5 (2002): 430–436.

Song, X. S., K. X. Gu, X. X. Duan, et al. "A myosin5 dsRNA that reduces the fungicide resistance and pathogenicity of *Fusarium asiaticum*". *Pestic Biochemistry and Physiology* 150 (2018): 1–9.

Sun, H., Z. Lang, L. Zhu, and D. Huang. "Optimized condition for protoplast isolation from maize, wheat and rice leaves". *Chinese Journal of Biotechnology* 29 (2013): 224–234.

Tiwari, S., S. L. Krishnamurthy, V. Kumar, et al. "Mapping QTLs for salt tolerance in rice (Oryza sativa l.) by bulked segregant analysis of recombinant inbred lines using 50K SNP chip". *PLos One* 11, no. 4 (2016): e0153610. doi: 10.1371/journal.pone.0153610.

Todaka, D., K. Nakashima, K. Shinozaki, and K. Yamaguchi-Shinozaki. "Toward understanding transcriptional regulatory networks in abiotic stress responses and tolerance in rice." *Rice* 5, no. 1 (2012): 6.

Upadhyay, S. K., J. Kumar, A. Alok, and R. Tuli "RNA-guided genome editing for target gene mutations in wheat". *G3: Genes Genome Genetics* 3 (2013): 2233–2238.

Usman, M. G., M. Y. Rafii, M. Y. Martini, O. A. Yusuff, M. R. Ismail, and G. Miah. "Molecular analysis of Hsp70 mechanisms in plants and their function in response to stress." *Biotechnology and Genetic Engineering Reviews* 33, no. 1 (2017): 26–39.

Wang, Y., X. Cheng, Q. Shan, et al. "Simultaneous editing of three homoeoalleles in hexaploid bread wheat confers heritable resistance to powdery mildew". *Nature Biotechnology* 32 (2014): 947.

Wang, S., S. Li, Q. Liu, et al. "The OsSPL16-GW7 regulatory module determines grain shape and simultaneously improves rice yield and grain quality". *Nature Genetics* 47 (2015a): 949–954.

Wang, S., Xiong, G., Hu, J., et al. "Copy number variation at the GL7 locus contributes to grain size diversity in rice". *Nature Genetics* 47 (2015b): 944–948.

Wang, F., C. Wang, P. Liu, C. Lei, W. Hao, Y. Gao, Y.-G. Liu, and K. Zhao. "Enhanced rice blast resistance by CRISPR/Cas9-targeted mutagenesis of the ERF transcription factor gene OsERF922." *PLoS One* 11, no. 4 (2016): e0154027.

Wang, K., B. Riaz, and X. Ye. "Wheat genome editing expedited by efficient transformation techniques: Progress and perspectives." *The Crop Journal* vol. 6, no. 1 (2018): 22–31.

Wayne, L. L., D. J. Gachotte, and T. A. Walsh. "Transgenic and genome editing approaches for modifying plant oils." In Kumar, S., Barone, P., and Smith, M. (eds), Transgenic Plants, pp. 367–394. Humana Press, New York (2019).

Zafar, S. A., S. Shan-e-Ali Zaidi, Y. Gaba, et al. "Engineering abiotic stress tolerance via CRISPR/Cas-mediated genome editing." *Journal of Experimental Botany* 71, no. 2 (2020): 470–479.

Zhang, G., X. Liu, Z. Quan, et al. "Genome sequence of foxtail millet (*Setaria italica*) provides insights into grass evolution and biofuel potential". *Nature Biotechnology* 30 (2012): 549–554.

Zhang, Y., Z. Liang, Y. Zong, et al. "Efficient and transgene-free genome editing in wheat through transient expression of CRISPR/Cas9 DNA or RNA". *Nature Communications* 7 (2016): 12617.

Zhang, Y., Y. Bai, G. Wu, et al. "Simultaneous modification of three homoeologs of Ta EDR 1 by genome editing enhances powdery mildew resistance in wheat". *Plant Journal* 91 (2017): 714–724.

Zhang, Z., L. Hua, A. Gupta, et al. "Development of an agrobacterium: Delivered CRISPR/Cas9 system for wheat genome editing". *Plant Biotechnology Journal* (2019). doi: 10. 1111/pbi.13088.

Zhao, F. J., and S. P. McGrath. "Biofortification and phytoremediation". *Current Opinion in Plant Biology* 12 (2009): 373–380.

Zhu, C., L. Bortesi, C. Baysal, et al. "Characteristics of genome editing mutations in cereal crops." *Trends in Plant Science* 22, no. 1 (2017): 38–52.

Zong, Y. et al. "Efficient C-to-T base editing in plants using a fusion of nCas9 and human APOBEC3A". *Nature Biotechnology* 36 (2018): 950–954.

6 Genome Editing for Improvement of Rice

Mayuri D. Mahalle
ICAR-National Institute for Plant Biotechnology

Suhas Gorakh Karkute
ICAR-National Institute for Plant Biotechnology
ICAR-Indian Institute of Vegetable Research

Umakanta Ngangkham
ICAR-Research Complex for North-Eastern Hill Region

Amolkumar U. Solanke
ICAR-National Institute for Plant Biotechnology

CONTENTS

6.1 INTRODUCTION

Rice (*Oryza sativa* L.) is a predominantly consumed cereal crop by the majority of the world population. Being the staple food crop and major source of energy, its value for humankind is very significant. In 2018–2019, the global rice production was recorded about 501.4 million tons by the U.S. Department of Agriculture (USDA) (https://www.ers.usda.gov/webdocs/publications/92853/rcs-19d.pdf?v=282.5). Globally, China ranks first followed by India in total rice area under cultivation. The crop is primarily a self-pollinated and exhibits two developmental stages: a long period of vegetative stage followed by a short period of reproductive stage comprising gametogenesis, sporogenesis, fertilization and grain developmental cascades (Chauhan et al., 2017). Reproductive stage is the critical phase in the lifecycle of rice, which determines the ultimate yield of the crop. There is a huge diversity of rice germplasm collections preserved at International Rice Research Institute (IRRI) holding <127,916 accessions of rice and around 4647 wild species, which serve as the best genomic resources for rice breeding programmes (https://www.irri.org/our-work/research/genetic-diversity/international-rice-genebank). The rice crop manifests an extensive adaptability to a wide range of agro-climatic conditions thriving well below the sea level to very high altitudes. However, the population explosion and shrinking agricultural land can cause a reduction in rice production in coming days. The global climate change and various abiotic and biotic stresses can further worsen its productivity. Rice is the first completely sequenced cereal crop having a genome size of 430 Mb, and it serves as a model plant for various cereal crops. This has facilitated in deciphering the genome function and loci involved in different plant developmental pathways, immune system, quality aspects etc. Rice plants deal with a number of environmental challenges like their susceptibility to drought, salt, cold, heat, submergence stress as well as lodging problems. The crop also suffers heavy yield losses as a result of biotic stresses like insect pests, fungal, bacterial, viral and several other pathogen encounters. These traits are multigenic and are influenced by various environmental factors. Several loci, pathways and mechanisms are involved in response to different stresses. In case of abiotic stresses, traits like osmotic adjustment, root and root-related traits, stomatal and leaf features, early flowering and grain yield are considered while selecting the tolerant genotypes. Similarly, rice grain quality is mainly influenced by their starch, protein, iron, mineral and vitamin contents. Rice improvement programmes have targeted all these traits, and various QTLs (quantitative trait loci)/genes linked to these traits have been identified and characterized so far (Vikram et al., 2011; Dai et al., 2007). Both traditional and molecular breeding practices have played a key role in the rice crop improvement. However, due to limiting factors like crossing barriers, requirement of time and labour, etc., it warrants the utilization of modern biotechnological tools. With the help of these tools, several genes have been cloned, transformed and functionally characterized for resistance to protect rice plants from these stresses. The advancements in new rice for Africa (NERICA) and hybrid rice are the successful influencers in the overall improvement of rice yields (Van Nguyen and Ferrero, 2006).

Modern biotechnological tools are essential parts of the crop improvement programmes to supplement the traditional and molecular breeding approaches. In recent past, genome editing techniques, which comprise sequence-specific nucleases (SSNs) mainly, zinc finger nucleases (ZFNs), transcription activator-like effector nucleases (TALENs) and clustered regularly interspaced short palindromic repeats/CRISPR-associated protein 9 (CRISPR/Cas9), have revolutionized and are extensively used for crop improvement. However, CRISPR/Cas9 method is ahead of all other editing technologies due to its several advantages like easy designing and implementation, more success rate, inexpensiveness and versatility. Genome editing techniques have been used on a large scale in a variety of cereal crops (Zhu et al., 2017), horticultural crops (Karkute et al., 2017), oilseed crops (Wayne et al., 2019) and other crops for improving the number of traits. The genome editing technology has the potential to utilize the genetic repertoire of plants for creating specific sequence diversity and thus improvement of desired traits. Different variants of Cas protein allow us to perform different kinds of applications such as site-directed mutations by creating blunt end cuts using Cas9, staggered cuts using Cas12a (Zetsche et al., 2015), base editing by using cytosine or adenine base editors (ABEs) (Komor et al., 2016), RNA knockdown using Cas13a (Abudayyeh et al., 2017) and RNA editing using Cas13b (Cox et al., 2017). The genome editing has been further advanced to a level where as many as eight genes can be targeted with a single construct utilizing endogenous tRNA processing system (Xie et al., 2015) to enable targeting multiple genes for their functional characterization. However, there is concern for legislative regulations of genetically modified organism and high chance of off-target modifications due to the Cas9–guide RNA (gRNA) expression cassette and selectable marker integration into the host genomic DNA by *Agrobacterium tumefaciens*-mediated transformation or particle bombardment (Toda et al., 2019). As an improvement to this technology, a DNA-free genome editing technique was developed where the ribonucleoproteins (RNPs) consisted of preassembled Cas9 protein and *in vitro* transcribed gRNA were directly delivered into the plant cells through PEG-mediated transfection or biolistics (Murovec et al., 2018). It offers several advantages such as selectable marker-free, less off-target modifications, not requiring for the manipulation of promoter and coding the sequence specific to host plants etc.

Being the most important cereal crop, rice is the prime target of genome editing for various traits like tolerance to biotic and abiotic stresses, and quality improvement. Fully characterized genome sequence enriched with resequencing of thousands of lines and standardized transformation and regeneration protocols in rice provides an additional benefit to effectively carry out genome editing. Although genome editing in rice has started with TALENs for producing disease resistance (Li et al., 2012), much of the work has been done after the entry of CRISPR/Cas9 technique. Various applications of genome editing tools in rice are discussed further in this chapter.

6.2 IMPROVING RESISTANCE TO BIOTIC STRESSES

Rice crop is found to be vulnerable to various biotic stresses like diseases due to bacteria, fungi and viruses, insect pests and herbicide toxicity. These stresses severely hamper the plant growth, development and seed quality, thereby reducing the productivity and yield. Highly efficient technologies like genomics and proteomics are being used in order to understand the regulatory networks involved in response of rice to various biotic stresses (Agrawal and Rakwal, 2011). Various biotechnological tools are used to develop resistance against these stresses, thereby reducing losses and improving yields. Several QTL/genes have been identified, cloned and transferred into susceptible rice cultivars, conferring resistance to various insects and pests. Recent emergence of genome editing technologies, specifically CRISPR/Cas9, has revolutionized the speed of functional characterization of genes that will ultimately lead to the development of novel varieties with enhanced resistance to various pests and diseases (Liang et al., 2016). This technology is predicted to aid in a varietal improvement, which is restricted due to the confined availability of natural and mutant strains. The technique can also aid in modulating various transcription factors to obtain an enhanced tolerance to various plant biotic stresses. This section mainly describes the molecular basis of biotic stress tolerance in rice and how CRISPR/Cas9 technology can be effectively applied in establishing resistance to these pests and pathogens in rice crop by highlighting a significant progress achieved during the past years.

6.2.1 Advancements in Molecular Basis of Biotic Stress Tolerance

In order to manipulate the stress response pathways in rice, it is essential to know the components playing an important role in such pathways. Several studies have contributed in identifying genes and regulatory network responsible for plant's defence against pathogens or insects. The crop counteracts the stress by down- and up-regulating various genes and transcription factors as well as by managing the plant regulatory pathways. Recent transcriptomics of rice plants by Cohen and Leach (2019) revealed the drastic reduction in photosynthesis and up-regulation of hormone-responsive elements like abscisic acid (ABA), jasmonic acid (JA) and salicylic acid (SA) pathways under biotic stress. Various promoter motifs like DREME (Bailey, 2011) and GOMO (Buske et al., 2010), which are involved in the regulation of stress response in rice plants, have also been identified. Plants respond to pathogens *via* reduced photosynthesis and stomatal closure (Bilgin et al., 2010; Melotto et al., 2006). The evidence is being provided in support to a model, which describes stomata as a part of intrinsic innate immune system which essentially serves as a blockade to various bacterial invasions (Bilgin et al., 2010). Moreover, the down-regulation of photosynthetic-responsive genes sharing the similar regulatory motifs like G-box (CACGTG), SORLIP (GCCAC) and T-box (ACTTG) forms an adaptive response of plants to pathogen rush (Melotto et al., 2006). The rice plants respond to pathogens by mostly relying on ethylene, JA and SA signalling. In contrast to abiotic stress, ABA mostly increases the susceptibility of plants to biotic stress exposure and very often responds antagonistically with SA (de Torres et al., 2009; Ding et al., 2016). Modulation of expression of transcription factors is one of the major responses towards biotic stress, which in turn regulates the expression of number of stress-responsive genes. These transcription factors are the eminent targets for manipulation to achieve superior stress resistance. However, very few transcription factors such as *bZIP*, *ERF*, *MYB*, *NAC* and *WRKY* are being studied in this aspect. *WRKY4* and *WRKY80* were reported to be responsible for an improved resistance to fungal sheath blight in rice (Peng et al., 2016). The WRKY80 gets attached to the WRKY4 W-Box promoter region, thus serving as a positive regulatory circuit for resistance against *Rhizoctonia solani*. Some WRKY transcription factors like *WRKY76*, *WRKY64*, *WRKY62*, *WRKY58* and *WRKY7* were significantly expressed in response to rice blast disease (Sureshkumar et al., 2019). *WRKY67* was also found to be up-regulated against both rice blast and bacterial blight, thus serving as an absolute target in improving tolerance to pathogen stress (Liu et al., 2018). *OsWRKY11* can enhance both pathogen defence and drought tolerance, thereby mediating a cross talk between biotic and abiotic stress responses (Lee et al., 2018). A variant of transcription factor, *Bsr-d1*, which is a C_2H_2 type, was identified using GWAS that confers non-race-specific rice blast resistance (Li et al., 2017). The variant is formed as a result of single-nucleotide change in the promoter of *bsr-d1* gene. This eventually turns down the gene expression *via* binding of repressive MYB transcription factor causing the inhibition of H_2O_2 and thus providing boosted resistance to rice blast. This gene can serve as a prime target for single-base editing approach engaging CRISPR/Cas technology. Another MADS-box transcription factor, i.e. *OsMADS26*, was conveyed to be down-regulated against two serious pathogens of rice, i.e. *Xanthomonas oryzae* and *Magnaporthe oryzae* (Khong et al., 2015). *OsMADS26* hence acts as an upstream regulator of stress-related genes functioning as a hub to regulate the response of rice to various biotic and abiotic stresses, and can be targeted using genome editing tool via the CRISPR/Cas9 system. Based on the basic understanding of stress-responsive pathways and genes involved in them, suitable strategies can be developed for improving the stress tolerance in rice. CRISPR/Cas tools have been effectively used lately for this purpose (Table 6.1), and the developments in this field are described in the section below.

TABLE 6.1
Application of Genome Editing Technology in Rice to Improve Biotic Stress Tolerance

Sr. No.	Trait/Function	Gene Targeted	Editing Approach	References
1	Rice blast resistance	*OsERF922*	CRISPR/Cas9	Wang et al. (2016)
2	Rice blast resistance	*Sec3A*	CRISPR/Cas9	Ma et al. (2018)
3	Bacterial leaf blight	*SWEET14*, *SWEET11*	CRISPR/Cas9	Jiang et al. (2013)
4	Bacterial leaf blight	*SWEET13*	CRISPR/Cas9	Zhou et al. (2015)
5	*Rice tungro spherical virus*	*eiF4g*	CRISPR/Cas9	Macovei et al. (2018)
6	Herbicide resistance	*ALS*	CRISPR/Cas9	Sun et al. (2016)
7	Herbicide resistance	*EPSPS*	Target-AID	Li et al. (2016)
8	Herbicide resistance	*ALS* (C-287 base)	Base editing using ncas9 and/or dCas9	Shimatani et al. (2018)
9	Herbicide resistance	*ACC* (C2186 base)	Adenine base editor using nCas9	Li et al. (2018)

6.2.2 Genome Editing to Improve Biotic Stress Tolerance in Rice

6.2.2.1 Development of Resistance against Fungal Diseases

Rice crop, during its life cycle, is infected with various fungal diseases like rice blast, sheath rot, sheath blight, scald diseases, smut, brown spot, rust, etc., resulting in the reduced yield as well as the grain quality of harvested crops. Likewise, certain mycotoxigenic fungi produce secondary metabolites called mycotoxins, which can lead to significant health issues in humans as well as animals when subjected to contaminated food/feed. Among all, rice blast caused by *Magnaporthe oryzae* is the most devastating disease causing 30%–50% reduction in the yield worldwide (Khush and Jena, 2009). One of the most adapted and economical approaches to manage the rice blast is the host plant resistance, and therefore, it is necessary to understand its mechanism. Large number of genes have been cloned and characterized in rice that play an important role in response to rice blast disease (Li et al., 2019). These genes include resistance (*R*) genes and defence regulator (DR) genes. Recently, 2688 *R* genes of NBS category have been identified from 11 species of rice (Rawal et al., 2018). DR genes are either positive or negative regulators of the defence response. These genes now serve as a potential targets for modification using CRISPR/Cas9 tool. One of the earliest genes targeted in order to enhance blast resistance in rice was *OsERF922*, a transcription factor acting as a negative regulator of blast resistance (Wang et al., 2016). Similarly, the disruption of *OsSEC3A* gene by CRISPR/Cas9 resulted in an intensified defence response against *Magnaporthe oryzae* through a rise in the levels of SA and related genes (Ma et al., 2018). This kind of improved resistance was devoid of any effects on other agronomic and yield traits demonstrating a useful approach to develop CRISPR/Cas9-mediated blast resistance in rice. The technique also helps in functional characterization of novel genes by site-directed mutagenesis. Recently, genes that are putatively responsible for resistance to blast are confirmed by CRISPR/Cas9-mediated gene targeting, e.g. *Ptr* gene (Zhao et al., 2018). *Ptr* is atypical resistance gene encoding a protein with four armadillo repeats and is responsible for broad-spectrum disease resistance. Alternatively, genome editing of fungus by targeting important genes responsible for causing disease and also for the functional characterization of putative genes involved in pathogenesis could lead to understanding of host–pathogen interactions during disease. The first use of genome editing tool was done for homologous recombination-mediated gene replacement in rice blast fungus *Pyricularia oryzae* using tailor-made TALENs (Arazoe et al., 2015). Recently, plasmid-free genome editing protocol based on CRISPR/Cas9 technology has been developed in *Magnaporthe oryzae*, where efficient gene edits were obtained by a transient expression of Cas9 complied with guide RNAs forming RNPs (Foster et al., 2018). Such findings will accelerate the precise genome alterations in *Magnaporthe oryzae* and can also be applied to other fungal species as well.

6.2.2.2 Development of Resistance against Bacterial Diseases

CRISPR/Cas9 can also serve as a versatile tool in managing bacterial pathogens. Rice crop is infected by various pathogenic bacteria like *Burkholderia plantarii, B. gladioli* and *B. glumae* leading to seedling blight, sheath rot and grain rot, respectively. These infestations cause a significant decline in crop yield in all the rice-growing countries. Bacterial leaf blight (BLB) caused by *Xanthomonas oryzae* is another major bacterial disease of rice causing severe crop loss. One of the most effective genes governing resistance to BLB is *Xa21* that has been used in several rice lines to develop BLB-resistant transgenic rice (Song et al., 1995). *Xa21* gene has also been transferred to a number of cultivated varieties of rice through marker-assisted breeding. In addition to transgenics and marker-assisted breeding which have in-built limitations of one or other kinds, genome editing technique has been utilized recently to develop resistance to bacterial diseases. Although there are limited reports of use of genome editing for this purpose, the technique has potential to develop strong resistance in crops against bacterial diseases. In rice, SWEET proteins play a role in plant–pathogen interaction and their expression is induced by effector protein named *PthXo2* produced by *X. oryzae*, ultimately causing host susceptibility. Mutations in SWEET protein result in BLB resistance in plants. In earlier experiments of genome editing, mutations in promoter of *OsSWEET14* gene were created using TALEN (Li et al., 2012). In the absence of SWEET14 protein, effectors of *X. oryzae* cannot bind to it and thus result in disease resistance. Taking into consideration the mechanism of BLB resistance in rice plants involving SWEET protein, strategies have been devised in order to use it for developing BLB-resistant rice plants.

In order to exploit the *PthXo2*-based disease susceptibility, null mutants of *OsSWEET13* were generated using CRISPR/Cas9 producing the BLB-resistant plants (Zhou et al., 2015). Recently, along with *SWEET13* and *SWEET14*, *SWEET11* was also found to be targeted by transcription activator-like effectors. Therefore, to achieve a broad-spectrum resistance to bacterial blight, promoters of all these three genes were targeted for site-directed mutagenesis using CRISPR/Cas9 (Oliva et al., 2019). Initially, mega varieties like IR64 and Ciherang-Sub1 that are growing on a large scale were used to mutate promoters of SWEET proteins, and observations of various agronomic and yield traits showed no difference between edited and non-edited plants except resistance to BLB. Similarly, *Os8N3* (*OsSWEET11*) gene was targeted in an attempt to enhance rice resistance to BLB (Kim et al., 2019). Genome-edited mutant plants showed higher resistance to bacterial blight without affecting agronomic traits. These results are highly promising as the genome-edited SWEET promoters and genes confer rice plants with broad-spectrum resistance to bacterial blight and can be implemented in other crops as well.

6.2.2.3 Development of Resistance against Viruses

Rice crop is damaged by a number of viruses like *rice tungro virus*, *rice yellow mottle virus*, *rice ragged stunt virus*, *rice bunchy stunt virus*, *rice grassy stunt virus* etc. Among them, *Rice tungro virus* causes a significant decline in yield damaging more than 350,000-ha area in major rice-growing countries (Azzam and Chancellor, 2002; Chancellor et al., 2006). Rice tungro disease (RTD) is induced when the *rice tungro spherical virus* (*RTSV*) and *rice tungro bacilliform virus* interact with each other. The resistance for *RTSV* is present in many traditional cultivars, which ultimately helped to control the infestation of this disease in fields. The *RTSV* tolerance is naturally a recessive trait provided by *eIF4G*, a translation initiation factor four gamma gene. In order to develop resistance in a susceptible variety IR64, mutations have been introduced in *eIF4G* gene using CRISPR/Cas9 system (Macovei et al., 2018). The resistance to virus depends on mutation site in *eIF4G*, *i.e.* in-frame mutations in residues of SVLFPNLAGKS (mainly NL), close to YVV residues, showed resistance. Some of the plants with truncated versions of eIF4G were not able to be maintained in homozygous conditions showing that the *eIF4G* function is essential for the normal growth and development of plants. Nevertheless, plants with novel alleles of *eIF4G* having *RSTV* resistance generated with genome editing techniques can be used in breeding programmes to develop elite rice lines resistant to *RSTV*.

In addition to targeting susceptible genes in plants, resistance against DNA viruses has been introduced in plants by developing transgenic plants expressing Cas9 and gRNA specific to viral genome (Ali et al., 2015). The strategy has been further extended to control RNA viruses by utilizing *FnCas9* and gRNA specific to viral genomes such as *cucumber mosaic virus* (CMV) or *tobacco mosaic virus* (Zhang et al., 2018). Recently, instead of Cas9, another enzyme FnCas13, which specifically cleaves RNA, has been utilized to develop transgenic plants (Zhang et al., 2019b). This is an efficient approach to induce stable RNA virus resistance in monocot plants. In general, CRISPR/Cas system can be effectively used to develop rice plants resistant to both RNA and DNA viruses.

6.2.2.4 Development of Herbicide Tolerance through Genome Editing

Weeds are one of the major biotic stresses responsible for significant yield losses in upland rice. Weed management is mainly done by manual and chemical measures; however, manual control is a costly affair as it requires huge labour. On the other hand, the use of herbicides is more effective and comparatively less costly. One of the limitations in using herbicide is that the broad-spectrum herbicides cannot be uniformly used for all crops. In order to use the broad-spectrum herbicides, herbicide resistance has been developed in several crops employing transgenic approach (Fartyal et al., 2018; Dong et al., 2017). However, the non-acceptance of transgenic crops for commercialization in several countries is a major barrier in their wide applicability. This necessitates the development of herbicide-resistant non-transgenic crops. Genome editing technology has been successfully used to produce transgene-free plants in a number of crops.

Molecular mechanism of herbicide tolerance in crops is based on mutations in the genes that are targets of herbicide molecules. *Acetolactate synthase 1* (*ALS1*) gene, which codes for a key enzyme involved in branched chain amino acid biosynthesis, is the prime target of herbicides such as chlorsulfuron and bispyribac solution (BS) (Mazur et al., 1987). Herbicide resistance has been successfully developed in soya bean using HR-mediated CRISPR/Cas9 editing of *ALS1* gene (Sun et al., 2016). The herbicide resistance was imparted due to two point mutations that reflected in W548L and S627I amino acid changes in protein. This report demonstrated the CRISPR/Cas9-mediated gene replacement using two gRNAs and a homologous recombination template. Several HR-mediated gene replacement experiments have been carried out in different plants, and the efficiency varies greatly depending on plant species. Nevertheless, the strategy successfully demonstrated the development of herbicide resistance in rice by creating site-directed point mutations in *ALS* gene and can be implemented in other crop species as well. Another enzyme 5 enolpyruvyl shikimate-3-phosphate synthase (EPSPS) consists of a conserved motif, which serves as a binding site for substrates like pyruvate and the glyphosate (herbicide). The motif is present in almost all plants, and substitution within two amino acids (T102I+P106S) causes herbicide resistance (Yu et al., 2015). Taking into account this basis of herbicide resistance, CRISPR/Cas9 tool has been used to replace the second exon in rice by targeting two gRNAs at first two introns and providing a suitable donor DNA template having desired mutations (Li et al., 2016a). In this study, the sgRNAs were targeted towards introns, and therefore, it was made sure that the splicing pattern is not altered *via* intronic mutations.

In addition to homologous recombination-based gene replacement, point mutations are being incorporated in plants with the help of recently developed base editors. Therefore, base editing is an effective way to develop herbicide resistance in plants by creating desired point mutations. Base editing has been successfully used in rice to edit *ALS* gene for C287T change, which conferred resistance against imazamox herbicide (Shimatani et al., 2018). This strategy employs a Cas9 fused with a base editor like *Petromyzon marinus* cytidine deaminase (PmCDA1). It is called as 'Target-AID', i.e. target-activation-induced cytidine deaminase. The process employs two Cas9 enzymes in mutated forms, *viz.* dCas9 exhibiting null activity and nCas9 conferring nickase activity combined to a base editor so as to achieve editing at a specific position *via* sgRNA. Another form of base editor is ABE, which comprises tRNA adenosine deaminase merged with nCas9 generating A-T to C-G changes governed by sgRNAs. This ABE system has been utilized in rice by creating fusion proteins of ABE varying on the site of adenosine deaminase and in the number and locus of nuclear localization sequence (Li et al., 2018). Using this, a point mutation reflecting in C2186R change in acetyl-coenzyme A carboxylase (ACC) was inserted that conferred resistance to herbicides like phenylpyrazoline, cyclohexanedione and aryloxyphenoxypropionate chemical groups in Zhonghua11 rice. All these gene replacement and base editing methods produce transgene-free herbicide-resistant plants and hence are not liable to controversies like transgenic crops.

6.3 IMPROVING RESISTANCE TO ABIOTIC STRESSES

Agriculture sector is severely affected due to uncertain rainfalls, deteriorating air, soil and water quality as a consequence of climate change. Rice is affected by various abiotic stresses like drought, high salinity, submergence, high and low temperatures, low nutrient availability, heavy metals and pollutants, mineral deficiencies, wind and mechanical injuries, which severely hamper rice production. Rice is a water-loving crop and scarcity of water hampers its yield, mostly during the reproductive stage of development (Swamy et al., 2017). On the other hand, it is highly sensitive to salt stress at early seedling stage. In order to make climate-resilient rice, understanding the mechanism of plant's response to various stresses plays a crucial role. Several studies have been successfully carried out so far to identify the key metabolic pathways, candidate genes, promoter motifs and QTLs involved in plant's response to abiotic stresses using transgenic and molecular breeding approaches. Abiotic stress regulation is a highly complex process involving diverse factors. This section mainly focuses on the most relevant studies done in the area of abiotic stress in rice and how genome editing technology can use this information effectively in order to reach the avid goal of high productivity under stressed conditions.

6.3.1 Molecular Mechanism and Candidate Genes Involved in Abiotic Stress Response

During abiotic stress, plant undergoes a series of morphophysiological changes, which include reduced photosynthesis, stomatal closure, increased root length, reduced leaf growth and increased reactive oxygen-scavenging activity. At the molecular level, plants regulate these changes by a series of coordinated responses producing various protein classes involving enzymes, functional proteins, molecular chaperones and transcription factors (Usman et al., 2017). Abiotic stresses are also regulated by hormones like ethylene, ABA and JA. Mainly, two types of components work in stress response: a signalling component that controls the expression of stress-responsive genes and a functional component that actually protects plant cells from the harmful impact of stress (Todaka et al., 2012a,b). Several genes have been identified so far, which are expressed in response to abiotic stress. The genes like *Asr1*, *calmodulin*, *catalase*, *dehydrins*, *lea*, *LIP9*, *LIP19*, *metallothionein*, *Myb* transcription factor, *RAB16*-C, *SalT*, *WS176* and *WS1724* are consistently observed to be up-regulated during abiotic stress in several studies. In addition to these genes, other important factors involved in abiotic stress response are aldehyde dehydrogenase, lipoxygenase, NAC6 transcription factor, O-methyltransferase, RING finger protein, sugar transporter, zinc finger proteins, etc. (Rabbani et al., 2003). Plants respond to different abiotic stresses in a coordinated manner as there is considerable cross talk between signalling pathways of response to various stresses (Shinozaki and Yamaguchi-Shinozaki, 2000; Seki et al., 2002). For instance, exposure of plants to drought, cold, high salinity and application of ABA induced 15 genes commonly, implying that they are regulated by the same or overlapping signalling pathways (Rabbani et al., 2003).

Rice has been extensively studied to unravel the abiotic stress response mechanism. Key regulatory transcription factors in abiotic stress tolerance are *OsDREB1s*, *OsDREB2s*, NAC (*NAM, ATAF, CUC*)-type TFs *OsNAC6*, *SNAC1*, *OsNAC5*,*OsNAC10* and *bZIPs* along with several other TFs. *OsSNAC3* transcription factor up-regulates during drought, salinity, high temperature and ABA stress, and confers stress tolerance by modulating ROS homeostasis (Lu et al., 2013). Thus, abiotic stress tolerance mechanism in plants is highly complex and maintains cellular homeostasis (Golldack et al., 2014; Mickelbart et al., 2015). Many of these genes have been utilized in rice breeding programmes to develop abiotic stress-tolerant rice genotypes using molecular breeding and transgenic approach. Since the inception of genome editing technology, it has been widely applied in rice for improving abiotic stress tolerance (Table 6.2).

6.3.2 Genome Editing to Improve Abiotic Stress Tolerance in Rice

6.3.2.1 Drought Stress

Genome editing is primarily used for creating mutations, and therefore, it is largely used for developing biotic stress resistance by mutating susceptible genes in plants. On the other hand, genome editing is mainly being used for the functional characterization of genes involved in abiotic stress response. Nevertheless, the technique has been significantly used for understanding as well as improving abiotic stress tolerance in rice. The main targets for genome editing to enhance abiotic stress tolerance are negative regulators playing a role in various cellular mechanisms like excessive ROS production, reduction of antioxidant activity, programmed cell death and disturbance of hormonal homeostasis (Zafar et al., 2020). One of the first experiments in the direction of development

TABLE 6.2
Application of CRISPR-Based Genome Editing to Impart Abiotic Stress Tolerance in Rice

Sr. No.	Trait/Function	Gene Targeted	References
1	Early-stage chloroplast development, chloroplast protection against heat stress	*HSA1*	Qiu et al. (2018)
2	Salt sensitivity	*OsBBS1*	Zeng et al. (2018)
3	Positive regulator in stomatal development/drought stress	*EPFL9*	Yin et al. (2017)
4	Intracellular H_2O_2 maintenance	*OsPRX2*	Mao et al. (2018)
5	Salt tolerance	*OsRR22*	Zhang et al. (2019a)
6	Cold tolerance	*OsAnn3*	Shen et al. (2017)
7	Cold tolerance	*OsMYB30*	Zeng et al. (2020)
8	Panicle length	*OsPIN5b*	Zeng et al. (2020)
9	Grain size	*Gs3*	Zeng et al. (2020)
10	Salt tolerance	*OsMIR528*	Zhou et al. (2017)
11	Drought tolerance	*SAPK2*	Lou et al. (2017)
12	Metal transporter (low Cd)	*OsNramp5*	Tang et al. (2017)

of abiotic stress tolerance in rice was a demonstration of multiple gene editing carried out to mutate 11 genes (Zhang et al., 2014). The target genes include *OsPMS3* (photoperiod-sensitive male sterile), *OsPDS* (albino), *OsMYB5* (MYB family transcription factor), *OsYSA* (albino young seedling), *OsEPSPS* (lethal), *OsMSH1* (pleiotrophic phenotype), *OsDERF1* (drought tolerance), *OsROC5* (abaxial leaf rolling), *OsMYB1* (MYB family transcription factor) and *OsSPP* (early seedling leaf chlorosis). This was an important experiment because abiotic stress tolerance is a complex trait and needs to manipulate multiple genes in order to develop stress tolerance in plants.

Closure of stomata is one of the major defence mechanisms of plants against the abiotic stress, thereby avoiding water loss and hence maintaining high tissue water potential. Therefore, one of the ways of development of abiotic stress tolerance in plants is to manipulate stomatal properties. This has been achieved by knocking out of *EPFL9* (epidermal pattern factor like-9) gene, a positive regulator of stomatal development, using CRISPR/Cas9 and CRISPR/Cpf technologies (Yin et al., 2017). The mutants showed an 8-fold reduction in stomatal density ultimately, improving tolerance to drought stress. Similarly, an overexpression of rice epidermal patterning factor *OsEPF1* resulted in drought-tolerant plants with substantially reduced stomatal density (Caine et al., 2019). These plants were able to conserve more water and showed higher survival under drought and high-temperature stress conditions.

6.3.2.2 Salt Stress

Apart from developing drought stress tolerance, genome editing has been successfully implemented in rice to understand the mechanism of salinity stress tolerance and also to enhance the same. The role of *OsPRX2* gene during salinity stress was characterized by knocking out using CRISPR/Cas9 and was shown to involve in maintaining intracellular hydrogen peroxide levels (Mao et al., 2018). The mutant plants show an increased level of ROS accumulation compared with control, and when these plants were deprived of potassium ions, an enhanced stomatal closure was observed. With the help of targeted mutagenesis using CRISPR/Cas9, the role of *OsRAV2* gene in salt stress tolerance was also elucidated (Duan et al., 2016). CRISPR/Cas9 technique has also been used to target *OsRR22* gene and the mutants exhibited higher level of salt stress tolerance compared to wild types without altering other agronomic traits (Zhang et al., 2019a). This signifies the importance of CRISPR/Cas9 technology in improving salt tolerance. Moreover, the technology can also pave the way to generate novel allelic variations in various transcription factors (e.g. *DREB2B*) as well as ion transporters (e.g. *NHX*, *HKT*) in order to develop tolerant varieties. Also, the allelic variations occurring naturally can be effectively exploited using GWAS as an eminent method of genetic breeding. Likewise, a GWAS done using 299 Indian wild-type rice accessions unravelled alleles tolerant to salt, *viz. HKT2;3* and *HKT1;5* (Mishra et al., 2016). From these SNPs, one SNP is associated with chlorophyll content and the other with shoot K^+ concentration. These variants can now be introduced in high-yielding varieties to expand a genetic pool and reduce yield losses due to salt stress. Alternatively, recently developed base editors can be effectively used to replace the SNPs of susceptible plants with these SNPs providing tolerance to the salt stress.

The microRNAs also play a significant role in gene regulation along with genes and transcription factors and thus can be targeted using gene editing. Chang et al. (2016) have obtained around 96% reduction in expression of miRNAs (miR-200c, miR141 and miR-17) by using the sgRNA that mainly targeted the biogenesis processing sites of miRNAs. Plants showed the stability in miRNA-knockdown phenotypes for a longer span. miRNA genes (*OsMIR528* and *OsMIR408*) as well as families of miRNA (miR820a/b/c and miR815a/b/c) were targeted in rice using CRISPR/Cas9 system (Zhou et al., 2017). Further, the reliability of marker systems in identifying the CRISPR mutants was also studied, and it has been found that single-strand conformation polymorphism (SSCP) is better than restriction fragment length polymorphism (RFLP). This experiment reported the role of *OsMIR528* as a positive regulator during salt stress. The study not only highlighted the function of miRNAs and complex cross talks in pathways but also provided a guideline on manipulating miRNAs using CRISPR/Cas9 system.

6.3.2.3 Heat Stress

Functional characterization of genes involved in abiotic stress tolerance has been carried out by genome editing in rice. CRISPR/Cas9-mediated mutation of *HSA1* (heat-sensitive albino 1) in rice resulted in sensitivity to heat stress and possessed faster greening phenotype compared with wild type at all temperatures (Qiu et al., 2018). This report established the role of *Hsa1* in the development of chloroplast at early stages as well as in providing protection to chloroplast during heat stress at subsequent stages of development in rice. Similarly, the role of *OsBBS1* gene in early senescence of leaf and sensitivity to salt stress has been depicted by using the similar method (Zeng et al., 2018).

6.3.2.4 Cold Stress

Rice crop is also influenced by cold stress affecting its growth and yield especially at seedling stage. Many cold-responsive genes like *OsSRFP1, COLD1* and *SGD1* have been isolated and cloned (Fang et al., 2015; Wang et al., 2017). Among the important transcription factors, *TIFY1b* is found to be involved in cold stress tolerance in rice. In order to explore its exact role and mechanism of tolerance, the *TIF1b* and its homologue *TIF1a* were edited by employing CRISPR/Cas9 technology (Huang et al., 2017). Functional characterization of rice annexin, *OsAnn3* gene, which is involved in plant development and protection from environmental stresses, was carried out by knocking out using CRISPR/Cas9 technique (Shen et al., 2017). The survival rate of these mutant plants was decreased, and the relative electrical conductivity was increased, suggesting the involvement of *OsAnn3* gene in cold stress tolerance. *OsMYB30* gene, which is a R2R3-type cold-responsive MYB gene, negatively affects the cold stress tolerance in rice by binding to β-amylase genes as transcription factor. In cold stress, this *OsMYB30* along with *OsJAZ9* works together and suppresses *BMY* gene expression ultimately affecting the maltose accumulation and starch degradation (Lv et al., 2017). CRISPR/Cas9-mediated targeting of *OsMYB30* gene could lead to an enhanced cold stress tolerance in rice. In the latest study, three genes, namely *OsMYB30* (a cold tolerance gene), *OsPIN5b* (a panicle length gene) and *GS3* (a grain size gene), were edited using CRISPR/Cas9 system by targeting two target sites of each gene with high efficiency (Zeng et al., 2020). As a result, some mutants showed an increased tolerance to cold stress, longer panicle length and bigger grain size (*osmyb30, ospin5b* and *gs3* mutants, respectively). The T_2 progeny mutants also showed higher yield and better tolerance to cold stress as compared to wild type. These studies again demonstrate that genome editing technology can be efficiently used to develop genotypes with significant tolerance to abiotic stresses.

6.3.2.5 Heavy Metal Stress

Application of genome editing has been extended to tackle the problem of heavy metal toxicity. Cadmium, a heavy metal, is highly toxic to the majority of living beings (Clemens et al., 2013), and its major dietary source in humans is rice. To reduce the accumulation of cadmium in rice, a metal transporter gene *OsNramp5* has been knocked out using CRISPR/Cas9 tool (Tang et al., 2017). The hydroponics culture of *osnramp5* mutants produced lesser cadmium in roots and shoots consistently, i.e. 0.05 mg/kg as compared to 0.33–2.90 mg/kg in wild type. The hybrid rice lines produced using these genome-edited plants performed well with regard to the reduced Cd content. Thus, the work demonstrated the effective use of genome editing technology in developing rice cultivars safer to particularly Cd pollution and to heavy metal stress in general.

6.4 IMPROVEMENT OF GRAIN QUALITY TRAITS IN RICE

Along with grain yield, its grain quality is highly important trait to be targeted for improvisation in rice crop. Grain quality is rightly a very significant attribute in rice as every region has different consumer preferences. It includes aspects like grain cooking qualities, micronutrient contents, aroma, taste and phytochemicals. Grain quality is a highly complex multigenic trait influenced by several molecular and environmental factors. These factors include ambient temperature during ripening, season effect, water regime, soil salinity, delayed threshing and drying of moist grains, etc. (Juliano, 1990). Various chemical pathways including genes and gene regulatory networks specifying the grain quality have been revealed so far by employing various techniques. The molecular breeding and functional genomics approach has also facilitated the generation of vital information about the molecular basis of grain quality. The plant breeding has definitely increased the yields of rice; however, the grain quality aspect is stagnated since a very long time. The genome editing techniques can play a significant role to accelerate the research in studying the different genes and pathways involved in controlling grain quality due to their well-known specific and simpler targeting capability.

6.4.1 Major Traits Associated with Grain Quality

Rice grain quality is mainly assessed based on traits like their milling properties, cooking quality, appearance and nutritive value (Yu et al., 2008). The important traits are discussed below.

i. Grain milling quality is mainly associated with the final yield and rate of broken kernel in milled rice, which is of prime concern to farmers and breeders. In order to evaluate the efficiency and quality of milling process, the three main parameters are considered, i.e. brown rice rate, milled rice rate and head rice rate. The rates indicate the grain ratios produced by different grades of milling.
ii. The grain appearance is linked to shape, size (length, width and thickness), chalkiness and translucency after milling process. Chalkiness of grains is caused during adverse climatic conditions leading to an incomplete filling of grains. It is related to size, shape and packing of amyloplast organelles required for the manufacture and storage of starch granules within the endosperm. Comparatively higher amylose content (AC) is observed in translucent rice kernels than in chalky grains (Rani and Bhattacharya, 1989; Lisle et al., 2000; Singh et al., 2003).
iii. Starch (amylose and amylopectin), protein content, gelatinization temperature (GT) and gel consistency (GC) are the important aspects of grain cooking quality. Hardiness and stickiness of rice grains after cooking largely depend on amylose and chain structure of amylopectin (Nakamura et al., 2006). Milling also reduces cooking time by decreasing the GT (Muramatsu et al., 2006), and improves chewiness, flavour, hardiness and adhesiveness of the cooked rice (Park et al., 2001; Saleh and Meullenet, 2007).

iv. On the other hand, the protein content is an important aspect in determining the nutritive value of grains. The milling process, though, enhances the appearance of grains, but at the cost of losing protein content on the outer part of the grains of brown rice.
v. The iron plays a very important role in both plant growth and human health. Rice is the important cereal crop; its Fe enrichment is the priority as most of it is present in the outer bran, which is removed while rice polishing.

6.4.2 Molecular Advancements in Rice Grain Quality

Molecular breeding had significantly contributed in identifying some major QTLs and genes for rice grain quality, further facilitating the development of functional markers for their selection. Being a highly complex trait, the grain quality is governed by several genes and highly influenced by environment. In order to increase its applicability in the rice grain quality improvement, deciphering key genes as well as negative regulators of grain quality (Q-genes) is equally important. Moreover, allelic variation in the existing rice germplasm has also an important role to play. The literature can thus serve as a valuable source of information in applying genome editing technology for improving the rice grain quality.

Brown rice recovery rate is influenced by two major QTLs, *qBRR3* and *qBRR5*, which were mapped on chromosomes three and five (Tan et al., 2001; Nevame et al., 2018), respectively. These QTLs also govern the grain width and length (Lou et al., 2009). Another major QTL *qHRR3* associated with head rice recovery and grain length is present on chromosome three (Jiang et al., 2013). It was concluded from the above studies that there is a strong relationship between grain shape and size with the percentage of head rice recovery. The *Waxy* (*Wx*) locus has been identified on chromosome six, which codes for granule-bound starch synthase linked to AC (Lapitan et al., 2009). The locus ultimately controls the GC, AC and GT. As discussed in Section 6.4.1, chalkiness also forms an integral part of grain quality negatively affecting milling, cooking, grain appearance, nutritional qualities and also head rice recovery (Nevame et al., 2018). Along with many internal and external factors, genes engaged in starch biosynthesis, starch granule structure and grain filling all restrict grain chalkiness (Bergman, 2019). Several QTLs governing chalkiness have also been fine-mapped. Despite many studies conducted, the mechanism involved in grain chalkiness remained elusive. Another major QTL *SSIIa* affects amylopectin structure and GT (Wang et al., 2007). Many genes related to starch biosynthesis like *AGPlar*, *BE3*, *SBE1*, *ISA* and *PUL* as well as genes responsible for starch synthesis like *SSI*, *SSIIa*, *SSIII-2* and *SSIV* were found to affect eating and cooking qualities (Bao et al., 2002). Another major QTL linked to protein content, *qPC1*, has been identified and functionally characterized in rice (Peng et al., 2014). A high-effect QTL in brown rice, *qCF5* controlling crude fat content, was identified on chromosome five (Qin et al., 2010). Iron biofortification is another important aspect to be considered while improving the rice grain quality. In order to enhance the Fe content in rice, various agronomic biofortification applications such as foliar application are practised but seem to be highly challenging. In an attempt to increase Fe content in rice grains, *OsVIT2* gene has been knocked down *via* T-DNA silencing (Bashir et al., 2013), which resulted in significantly increased iron content. Fe-enriched rice can be developed by editing such related genes involved in iron homeostasis by altering their expression. β-Carotene, the vitamin A precursor, is another important nutrient but is deficient in rice grains. β-Carotene has been enhanced in golden rice by introducing genes like *phytoene synthase* and *phytoene desaturase* (Paine et al., 2005). These genes and QTLs can thus serve as an effective choice for genome editing *via* CRISPR/Cas, by implementing the high transformation efficiency and extremely diverse genetic reservoir of rice crop.

6.4.3 Genome Editing to Improve Grain Quality Traits

Although they are highly important, quality traits in rice are the least studied as compared to yield traits and resistance to various stresses. Plentiful of information about the molecular basis of grain quality traits in rice facilitates genome editing for crop improvement. Genome editing has been

TABLE 6.3
Quality Trait Improvement in Rice via Application of CRISPR Technology

Sr. No.	Trait/Function	Gene Targeted	References
1	Amylose content	*SBEIIb*	Sun et al. (2017)
2	Amylose synthesis/sticky rice	*Wx*	Yunyan et al. (2019)
3	Amylopectin synthesis	*ISA1*	Shufen et al. (2019)
4	Carotenoid catabolism	*OsCYP97A4, OsDSM2, OsCCD4a, OsCCD4b* and *OsCCD7*	Yang et al. (2017)
5	Aroma	*OsPDS, OsBADH2, Oso2g23823, OsMPK2*	Shan et al. (2014)

applied to improve various aspects of grain quality, and this section briefly highlights it (Table 6.3). CRISPR/Cas9 system has been used in rice to target genes like *OsMPK2*, *OsBADH2*, *OsPDS* and *Os02g23823A* for introgression of mutations using single sgRNA (Shan et al., 2014). Targeted knock-out of *OsBADH2* gene in rice results in enhanced fragrance, and it has also been demonstrated by using TALENs (Shan et al., 2015). CRISPR/Cas9-based editing has been further used for improving the AC in rice by mutating *starch-branching enzyme* (*SBE*), which plays a vital role in regulating the physical and fine structure of rice grains (Sun et al., 2017). Both the alleles *SBEI* and *SBEIIb* were targeted and *sbeII* mutants had showed a higher amount of debranched amylopectin, ultimately causing a significant rise in the content of AC and RS by 25% and 9.8%, respectively. This leads to a change in fine structure as well as nutritional aspects of starch. In order to modify the starch profile of rice seeds, two elite lines, *Huaidao5* (HD5) and *Suken118* (SK118), from *japonica* group were subjected to CRISPR/Cas9-mediated genome editing to develop knockout lines of *Wx* gene involved in amylose synthesis (Yunyan et al., 2019). In this way, transgene-free homozygous T_1 lines were generated, which were all opaque and white-like sticky rice consisting of merely 2.6%–3.2% of amylose without an alteration in rice quality. The study had ultimately produced two elite sticky rice varieties. The isoamylase-debranching enzyme *isoamylase1* (*ISA1*), which plays a major role in the synthesis of amylopectin, was manipulated using CRISPR/Cas9 technology, and the edited mutants showed altered seed sizes, shrunken endosperm and significantly reduced grain weight (Shufen et al., 2019). These mutant cells showed abnormal amyloplast and starch granules. The *cr-isa-1-1* mutants also consisted of significantly lesser sugar content and starch GC. Moreover, the starch chain length distribution as well as GT was also altered in cr-isa-1-1 mutants. The mutants also showed a reduced level of transcripts of genes related to starch synthesis. Thus, the *ISA1* gene can serve as a potential target for CRISPR/Cas9-mediated genome editing. Another quality trait related to endosperm carotene content has also been targeted for modification by knocking out the genes like *OsCCD7*, *OsCCD4b*, *OsCCD4a*, *OsDSM2* and *OsCYP97A4* involved in carotenoid catabolism (Yang et al., 2017). However, mutants did not show any significant increase in the carotenoid accumulation. Nevertheless, such approach can be further investigated and used for the manipulation of carotenoid content.

6.5 GENOME EDITING FOR IMPROVEMENT OF YIELD

The prime goal of any agricultural research is to fulfil the food demand of rapidly increasing population by increasing crop productivity. The ultimate aim of all the breeding programmes is to enhance the yield by reducing the effects of environmental stress as well as by increasing the water- and nutrient-use efficiency. For the yield improvement in rice, several traits like the number of productive tillers per plant, number of grains per panicle, grain length, width and weight, etc. have been subjected to extensive research. Large number of genes and QTLs governing various traits directly or indirectly contributing to the yield of rice have been identified and utilized in breeding programmes. Genome editing techniques can further strengthen the breeding programmes by validating the function and role of these identified genes/QTLs (Table 6.4). For instance, one such study using

TABLE 6.4
Enhancements of Yield Traits through Application of CRISPR/Cas9 Technology

Sr. No.	Trait/Function	Gene Targeted	References
1	Grain length and width	*OsSNB*	Ma et al. (2019)
2	Grain number	*Gn1a*	Li et al. (2016b)
3	Panicle architecture	*DEP1*	Li et al. (2016b)
4	Plant architecture	*IPA1*	Li et al. (2016b)
5	Grain size	*GS3*	Li et al. (2016b)
6	Semi-dwarf rice gene	*SD1*	Huang et al. (2018) and Hu et al. (2019)
7	Grain size, yield, nitrogen assimilation and panicle branching	*OsGRF4* and *OsGRF6*	Kim and Tsukaya (2015)

association mapping approach employing 996,722 SNP markers and 270 rice accessions leads to the identification of some novel and important QTLs linked to grain length and width (Ma et al., 2019). After sequencing and haplotype screening in 128 diverse accessions from different geographical backgrounds, eight haplotypes were identified. One haplotype, Hap3, showed the maximum grain width in contrast to other haplotypes based on which OsSNB_Indel2, a functional marker, was designed. Further, the *OsSNB* gene has been identified from the QTL *qGW7* negatively regulating the grain size. CRISPR/Cas9-generated knockout mutants of this gene showed the increased grain length and width, thus further validating its function. Similarly, four yield governing genes, *viz. Gn1a (Grain number 1a), DEP1, IPA1* and *GS3* from rice cultivar *Zhonghua*, were reassessed in rice using CRISPR/Cas9 system (Li et al., 2016b). These genes are known to regulate the grain number, panicle architecture, plant architecture and grain size in rice crop, respectively. The T_2 mutant lines of *gn1a*, *gs3* and *dep1* showed enhanced grain number, larger grain size and dense, erect panicles, respectively. Moreover, *dep1* and *gs3* mutants exhibit semi-dwarf and long awn grain phenotypes, respectively. Depending upon the changes prompted in the target region of *OsmiR156*, the *ipa1* mutants revealed two distinct phenotypes: some with fewer tillers and another with more tillers. The *DEP1* and *Gn1a* genes were again targeted for editing using CRISPR/Cas9 technology (Huang et al., 2018), and the sequence analysis of the target region leads to the identification of several mutants. The phenotypic study of three mutant alleles of *DEP1* and one of *Gn1a* conferred superior yields compared to usual high-yielding alleles. These studies demonstrate that the important regulators of multiple traits can be altered in single cultivar using CRISPR/Cas9 technology. Thus, the genome editing technology can be exploited in dissecting the complex regulatory network of genes in the genetic background of the same cultivar and stacking the cultivated varieties with desirable traits.

Plant architecture is also a key determinant in grain yield, which is controlled by several environmental and developmental factors. The green revolution (GR) used random mutagenesis in order to couple natural variants for the key plant architectural genes in cereal crops yielding varieties with short height and more tillers subsequently improving crop yields (Sakamoto and Matsuoka, 2004). Strigolactone hormones (SGs) are the important determinants in the plant architecture known to hinder shoot branching. The CRISPR/Cas9 technology was used to disrupt a key enzyme in the biosynthesis of SG, i.e. *carotenoid cleavage dioxygenase7* (CCD7) (Butt et al., 2018), and mutants displayed strikingly increased tiller number with dwarf phenotype. Thus, CRISPR/Cas9 technology can be exploited in targeting plant architectural genes and understanding the molecular foundation in architecture-related genes.

A complex trait like yield is governed by several genes, each contributing to a very little phenotypic effect. Hence, the identification of such loci is quiet challenging, labour-intensive and time-demanding, which is usually done using QTL mapping and association mapping studies.

So another method that combines pedigree analysis with whole-genome sequencing along with CRISPR/Cas9 was recently tried and tested (Huang et al., 2018). The pedigree study comprises 30 cultivars, including IR8 and their descendants and ancestors plus other allied cultivars to pinpoint high-yield genes. The sequencing results identified 28 blocks on chromosomes, which were common in all the high-yielding cultivars. From these blocks, six known function genes, including *sd1* gene with 123 genetic loci with unknown function, were identified and 57 genes from these loci were knocked out using CRISPR/Cas9. The analysis of the mutants showed that the majority of these mutants had an effect on phenotypic traits linked to rice production. The knockout mutants showed a significant effect on the height, a crucial GR trait. Some knockouts in gene *glycosyl transferase* affected the plant growth and panicle architecture. Additionally, certain GR-related genes were grouped in one block although the sequences were unrelated, suggesting grouping of genes with related functions. The strategy also has given insights on many high-productivity genes in rice.

Vascular bundle associated with panicle neck has a substantial role in rice production that varies in both *japonica* and *indica* species. Therefore, researchers have been exploiting this trait for improving the crop yield. *Indica* pedigree affects the ratio of large vascular bundles (LVBs) to small vascular bundles (SVBs), and it governs the number of LVBs but not the number of SVBs (Fei et al., 2019). The locus *qLVB9* showed a pleiotrophic effect for both LVBs and SVBs, which was similar to *DENSE AND ERECT PANICLE 1* (*DEP1*). Its function was further confirmed by creating truncated alleles of *dep1* using CRISPR/Cas9 technology. The truncated allele leads to a rise in LVBs, and the resulting LVBs were more closely related to *indica* pedigree. In addition, the RNA sequencing also revealed that the truncated *dep1* down-regulates the AP2-like gene family, and *DEP1* and AP2-like genes double mutants showed the reduced ABA level and insensitivity to the treatment of exogenous ABA. Thus, the QTL analysis along with CRISPR/Cas technology could speed up the study in understanding the function of both *DEP1* and AP2-like genes in ABA-response mechanism.

Genetic diversity is a primary requirement in any crop breeding programme. Though tremendous genetic resources are available in rice, only few unique landraces are used frequently in the genetic background of modern cultivars. This has created a genetic bottleneck and also escalated permanent erosion of many valuable landraces (Van de Wouw et al., 2010; Nourollah, 2016). In order to develop superior varieties promising better yields, good quality and resistance to various abiotic stresses, it is highly essential to enlarge the genetic diversity in rice crop. Various germplasm resources are still unexplored, and certain lesser known traditional rice accessions are found to exhibit a higher yield under drought stress (Mahalle et al., 2021). Identifying favourable alleles in such landraces through association mapping and targeting them through genome editing can yield better results. Genome editing with the help of CRISPR/Cas9 has been implemented to edit *SD1* gene in elite rice landraces *TeTePu* (TTP) and *Kasalath* for reducing the genetic erosion (Hu et al., 2019). These landraces are known to harbour several disease- and insect-resistant genes as well as genes for low phosphate tolerance. Introduction of *SD1* mutations in these lines developed short plants with better lodging resistance and resulted in higher yields in field trials compared to wild-type progenitors. Such studies enlightened the applicability of CRISPR/Cas9 in preventing genetic erosion of modern rice cultivars.

Among the many regulatory genes, miRNAs are known as master regulators and control almost all the traits such as growth, development and tolerance to various stresses (Secco et al., 2013; Jiao et al., 2010; Zhang et al., 2013). The miRNA, *miR396*, is a *GROWTH REGULATING FACTOR* (*OsGRFs*) directly involved in the regulation of rice yield as well as nitrogen assimilation in rice crop (Kim and Tsukaya, 2015). Targeting *OsGRF4* and *OsGRF6* via overexpression of *miR396* increases the grain size and panicle branching. In order to assess the role of *miR396* genes of rice, CRISPR/Cas9 is applied to knock out miR396ef (*MIR396e* and *MIR396f* isoforms) (Zhang et al., 2020). The mutants of *mir396ef* survived deficient nitrogen conditions and displayed the greater grain yield as well as the prominent above-ground mass. Disruption of *OsGRF8*, which is another target of miR396, shows the increased grain size and longer panicles. The study hence suggested the role of miR396ef-GRxF4/6/8-GIF1/2/3 modules in the regulation of rice grain and panicle development. The miR396ef can serve as a convincing target in genome editing to breed environment-favourable varieties of rice that can resist low nitrogen fertilization. Thus, it opens up the new strategies to improve the yield of crops where genome editing can be efficiently used to target miRNAs, which play an important role in plant developmental pathways.

6.6 CONCLUSION

Rice is the most important cereal crop, which has been domesticated for a thousand of years ago, and due to its selective and artificial breeding during domestication processes, many useful genetic resources have been devoid in the present cultivated rice genotypes, resulting in more vulnerable towards abiotic and abiotic stresses with deficiencies of several micronutrients. Conventionally, available rice germplasm and induced mutation have been utilized to identify novel genes/QTL and development of tolerant/resistant rice cultivars. However, such approaches involve the generation of new genetic combinations by sexual hybridization and the subsequent application of phenotypically selection methods yielding many undesirable phenotypes, and remain time-taking, tedious and costly. Therefore, a new biotechnological approach is needed to maximize the probability of success to develop desired genotypes, and it can be achieved through genome editing, which has been widely used in the studies of functional genomics, reverse genetics, genome engineering and targeted transgene integration experiments. The unique advantages of this technology over conventional breeding are efficiency, precision and wide applicability to living organisms. Genome editing causes site-specific alterations in the genome of organisms, thereby generating desired insertions, deletions or substitutions to achieve the desired phenotype without affecting other agronomic traits. Unlike other genome modifications/editing technologies, *viz* ZFNs, TALENs etc., CRISPR/Cas9 depends on small RNA for sequence-specific cleavage, which is required to generate sequence specificity and easily applicable, providing a robust tool for genetic improvement of important complex traits. Thus, this technology along with the conventional molecular breeding would help in the genetic improvement of rice within a limited period of time to increase the productivity of rice and quality of rice grains, thereby ensuring the world food security.

ACKNOWLEDGEMENT

We acknowledge SERB (EMR/2017/003036) and ICAR-National Institute for Plant Biotechnology, New Delhi for financial and logistic support.

REFERENCES

Abudayyeh, O. O., J. S. Gootenberg, P. Essletzbichler, et al. "RNA targeting with CRISPR–Cas13." *Nature* 550, no. 7675 (2017): 280–284.

Agrawal, G. K. and R. Rakwal. "Rice proteomics: A move toward expanded proteome coverage to comparative and functional proteomics uncovers the mysteries of rice and plant biology." *Proteomics* 11, no. 9 (2011): 1630–1649.

Ali, Z., A. Abulfaraj, A. Idris, S. Ali, M. Tashkandi, and M. M. Mahfouz. "CRISPR/Cas9-mediated viral interference in plants." *Genome Biology* 16, no. 1 (2015): 238.

Arazoe, T., T. Ogawa, K. Miyoshi, et al. "Tailor-made TALEN system for highly efficient targeted gene replacement in the rice blast fungus." *Biotechnology and Bioengineering* 112, no. 7 (2015): 1335–1342.

Azzam, O., and T. C. B Chancellor. "The biology, epidemiology, and management of rice tungro disease in Asia." *Plant Disease* 86, no. 2 (2002): 88–100.

Bailey, T. L. "DREME: Motif discovery in transcription factor ChIP-seq data." *Bioinformatics* 27, no. 12 (2011): 1653–1659.

Bao, J. S., Y. R. Wu, B. Hu, P. Wu, H. R. Cui, and Q. Y. Shu. "QTL for rice grain quality based on a DH population derived from parents with similar apparent amylose content." *Euphytica* 128, no. 3 (2002): 317–324.

Bashir, K., R. Takahashi, S. Akhtar, Y. Ishimaru, H. Nakanishi, and N. K. Nishizawa. "The knockdown of OsVIT2 and MIT affects iron localization in rice seed." *Rice* 6, no. 1 (2013): 1–7.

Bergman, C. J. "Rice end-use quality analysis." In Rice, pp. 273–337. AACC International Press: St. Paul, MN (2019).

Bilgin, D. D., J. A. Zavala, J. I. N. Zhu, S. J. Clough, D. R. Ort, and E. H. DeLUCIA. "Biotic stress globally downregulates photosynthesis genes." *Plant, Cell & Environment* 33, no. 10 (2010): 1597–1613.

Buske, F. A., M. Bodén, D. C. Bauer, and T. L. Bailey. "Assigning roles to DNA regulatory motifs using comparative genomics." *Bioinformatics* 26, no. 7 (2010): 860–866.

Butt, H., M. Jamil, J. Y. Wang, S. Al-Babili, and M. Mahfouz. "Engineering plant architecture via CRISPR/Cas9-mediated alteration of strigolactone biosynthesis." *BMC Plant Biology* 18, no. 1 (2018): 1–9.

Caine, R. S., X. Yin, J. Sloan, et al. "Rice with reduced stomatal density conserves water and has improved drought tolerance under future climate conditions." *New Phytologist* 221, no. 1 (2019): 371–384.

Chancellor, T. C. B., J. Holt, S. Villareal, E. R. Tiongco, and J. Venn. "Spread of plant virus disease to new plantings: A case study of rice tungro disease." *Advances in Virus Research* 66 (2006): 1–29.

Chang, H., B. Yi, R. Ma, X. Zhang, H. Zhao, and Y. Xi. "CRISPR/cas9, a novel genomic tool to knock down microRNA in vitro and in vivo." *Scientific Reports* 6, no. 1 (2016): 1–12.

Chauhan, B. S., K. Jabran, and G. Mahajan, eds. *Rice Production Worldwide*, Vol. 247. Springer International Publishing: Berlin, Heidelberg (2017).

Clemens, S., M. G. M. Aarts, S. Thomine, and N. Verbruggen. "Plant science: The key to preventing slow cadmium poisoning." *Trends in Plant Science* 18, no. 2 (2013): 92–99.

Cohen, S. P., and J. E. Leach. "Abiotic and biotic stresses induce a core transcriptome response in rice." *Scientific Reports* 9, no. 1 (2019): 1–11.

Cox, D. B. T., J. S. Gootenberg, O. O. Abudayyeh, et al. "RNA editing with CRISPR-Cas13." *Science* 358, no. 6366 (2017): 1019–1027.

Dai, L., X. Liu, Y. Xiao, and G. Wang. "Recent advances in cloning and characterization of disease resistance genes in rice." *Journal of Integrative Plant Biology* 49, no. 1 (2007): 112–119.

de Torres Zabala, M., M. H. Bennett, W. H. Truman, and M. R. Grant. "Antagonism between salicylic and abscisic acid reflects early host–pathogen conflict and moulds plant defence responses." *The Plant Journal* 59, no. 3 (2009): 375–386.

Ding, Y., D. Matthew, and Z. Mou. "Abscisic acid promotes proteasome-mediated degradation of the transcription coactivator NPR 1 in Arabidopsis thaliana." *The Plant Journal* 86, no. 1 (2016): 20–34.

Dong, Y., X. Jin, Q. Tang, et al. "Development and event-specific detection of transgenic glyphosate-resistant rice expressing the G2-EPSPS gene." *Frontiers in Plant Science* 8 (2017): 885.

Duan, Y., J. Li, R. Qin, et al. "Identification of a regulatory element responsible for salt induction of rice OsRAV2 through ex situ and in situ promoter analysis." *Plant Molecular Biology* 90, no. 1–2 (2016): 49–62.

Fang, H., Q. Meng, J. Xu, et al. "Knock-down of stress inducible OsSRFP1 encoding an E3 ubiquitin ligase with transcriptional activation activity confers abiotic stress tolerance through enhancing antioxidant protection in rice." *Plant Molecular Biology* 87, no. 4–5 (2015): 441–458.

Fartyal, D., A. Agarwal, D. James, et al. "Developing dual herbicide tolerant transgenic rice plants for sustainable weed management." *Scientific Reports* 8, no. 1 (2018): 1–12.

Fei, C., X. Geng, Z. Xu, and Q. Xu. "Multiple areas investigation reveals the genes related to vascular bundles in rice." *Rice* 12, no. 1 (2019): 17.

Foster, A. J., M. Martin-Urdiroz, X. Yan, H. S. Wright, D. M. Soanes, and N. J. Talbot. "CRISPR-Cas9 ribonucleoprotein-mediated co-editing and counter selection in the rice blast fungus." *Scientific Reports* 8, no. 1 (2018): 1–12.

Golldack, D., C. Li, H. Mohan, and N. Probst. "Tolerance to drought and salt stress in plants: Unravelling the signaling networks." *Frontiers in Plant Science* 5 (2014): 151.

Hu, X., Y. Cui, G. Dong, et al. "Using CRISPR-Cas9 to generate semi-dwarf rice lines in elite landraces." *Scientific Reports* 9, no. 1 (2019): 1–7.

Huang, X., X. Zeng, J. Li, and D. Zhao. "Construction and analysis of tify1a and tify1b mutants in rice (Oryza sativa) based on CRISPR/Cas9 technology." *Journal of Agricultural Biotechnology* 25, no. 6 (2017): 1003–1012.

Huang, L., R. Zhang, G. Huang, et al. "Developing superior alleles of yield genes in rice by artificial mutagenesis using the CRISPR/Cas9 system." *The Crop Journal* 6, no. 5 (2018): 475–481.

Jiang, W., D. Bikard, D. Cox, F. Zhang, and L. A. Marraffini. "RNA-guided editing of bacterial genomes using CRISPR-Cas systems." *Nature Biotechnology* 31, no. 3 (2013): 233.

Jiang, X., P. F. O'Reilly, H. Aschard, et al. "Genome-wide association study in 79,366 European-ancestry individuals informs the genetic architecture of 25-hydroxyvitamin D levels." *Nature Communications* 9, no. 1 (2018): 1–12.

Jiao, Y., Y. Wang, D. Xue, et al. "Regulation of OsSPL14 by OsmiR156 defines ideal plant architecture in rice." *Nature Genetics* 42, no. 6 (2010): 541.

Juliano, B. O. "Rice grain quality: Problems and challenges." *Cereal Foods World* 35, no. 2 (1990): 245–253.

Karkute, S. G., A. K. Singh, O. P. Gupta, P. M. Singh, and B. Singh. "CRISPR/Cas9 mediated genome engineering for improvement of horticultural crops." *Frontiers in Plant Science* 8 (2017): 1635.

Khong, G., P. K. Pati, F. Richaud, et al. "OsMADS26 negatively regulates resistance to pathogens and drought tolerance in rice." *Plant Physiology* 169, no. 4 (2015): 2935–2949.

Khush, G. S., and K. K. Jena. "Current status and future prospects for research on blast resistance in rice (Oryza sativa L.)." In *Advances in Genetics, Genomics and Control of Rice Blast Disease*, edited by W. Xiaofan and B. Valent, pp. 1–10. Springer: Dordrecht (2009).

Kim, J. H., and H. Tsukaya. "Regulation of plant growth and development by the growth-regulating factor and GRF-interacting factor duo." *Journal of Experimental Botany* 66, no. 20 (2015): 6093–6107.

Kim, Y., H. Moon, and C. Park. "CRISPR/Cas9-targeted mutagenesis of Os8N3 in rice to confer resistance to Xanthomonas oryzae pv. oryzae." *Rice* 12, no. 1 (2019): 1–13.

Komor, A. C., Y. B. Kim, M. S. Packer, J. A. Zuris, and D. R. Liu. "Programmable editing of a target base in genomic DNA without double-stranded DNA cleavage." *Nature* 533, no. 7603 (2016): 420–424.

Lapitan, V. C., E. D. Redona, T. Abe, and D. S. Brar. "Mapping of quantitative trait loci using a doubled-haploid population from the cross of Indica and Japonica cultivars of rice." *Crop Science* 49, no. 5 (2009): 1620–1628.

Lee, H., J. Cha, C. Choi, et al. "Rice WRKY11 plays a role in pathogen defense and drought tolerance." *Rice* 11, no. 1 (2018): 1–12.

Li, T., B. Liu, M. H. Spalding, D. P. Weeks, and B. Yang. "High-efficiency TALEN-based gene editing produces disease-resistant rice." *Nature Biotechnology* 30, no. 5 (2012): 390.

Li, J., X. Meng, Y. Zong, et al. "Gene replacements and insertions in rice by intron targeting using CRISPR–Cas9." *Nature Plants* 2, no. 10 (2016a): 1–6.

Li, M., X. Li, Z. Zhou, et al. "Reassessment of the four yield-related genes Gn1a, DEP1, GS3, and IPA1 in rice using a CRISPR/Cas9 system." *Frontiers in Plant Science* 7 (2016b): 377.

Li, W., Z. Zhu, M. Chern, et al. "A natural allele of a transcription factor in rice confers broad-spectrum blast resistance." *Cell* 170, no. 1 (2017): 114–126.

Li, C., Y. Zong, Y. Wang, et al. "Expanded base editing in rice and wheat using a Cas9-adenosine deaminase fusion." *Genome Biology* 19, no. 1 (2018): 59.

Li, W., M. Chern, J. Yin, J. Wang, and X. Chen. "Recent advances in broad-spectrum resistance to the rice blast disease." *Current Opinion in Plant Biology* 50 (2019): 114–120.

Liang, G., H. Zhang, D. Lou, and D. Yu. "Selection of highly efficient sgRNAs for CRISPR/Cas9-based plant genome editing." *Scientific Reports* 6, no. 1 (2016): 1–8.

Lisle, A. J., M. Martin, and M. A. Fitzgerald. "Chalky and translucent rice grains differ in starch composition and structure and cooking properties." *Cereal Chemistry* 77, no. 5 (2000): 627–632.

Liu, Q., X. Li, S. Yan, et al. "OsWRKY67 positively regulates blast and bacteria blight resistance by direct activation of PR genes in rice." *BMC Plant Biology* 18, no. 1 (2018): 1–13.

Lou, D., H. Wang, G. Liang, and D. Yu. "OsSAPK2 confers abscisic acid sensitivity and tolerance to drought stress in rice." *Frontiers in Plant Science* 8 (2017): 993.

Lou, J., L. Chen, G. Yue, et al. "QTL mapping of grain quality traits in rice." *Journal of Cereal Science* 50, no. 2 (2009): 145–151.

Lu, W., X. Chu, Y. Li, C. Wang, and X. Guo. "Cotton GhMKK1 induces the tolerance of salt and drought stress, and mediates defence responses to pathogen infection in transgenic Nicotiana benthamiana." *PLoS One* 8, no. 7 (2013): e68503.

Lv, Y., M. Yang, D. Hu, et al. "The OsMYB30 transcription factor suppresses cold tolerance by interacting with a JAZ protein and suppressing β-amylase expression." *Plant Physiology* 173, no. 2 (2017): 1475–1491.

Ma, J., J. Chen, M. Wang, et al. "Disruption of OsSEC3A increases the content of salicylic acid and induces plant defense responses in rice." *Journal of Experimental Botany* 69, no. 5 (2018): 1051–1064.

Ma, X., F. Feng, Y. Zhang, et al. "A novel rice grain size gene OsSNB was identified by genome-wide association study in natural population." *PLoS Genetics* 15, no. 5 (2019): e1008191.

Macovei, A., N. R. Sevilla, C. Cantos, et al. "Novel alleles of rice eIF4G generated by CRISPR/Cas9-targeted mutagenesis confer resistance to Rice tungro spherical virus." *Plant Biotechnology Journal* 16, no. 11 (2018): 1918–1927.

Mahalle, M. D., P. C. Dey, S. K. Chetia, et al. "Association mapping for yield traits under drought stress in Autumn rice germplasm collection of Assam." *Journal of Plant Biochemistry and Biotechnology* 30 (2021): 26–36.

Mao, X., Y. Zheng, K. Xiao, et al. "OsPRX2 contributes to stomatal closure and improves potassium deficiency tolerance in rice." *Biochemical and Biophysical Research Communications* 495, no. 1 (2018): 461–467.

Mazur, B. J., C. Chui, and J. K. Smith. "Isolation and characterization of plant genes coding for acetolactate synthase, the target enzyme for two classes of herbicides." *Plant Physiology* 85, no. 4 (1987): 1110–1117.

Melotto, M., W. Underwood, J. Koczan, K. Nomura, and S. Y. He. "Plant stomata function in innate immunity against bacterial invasion." *Cell* 126, no. 5 (2006): 969–980.

Mickelbart, M. V., P. M. Hasegawa, and J. Bailey-Serres. "Genetic mechanisms of abiotic stress tolerance that translate to crop yield stability." *Nature Reviews Genetics* 16, no. 4 (2015): 237–251.

Mishra, S., B. Singh, K. Panda, et al. "Association of SNP haplotypes of HKT family genes with salt tolerance in Indian wild rice germplasm." *Rice* 9, no. 1 (2016): 15.

Muramatsu, Y., A. Tagawa, E. Sakaguchi, and T. Kasai. "Water absorption characteristics and volume changes of milled and brown rice during soaking." *Cereal Chemistry* 83, no. 6 (2006): 624–631.

Murovec, J., K. Gucek, B. Bohanec, M. Avbelj, and R. Jerala. "DNA-free genome editing of Brassica oleracea and B. rapa protoplasts using CRISPR-Cas9 ribonucleoprotein complexes." *Frontiers in Plant Science* 9 (2018): 1594.

Nakamura, Y., A. Sato, and B. O. Juliano. "Short-chain-length distribution in debranched rice starches differing in gelatinization temperature or cooked rice hardness." *Starch-Starke* 58, no. 3–4 (2006): 155–160.

Nevame, A. Y. M., R. M. Emon, M. A. Malek, et al. "Relationship between high temperature and formation of chalkiness and their effects on quality of rice." *BioMed Research International* (2018): 1653721–1653739.

Nourollah, A. "Genetic diversity, genetic erosion, and conservation of the two cultivated rice species (Oryza sativa and Oryza glaberrima) and their close wild relatives." In *Genetic Diversity and Erosion in Plants*, edited by M. R. Ahuja and S. M. Jain, pp. 35–73. Springer: Cham (2016).

Oliva, R., C. Ji, G. Atienza-Grande, et al. "Broad-spectrum resistance to bacterial blight in rice using genome editing." *Nature Biotechnology* 37, no. 11 (2019): 1344–1350.

Paine, J. A., C. A. Shipton, S. Chaggar, et al. "Improving the nutritional value of Golden Rice through increased pro-vitamin A content." *Nature Biotechnology* 23, no. 4 (2005): 482.

Park, J. K., S. S. Kim, and K. O. Kim. "Effect of milling ratio on sensory properties of cooked rice and on physicochemical properties of milled and cooked rice." *Cereal Chemistry* 78, no. 2 (2001): 151–156.

Peng, B., H. Kong, Y. Li, et al. "OsAAP6 functions as an important regulator of grain protein content and nutritional quality in rice." *Nature Communications* 5, no. 1 (2014): 1–12.

Peng, X., H. Wang, J. Jang, et al. "OsWRKY80-OsWRKY4 module as a positive regulatory circuit in rice resistance against Rhizoctonia solani." *Rice* 9, no. 1 (2016): 1–14.

Qin, Y., S. Kim, X. Zhao, et al. "QTL detection and MAS selection efficiency for lipid content in brown rice (Oryza sativa L.)." *Genes & Genomics* 32, no. 6 (2010): 506–512.

Qiu, Z., S. Kang, L. He, et al. "The newly identified heat-stress sensitive albino 1 gene affects chloroplast development in rice." *Plant Science* 267 (2018): 168–179.

Rabbani, M. A., K. Maruyama, H. Abe, et al. "Monitoring expression profiles of rice genes under cold, drought, and high-salinity stresses and abscisic acid application using cDNA microarray and RNA gel-blot analyses." *Plant Physiology* 133, no. 4 (2003): 1755–1767.

Rani, M. R. S., and K. R. Bhattacharya. "Slurry viscosity as a possible indicator of rice quality." *Journal of Texture Studies* 20, no. 2 (1989): 139–149.

Rawal, H. C., S. V. A. Mithra, K. Arora, et al. "Genome-wide analysis in wild and cultivated Oryza species reveals abundance of NBS genes in progenitors of cultivated rice." *Plant Molecular Biology Reporter* 36, no. 3 (2018): 373–386.

Sakamoto, T., and M. Matsuoka. "Generating high-yielding varieties by genetic manipulation of plant architecture." *Current Opinion in Biotechnology* 15, no. 2 (2004): 144–147.

Saleh, M. I., and J. Meullenet. "Effect of moisture content at harvest and degree of milling (based on surface lipid content) on the texture properties of cooked long-grain rice." *Cereal Chemistry* 84, no. 2 (2007): 119–124.

Secco, D., M. Jabnoune, H. Walker, H. Shou, P. Wu, Y. Poirier, and J. Whelan. "Spatio-temporal transcript profiling of rice roots and shoots in response to phosphate starvation and recovery." *The Plant Cell* 25, no. 11 (2013): 4285–4304.

Seki, M., M. Narusaka, J. Ishida, et al. "Monitoring the expression profiles of 7000 Arabidopsis genes under drought, cold and high-salinity stresses using a full-length cDNA microarray." *The Plant Journal* 31, no. 3 (2002): 279–292.

Shan, Q., Y. Wang, J. Li, and C. Gao. "Genome editing in rice and wheat using the CRISPR/Cas system." *Nature Protocols* 9, no. 10 (2014): 2395.

Shan, Q., Y. Zhang, K. Chen, K. Zhang, and C. Gao. "Creation of fragrant rice by targeted knockout of the Os BADH 2 gene using TALEN technology." *Plant Biotechnology Journal* 13, no. 6 (2015): 791–800.

Shen, C., Z. Que, Y. Xia, et al. "Knock out of the annexin gene OsAnn3 via CRISPR/Cas9-mediated genome editing decreased cold tolerance in rice." *Journal of Plant Biology* 60, no. 6 (2017): 539–547.

Shimatani, Z., S. Kashojiya, M. Takayama, et al. "Targeted base editing in rice and tomato using a CRISPR-Cas9 cytidine deaminase fusion." *Nature Biotechnology* 35, no. 5 (2017): 441–443.

Shimatani, Z., U. Fujikura, H. Ishii, R. Terada, K. Nishida, and A. Kondo. "Herbicide tolerance-assisted multiplex targeted nucleotide substitution in rice." *Data in Brief* 20 (2018): 1325–1331.

Shinozaki, K., and K. Yamaguchi-Shinozaki. "Molecular responses to dehydration and low temperature: Differences and cross-talk between two stress signaling pathways." *Current Opinion in Plant Biology* 3, no. 3 (2000): 217–223.

Shufen, C., C. Yicong, F. Baobing, et al. "Editing of rice isoamylase gene ISA1 provides insights into its function in starch formation." *Rice Science* 26, no. 2 (2019): 77–87.

Singh, N., N. S. Sodhi, M. Kaur, and S. K. Saxena. "Physico-chemical, morphological, thermal, cooking and textural properties of chalky and translucent rice kernels." *Food Chemistry* 82, no. 3 (2003): 433–439.

Song, W., G. Wang, L. Chen, et al. "A receptor kinase-like protein encoded by the rice disease resistance gene, Xa21." *Science* 270, no. 5243 (1995): 1804–1806.

Sun, Y., X. Zhang, C. Wu, et al. "Engineering herbicide-resistant rice plants through CRISPR/Cas9-mediated homologous recombination of acetolactate synthase." *Molecular Plant* 9, no. 4 (2016): 628–631.

Sun, Y., G. Jiao, Z. Liu, et al. "Generation of high-amylose rice through CRISPR/Cas9-mediated targeted mutagenesis of starch branching enzymes." *Frontiers in Plant Science* 8 (2017): 298.

Sureshkumar, V., B. Dutta, V. Kumar, et al. "RiceMetaSysB: A database of blast and bacterial blight responsive genes in rice and its utilization in identifying key blast-resistant WRKY genes." *Database* 2019 (2019): 1–13.

Swamy, B. P. M., N. A. Shamsudin, S. N. A. Rahman, et al. "Association mapping of yield and yield-related traits under reproductive stage drought stress in rice (Oryza sativa L.)." *Rice* 10, no. 1 (2017): 21.

Tan, Y. F., M. Sun, Y. Z. Xing, et al. "Mapping quantitative trait loci for milling quality, protein content and color characteristics of rice using a recombinant inbred line population derived from an elite rice hybrid." *Theoretical and Applied Genetics* 103, no. 6–7 (2001): 1037–1045.

Tang, L., B. Mao, Y. Li, et al. "Knockout of OsNramp5 using the CRISPR/Cas9 system produces low Cd-accumulating Indica rice without compromising yield." *Scientific Reports* 7, no. 1 (2017): 1–12.

Toda, E., N. Koiso, A. Takebayashi, et al. "An efficient DNA-and selectable-marker-free genome-editing system using zygotes in rice." *Nature Plants* 5, no. 4 (2019): 363.

Todaka, D., K. Nakashima, K. Maruyama, et al. "Rice phytochrome-interacting factor-like protein OsPIL1 functions as a key regulator of internode elongation and induces a morphological response to drought stress." *Proceedings of the National Academy of Sciences* 109, no. 39 (2012a): 15947–15952.

Todaka, D., K. Nakashima, K. Shinozaki, and K. Yamaguchi-Shinozaki. "Toward understanding transcriptional regulatory networks in abiotic stress responses and tolerance in rice." *Rice* 5, no. 1 (2012b): 6.

Usman, M. G., M. Y. Rafii, M. Y. Martini, O. A. Yusuff, M. R. Ismail, and G. Miah. "Molecular analysis of Hsp70 mechanisms in plants and their function in response to stress." *Biotechnology and Genetic Engineering Reviews* 33, no. 1 (2017): 26–39.

Van de Wouw, M., C. Kik, T. Hintum, R. Treuren, and B. Visser. "Genetic erosion in crops: Concept, research results and challenges." *Plant Genetic Resources* 8, no. 1 (2010): 1–15.

Van Nguyen, N., and A. Ferrero. "Meeting the challenges of global rice production." *Paddy and Water Environment* 4, no. 1 (2006): 1–9.

Vikram, P., B. P. M. Swamy, S. Dixit, et al. "qDTY 1.1, a major QTL for rice grain yield under reproductive-stage drought stress with a consistent effect in multiple elite genetic backgrounds." *BMC Genetics* 12, no. 1 (2011): 89.

Wang, F., C. Wang, P. Liu, et al. "Enhanced rice blast resistance by CRISPR/Cas9-targeted mutagenesis of the ERF transcription factor gene OsERF922." *PLoS One* 11, no. 4 (2016): e0154027.

Wang, L. Q., W. J. Liu, Y. Xu, et al. "Genetic basis of 17 traits and viscosity parameters characterizing the eating and cooking quality of rice grain." *Theoretical and Applied Genetics* 115, no. 4 (2007): 463–476.

Wang, D., Y. Wang, W. Long, et al. "SGD1, a key enzyme in tocopherol biosynthesis, is essential for plant development and cold tolerance in rice." *Plant Science* 260 (2017): 90–100.

Wayne, L. L., D. J. Gachotte, and T. A. Walsh. "Transgenic and genome editing approaches for modifying plant oils." In *Transgenic Plants*, edited by S. Kumar, P. Barone, and M. Smith, pp. 367–394. Humana Press: New York (2019).

Xie, K., B. Minkenberg, and Y. Yang. "Boosting CRISPR/Cas9 multiplex editing capability with the endogenous tRNA-processing system." *Proceedings of the National Academy of Sciences* 112, no. 11 (2015): 3570–3575.

Yang, X., L. Chen, J. He, and W. Yu. "Knocking out of carotenoid catabolic genes in rice fails to boost carotenoid accumulation, but reveals a mutation in strigolactone biosynthesis." *Plant Cell Reports* 36, no. 10 (2017): 1533–1545.

Yin, X., A. K. Biswal, J. Dionora, et al. "CRISPR-Cas9 and CRISPR-Cpf1 mediated targeting of a stomatal developmental gene EPFL9 in rice." *Plant Cell Reports* 36, no. 5 (2017): 745–757.

Yu, T., W. Jiang, T. Ham, et al. "Comparison of grain quality traits between japonica rice cultivars from Korea and Yunnan Province of China." *Journal of Crop Science and Biotechnology* 11, no. 2 (2008): 135–140.

Yu, Q., A. Jalaludin, H. Han, M. Chen, R. D. Sammons, and S. B. Powles. "Evolution of a double amino acid substitution in the 5-enolpyruvylshikimate-3-phosphate synthase in Eleusine indica conferring high-level glyphosate resistance." *Plant Physiology* 167, no. 4 (2015): 1440–1447.

Yunyan, F., Y. Jie, W. Fangquan, et al. "Production of two elite glutinous rice varieties by editing wx gene." *Rice Science* 26, no. 2 (2019): 118–124.

Zafar, S. A., S. S. Zaidi, Y. Gaba, et al. "Engineering abiotic stress tolerance via CRISPR/Cas-mediated genome editing." *Journal of Experimental Botany* 71, no. 2 (2020): 470–479.

Zeng, D., C. Yang, R. Qin, et al. "A guanine insert in OsBBS1 leads to early leaf senescence and salt stress sensitivity in rice (Oryza sativa L.)." *Plant Cell Reports* 37, no. 6 (2018): 933–946.

Zeng, Y., J. Wen, W. Zhao, Q. Wang, and W. Huang. "Rational improvement of rice yield and cold tolerance by editing the three genes OsPIN5b, GS3 and OsMYB30 with the CRISPR-Cas9 system." *Frontiers in Plant Science* 10 (2020): 1663.

Zetsche, B., J. S. Gootenberg, O. O. Abudayyeh, et al. "Cpf1 is a single RNA-guided endonuclease of a class 2 CRISPR-Cas system." *Cell* 163, no. 3 (2015): 759–771.

Zhang, Y., Y. Yu, C. Wang, et al. "Overexpression of microRNA OsmiR397 improves rice yield by increasing grain size and promoting panicle branching." *Nature Biotechnology* 31, no. 9 (2013): 848.

Zhang, H., J. Zhang, P. Wei, et al. "The CRISPR/C as9 system produces specific and homozygous targeted gene editing in rice in one generation." *Plant Biotechnology Journal* 12, no. 6 (2014): 797–807.

Zhang, T., Q. Zheng, X. Yi, et al. "Establishing RNA virus resistance in plants by harnessing CRISPR immune system." *Plant Biotechnology Journal* 16, no. 8 (2018): 1415–1423.

Zhang, A., Y. Liu, F. Wang, et al. "Enhanced rice salinity tolerance via CRISPR/Cas9-targeted mutagenesis of the OsRR22 gene." *Molecular Breeding* 39, no. 3 (2019a): 47.

Zhang, T., Y. Zhao, J. Ye, et al. "Establishing CRISPR/Cas13a immune system conferring RNA virus resistance in both dicot and monocot plants." *Plant Biotechnology Journal* 17, no. 7 (2019b): 1185–1187.

Zhang, J., Z. Zhou, J. Bai, et al. "Disruption of MIR396e and MIR396f improves rice yield under nitrogen-deficient conditions." *National Science Review* 7, no. 1 (2020): 102–112.

Zhao, H., X. Wang, Y. Jia, et al. "The rice blast resistance gene Ptr encodes an atypical protein required for broad-spectrum disease resistance." *Nature Communications* 9, no. 1 (2018): 1–12.

Zhou, J., Z. Peng, J. Long, et al. "Gene targeting by the TAL effector PthXo2 reveals cryptic resistance gene for bacterial blight of rice." *The Plant Journal* 82, no. 4 (2015): 632–643.

Zhou, J., K. Deng, Y. Cheng, et al. "CRISPR-Cas9 based genome editing reveals new insights into microRNA function and regulation in rice." *Frontiers in Plant Science* 8 (2017): 1598.

Zhu, C., L. Bortesi, C. Baysal, et al. "Characteristics of genome editing mutations in cereal crops." *Trends in Plant Science* 22, no. 1 (2017): 38–52.

7 Genome Editing Using CRISPR/Cas Systems in Legumes

Shallu Thakur, Neetu Kushwah, Sudhir Kumar, Prateek Singh, Meenal Rathore, Narendra Pratap Singh, and Alok Das
ICAR-Indian Institute of Pulses Research

CONTENTS

7.1 LEGUMES

Legumes are members of *Fabaceae* or *Leguminosae* family of flowering plants comprising ~751 genera and 19,000 species, and generally refer to the type of fruit characteristics of most of the species (Christenhusz and Byng, 2016). Legumes refer to dry fruits or pods that develop from the female reproductive organ of flower and on maturity, dehisce on two sides releasing the seeds attached to one or both the seams. Legumes are grown primarily for human consumption (protein-rich seeds), aesthetics, livestock feed, as soil-enhancing green manure and industrial uses (gums, dyes etc). The term 'legume' is derived from Latin word *legumen*, which means *anything that can be gathered*, and the legumes were among the first plants cultivated by humankind. All legumes have five-petaled flower in which the ovary is attached to receptacle above the attachment of other floral parts, and that subsequently ripens to form a *pod*. Legumes are classified into six subfamilies, based on the shape of the flowers: *Cercidoideae, Detarioideae, Duparquetioideae, Dialioideae, Caesalpinioideae* and *Faboideae* or *Papilionoideae*, and comprise a wide variety of growth forms, including trees, shrubs, perennial or annual herbs and even vines or lianas (LPWG, 2017). Legumes are versatile plants and are distributed in major biomes of the world: tropical forest, temperate forest, grass and succulent (Schrire, 2005). Most of the members of this family host bacteria in their roots within structures called *root nodules* for a symbiotic relationship. Commonly referred to as *Rhizobia*, these bacteria have the ability to utilize nitrogen gas (N_2) out of atmosphere and convert into a plant-usable form of nitrogen (NO_3 or NH_3), whereas the bacteria derive carbohydrates from the plants. This process, called biological nitrogen fixation (BNF), is essential for the biosynthesis of nitrogen-containing organic compounds and also adds nitrogen content to soil for subsequent crop production (Geddes and Oresnik, 2016).

Legumes thus hold an important place in *agro*-ecosystem around the world, and creation and judicious utilization of genetic diversity in these valuable legumes shall address the specific needs and sustain agriculture. For thousands of years, efforts have been directed towards the utilization of genetic diversity emanated from spontaneous mutation and recombination. Due to the inherent random nature of recombination and mutagenesis, improvement of germplasm is a time-taking process, and currently, exploring *wild* species is an active area of research. These approaches can be accelerated by increasing the selection efficiency using the marker-assisted selection (MAS) and genomic selection (GS) and speed breeding techniques (Collard and Mackill, 2008; Desta and Ortiz, 2014; Watson et al., 2018). Currently, mutation breeding by gamma radiation is extensively employed to achieve genetic variability and select phenotypes based on the target trait. In the last two decades, genetic engineering approaches were available that allowed traits from unrelated species to become accessible. Notably, both mutation breeding and genetic engineering approaches are mostly non-specific and random in nature.

7.2 GENOME EDITING SYSTEMS

Attempts for site-specific modification in the genome have been a long-standing goal of plant researchers. This can only be affected by DNA double-strand breaks (DSBs) at the predetermined sites that can be subsequently repaired by non-homologous end-joining (NHEJ), homology-dependent repair (HDR) or microhomology-mediated end-joining (MMEJ) (Bleuyard et al., 2006; Symington and Gautier, 2011).

Homing endonucleases (HEs), or meganucleases, are sequence-specific endonucleases, discovered in microbes, phage and eukaryotic organelles, that drive the mobility and persistence of their own genes. Genes encoding HEs are found in self-splicing elements like introns (groups I & II) and *inteins*, and elicit the recombination process called homing (Stoddard, 2014). Several tens and hundreds of HEs have been identified but their cleavage sites in a genic sequence are obfuscating. HE recognition sites do not normally exist in plant genomes, and hence, novel strategies were devised to specifically recognize a given genomic sequence. Leveraging on the natural degeneracy of HE, I-CreI recognition sequence, strategic designs and screening approaches were used to create I-CreI derivatives with specific recognition DNA sequences (Smith et al., 2006). Employing modified I-CreI capable of recognizing maize liguleless locus and generating mutations with 2- to 220-bp InDels at the expected cleavage site was reported (Gao et al., 2010). However, the lack of modular DNA binding domain and degeneracy of binding with more possible off-targets, together with the lack of customizable reagents and principles for designing a vector that recognizes a specific DNA sequence, rendered HEs less useful for routine genome engineering applications.

Zinc finger nucleases (ZFNs) are another group of endonucleases that use designer zinc fingers to specifically bind a DNA sequence enabling the linked FokI endonuclease to generate a DSB in the predetermined sequence (Durai et al., 2005). Individual zinc finger recognizes and binds to the specific nucleotide triplet, and fingers are designed to assemble into groups to bind to specific DNA sequences. ZFNs work in duplex since FokI nuclease subunits have to dimerize to cleave the specific DNA. Mutations, deletions, insertions or targeted gene integrations at the recognition sites were achieved in a wide range of plant species using this technology (Lloyd et al., 2005; Wright et al., 2005; De Pater et al., 2009). Design of ZFNs remains technically challenging and complicated process, and with low efficacy (Zhang et al., 2018).

Transcription activator-like effector nucleases (TALENs) are another group of engineered endonucleases with an increased versatility to bind practically any DNA sequence. TALENs are created by fusing transcriptional activator-like effector (TALE) repeats from *Xanthomonas* and the FokI restriction endonuclease. Individual TALE targets a single nucleotide, with a flexible target design, thus increasing the repertoire of potential target sites compared to ZFNs (Cermak et al., 2011; Sun and Zhao, 2013; Chen and Gao, 2013). The first successful report of TALEN-mediated genome editing was observed in rice, where the disease-susceptibility gene *Os*SWEET14 was disrupted resulting in bacterial blight-resistant rice (Li et al., 2012). A range of mutations were observed when endogenous genes in the model plant, Arabidopsis, were targeted with multiple TALENs, some of which were heritable to progenies of subsequent generation (Christian et al., 2013). Another successful report of modification of two fatty acid desaturase 2 (FAD2) genes using TALENs was to obtain mutants with desired fatty acid profiles viz. oleic acid (80%) and linoleic acid (4%) (Haun et al., 2014). There are umpteen reports of potential of TALEN technology for trait improvement. However, the designing of specific TALE repeats poses a challenge, and the realized efficiency of gene targeting is variable (Zhang et al., 2018).

The CRISPR/Cas system, comprising clustered regularly interspaced short palindromic repeat (CRISPR) repeat-spacer arrays and (CRISPR-associated) Cas proteins, is an adaptive immune system discovered in bacteria and archaea. The system provides defence against phages and other elements by specifically cleaving the invader's nucleic acid sequence. Type II-A CRISPR/Cas9 originally from the bacteria, *Streptococcus pyogenes*, is the first reported system demonstrated to specifically cleave DNA *in vitro* and in plant cells (Jinek et al., 2012; Shan et al., 2013). Type II-A CRISPR-Cas systems were also identified from other species like *Streptococcus thermophilus*, *Staphylococcus aureus*, *Neisseria meningitidis*, *Campylobacter jejuni* etc. Basically, the editing system has been modified/repurposed and commonly has two components: the Cas9 nuclease (protein) (1000–1600 a.a) and a single guide RNA (sgRNA) [including artificial fusion of crRNA and trans-activating crRNA] (~100 nt length) for most uses. Further, a 20-bp target sequence followed by a protospacer-adjacent motif (PAM) NGG facilitates cleavage in the targeted site between third and fourth nucleotides upstream of the PAM, producing *blunt-ended* dsDNA break. Mechanistically, the recognition of PAM by Cas9-gRNA complex initiates DNA unwinding and RNA-DNA rewinding towards heteroduplex formation (Doudna and Charpentier, 2014; Sternberg et al., 2014). Type II-A Cas9s affect high genome editing efficiency; however, off-target cleavage at unintended sites was reported leading to engineering of other variants like type II-B and type II-C. Recently, a plethora of nucleases were being identified and repurposed for usage in plants like CRISPR-Cas12 (Type V CRISPR-Cas system) and CRISPR-Cas13 (Type VI CRISPR-Cas system) (Zetsche et al., 2015). Type V CRISPR-Cas system isolated from *Francisella novicida*, *Acidaminococcus* sp., *Prevotella* sp., *Lachnospiraceae* sp. etc has two subtypes: Cpf1 (type V-A) or C2c1 (type V-B) of sizes (1100–1300 a.a) and a total guide length of 42–44 nt that produces 5-nt 5′ *overhang ds breaks*. Cas12 has also been engineered for epigenome editing and detecting tiny amount of DNA in mixture. In contrast, Cas13 targets ssRNA and has been exploited in altering gene expression without altering genome sequence (Yan et al., 2019). Recently, the application of Cpf1 in plant genome editing has been reviewed extensively (Alok et al., 2020). Unlike HEs, ZFNs and TALENs that rely on protein domains only to recognize specific DNA sequences, CRISPR uses a short stretch of nucleotides for the precise genome editing.

CRISPR/Cas gene editing is the currently most widely used tool used to introduce a variety of genomic modifications, including knockout of negative elements, knock-in and replacements, base editing, prime editing, tuning of gene regulation, accelerating domestication and most importantly reverse genetic tool to ascertain the function of genes. Two key steps in genome editing are a) delivery of editing reagents to plant cells and b) production of editing lines and their analyses. Editing reagents DNA, RNA and ribonucleoproteins (RNPs) can be delivered into plant cells by a wide range of techniques, including *Agrobacterium*-mediated genetic transformation (T-DNA) (*A tumefaciens* and *A. rhizogenes*), protoplast transfection, particle bombardment and nanoparticle-mediated delivery. Generally, protoplast transfection is used for the transient expression in plant cells, whereas *Agrobacterium*-mediated transformation, particle bombardment and nanoparticle-routed delivery are the major delivery methods for the production of edited plants. T-DNA baggage-free edited lines can be generated based on subsequent segregation in progenies or the use of RNPs in protoplasts and their subsequent regeneration into whole plants.

7.3 GENOME EDITING IN LEGUMES

Reports of genome editing using CRISPR/Cas system in five legume species, *viz*. birdsfoot trefoil (*Lotus japonicus*), barrel medic (*Medicago truncatula*), alfalfa (*Medicago sativa*), soya bean (*Glycine max*) and cowpea (*Vigna unguiculata*), are discussed. The majority of the genome editing reports are from soya bean where both exogenous and endogenous loci were reportedly edited. We have also included some common web-based tools for designing reagents particularly guide RNA (gRNA) for editing of legumes.

7.3.1 Genome Editing in Birds Foot Trefoil (*Lotus japonicus*)

The first report of CRISPR/Cas9-mediated editing in *Lotus japonicus via. Agrobacterium rhizogenes*-mediated hairy root transformation of SNF (symbiotic nitrogen fixation)-related genes was documented in the study by Wang et al. (2016). Initially, a designer sgRNA targeting *SYMRK* (symbiosis receptor-like kinase) locus was designed, and deleted mutant (2-bp deletion near the PAM region) was achieved in selected edited lines with approx. 35% mutagenic efficiency. Further, sgRNAs targeting three leghaemoglobin homologs (*LjLb1, LjLb2* and *LjLb3*) were designed for generating multigene knockouts. Edited transgenic plants exhibited white nodules, with at least double *LjLbs* disrupted in each plant. A nodule-specific *LjLb2* promoter was effective in gene editing in nodules by hairy root transformation. Notably, triple-mutant knockouts were also obtained by stable transformation using two sgRNAs. Hence, CRISPR/Cas9 system was shown to facilitate functional analyses of endogenous plant genes (either singly or homologs). Subsequently, RNA-sequencing (RNA-seq) analysis of the nodules of triple mutants indicated the overproduction of superoxide radicals and hydrogen peroxide, possibly linked to the activation of NADPH oxidases and changes in superoxide dismutase isoforms expression. Further, the mutant nodules exhibited a major ultrastructural alterations, including vacuolization, poly-β-hydroxybutyrate accumulation and mitochondria disruption. Lb-deficient nodules had *ca*. 30- to 50-fold less haeme but the similar transcript levels of haeme biosynthetic genes, suggesting a post-translational regulatory mechanism. Lbs act synergistically in nodules, and the lack of Lbs results in early nodule senescence (Wang et al., 2019).

Besides symbiotic nodulation, *Lotus japonicus* serves as an important model legume plant in secondary (specialized) metabolism. Triterpenoid composition indicated the presence of C-28-oxidized triterpenoids (ursolic, betulinic and oleanolic acids) and soyasapogenols (soyasapogenols B, A and E) in a tissue-dependent manner. A candidate cytochrome P450 (CYP716A51) is predicted to play an important role in triterpenoid C-28 oxidation activity and validated by the loss of function of the *cyp716a51* mutant in hairy roots generated by genome editing technique. The knockout mutant lines produced no C-28 oxidized triterpenoids, indicating their role in triterpenoid biosynthesis pathway (Suzuki et al., 2019).

7.3.2 Genome Editing in Barrel Medic (*Medicago truncatula*)

The first report of genome editing in *Medicago truncatula* was from University of Minnesota, USA (Michno et al., 2015). The report was basically to study whether CRISPR/Cas9 system could cause mutations in the GUS transgene of transgenic *M. truncatula* cells, generated earlier. The control roots (vector with no sgRNA) exhibited GUS staining, while some edited roots transformation (vector with sgRNA) showed the loss of GUS staining. An enrichment PCR CAPS assay consisted of using a restriction endonuclease, *Mae*II, designed 3–5 bp adjacent to the PAM region, was conducted. Electrophoresis of the CAPS bands indicated the presence of templates recalcitrant to *Mae*II digestion, indicating the generation of new mutations at the targeted locus. Cloning and sequencing confirmed the successful mutagenesis at the target GUS locus (frameshift mutations).

In another report, *Medicago truncatula* phytoene desaturase (*Mt*PDS) was used as the target gene based on sgRNA specific to the second exon of the gene that was designed using CRISPR-P software. Both monoallelic and biallelic mutants of the loci were reported with albino phenotype in few lines suggesting the disruption of the *Mt*PDS gene. Genotyping studies on these edited lines using the specific primers indicated mutations of the *Mt*PDS gene. Off-target effect of the CRISPR/Cas9 construct targeting the MtPDS gene indicated the two potential off-target sites (CRISPR-P), but did not exhibit mutations (Meng et al., 2017).

Further, Curtin and colleagues demonstrated the synergistic use of different platforms (ZF, TALEN and CRISPR) to generate useful mutant lines for the study of small RNA processing in legume crops. Processing of dsRNA precursors into small RNAs is an important step in gene regulation and expression. Mutations in *M. truncatula* Hua enhancerl (*Mt*Hen1) gene were attempted in two different sites in MtHen1 gene (Medtr4g094545). Twelve T_0 plants were recovered after genetic transformation, and four lines exhibited the digestion-resistant amplification products indicative of germ-line heterozygous mutations. Progeny screening indicated 9-bp deletion and insertion that did not disrupt the gene function at reduced mutagenesis efficiencies, suggesting that frameshift knockout of Hen1 may be embryo lethal (Curtin et al., 2018).

7.3.3 Genome Editing in Alfalfa (*Medicago sativa*)

CRISPR/Cas9 technology was successfully reported in tetraploid alfalfa (*M. sativa*). Two sgRNAs specific to the squamosa promoter-binding protein-like 9 (SPL9) gene in alfalfa target sequence were designed using CRISPR-P, and the edited lines were regenerated. Droplet digital PCR (ddPCR)-based screening of large populations of edited plants was employed to identify the integration of the transgene construct. Selected plants were analysed by restriction enzyme digestion, and PCR amplification products of positive plants were subcloned and Sanger-sequenced to identify genome editing. Sequence information indicated deletion, insertion and substitution mutations in SPL9 locus (Gao et al., 2018).

In another study, gene editing in the ubiquitin E2-conjugating enzyme PHO2, involved in P signalling and P homeostasis in alfalfa, was reported so that plants hyperaccumulate phosphate with the purpose to reduce soil P levels and reclaim P for use as a fertilizer. Initial attempts to transform alfalfa cv. RegenSY with CRISPR/Cas9/Cys4 reagent targeting all three genes were made, and a total of 67 verified transgenic plants were screened by acrylamide gel shift assays, cloning and sequencing to identify plants with mutations. Mutations ranging from a 1-bp insertion to a 25-bp deletion were identified, and some plants had multiple targets hit. Another attempt at CRISPR/Cas9 mutation utilized a cassette vector system with either the tRNA or Cys4 splicing system, and exonuclease components were made. TaqMan probes were designed to identify plants with changes in the target sites and were subsequently verified by restriction digestions, cloning and sequencing (Samac et al., 2019).

7.3.4 Genome Editing in Cowpea (*Vigna unguiculata*)

CRISPR/Cas system was also successfully demonstrated in cowpea (*Vigna unguiculata*) to edit the SNF gene. gRNAs specific to SYMRK specifically exon 2 (corresponding to partial Malectin-like domain) resulted in mutagenic efficiency (up to 67%) in hairy root-transformed lines. Root nodule formation was completely blocked in the confirmed edited lines. Mutations ranging from small- and large-scale insertion and deletion near the PAM region of the respective gRNA were observed (Ji et al., 2019).

7.3.5 Genome Editing in Soya bean (*Glycine max*)

Genome editing was successfully reported in soya bean for targeted mutagenesis, gene integration and gene editing employing both endogenous and exogenous gene/loci leveraging on single and multiplexing strategy. Recently, DNA-free genome editing tool comprising Cpf1–crRNA complex was also effectively reported in soya bean. Currently, there are more than a dozen research reports of gene editing in soya bean, and significant findings of the reports are highlighted and discussed. A recent review also highlights the application of genome editing technology in soya bean improvement (Bao et al., 2020). Detailed protocols of soya bean genome editing using CRISPR/Cas9 are also available (Liu et al., 2019).

The first report of genome editing in soya bean was from University of Georgia, Athens, USA, demonstrating a green fluorescent protein (GFP) knockout and also modifying nine endogenous loci (Jacobs et al., 2015). Small deletions, SNPs and short insertions were observed in the edited progenies of soya bean. Homoeologous genes from soya bean were also successfully targeted singly and in unison, demonstrating the ability of CRISPR/Cas9. Besides, hairy roots and somatic embryos were also edited, and plants produced harboured stable and heritable mutations, with the frequency increasing with culture time. A novel in-fusion cloning strategy was also developed to ease the generation of CRISPR/Cas9 reagents, with potential applications in any organism.

In another report, CRISPR/Cas9 system was used to generate and estimate genome editing using a sgRNA targeting *bar* gene and another set of six sgRNAs specific to two endogenous soya bean genes (GmFEI2 and GmSHR). Both the components, single gRNA and Cas9 cassettes, were re-assembled in a single vector to enhance the transformation efficiency, and as expected, mutations were detected for targeted genes in soya bean hairy roots. Using only one customized sgRNA can simultaneously edit two endogenous soya bean genes. Besides amenability to affect editing, the method also demonstrated the functional genomic studies in soya bean roots (Cai et al., 2015).

In another breakthrough report, *Streptococcus pyogenes* Cas9-gRNA was shown to generate targeted mutagenesis, gene integration and gene editing. To date, this is the only report of HDR in soya bean employing suitable CRISPR-Cas9 reagents. Targeted mutagenesis of two loci, *viz.* DD20 and DD43, resulted in deletions/insertions, with frequencies of 59% and 76%, respectively. Targeted gene integrations could be detected by 'border-specific PCR' analysis for both sites. Mendelian segregation pattern and clean homozygous T_1 plants were reported for one event with a precise insertion at the DD43 locus. *In planta* gene editing was also successfully reported towards a directed mutation of acetolactate synthase 1 gene (P178S) (Li et al., 2015).

In another study, an online web tool (http://stuparcrispr.cfans.umn.edu/CRISPR/) for a quicker identification of target loci based on soya bean gene model was reported. The tool identifies the number of potential target sites, including restriction endonuclease sites, facilitating the detection of newly generated mutations. Using the developed tool, gRNAs targeted towards glutamine synthase (GS1) and chalcone-flavanone isomerase (CHI20) were designed, and analyses of transformed hairy root revealed mutations that disrupted the predetermined MspA1 restriction site. Transformed hairy roots stained blue (the presence of GUS gene in transformed vector), and 2-bp deletion was identified in the three selected samples (Michno et al., 2015).

In another study, sgRNA were designed targeting three soya bean genes, Glyma06g14180, Glyma08g02290 and Glyma12g37050, driven by soya bean U6–10 and *Arabidopsis* U6–26 promoters. Mutations of all three genes were detected in protoplasts with higher efficiency for soya bean-derived U6 promoters. Further, transformed hairy roots were obtained with mutation frequencies ranging from 3.2% to 9.7% (for At-derived promoter) and 14.7% to 20.2% (for Gm-derived promoter) vector. Off-target effects of the designed sgRNAs were also detected in the study (Sun et al., 2015).

In another study, comparative studies of these two technologies TALEN and CRISPR/Cas9 for two soya bean genome editing targets, GmPDS11 and GmPDS18, were reported. Higher targeting efficiency range of 17.5% –21.1% for TALENs and 11.7%–18.1% for CRISPR/Cas9 was reported. Further, CRISPR/Cas9 employing the soya bean U6 promoter was found more efficient tool for simultaneous editing of multiple homeo-alleles (Du et al., 2016).

In another significant study, *Rj4* gene that restricts nodulation by several strains of *Bradyrhizobium elkanii* was studied. Their symbiotic association has lower nitrogen fixation efficiency, but the bacterial strains are highly competitive for nodulation particularly for cultivars harbouring an Rj4 allele. Validating that *Rj4* is Glyma.01g165800-D – a duplicated and unique version of

Glyma.01g165800 – that has evolved over time to control the symbiotic specificity in soya bean genotypes, is a remarkable example of how CRISPR/Cas9 tool can be employed in functional validation (Tang et al., 2016).

Kim and colleagues from South Korea were able to demonstrate a significant breakthrough by employing ribonucleoprotein complex (Cpf1-crRNA) to edit FAD2 locus in soya bean. Concomitant delivery of recombinant Cpf1 proteins with *in vitro* transcribed FAD2-specific crRNAs into soya bean protoplasts was reported in the study. Targeted deep sequencing analyses corroborate that specific mutations were successfully induced in FAD2 paralogues. Further, no significant off-target mutations were detected in the soya bean genome (Kim et al., 2017).

Understanding flowering biology is an important trait for breeders and effective strategy for wider adaptability and climate resilience breeding. Cai and colleagues employed CRISPR/Cas9 system to induce changes in GmFT2a, an important component in the flowering pathway. Sequencing analysis indicated the site-directed mutations at all targeted sites and homozygous GmFT2a mutants exhibited late flowering under both long-day and short-day conditions. Further, the T-DNA elements could also be segregated out in subsequent generations (Cai et al., 2018a). In another study, the same group also demonstrated the use of multiple gRNAs for the deletion of large DNA fragments in soya bean from two target loci, *viz. GmFT2a* (Glyma16g26660) and *GmFT5a* (Glyma16g04830), and its heritability. Deletion fragment size ranged from 599 to >4.5 kb (*GmFT2a*) and 1069 to 1161 bp (*GmFT5a*). The demonstrated amenability shall benefit functional genomics and chromosome engineering or customized genetic breeding (Cai et al., 2018b). CRISPR-Cas9-mediated ft2a/ft5a double mutants exhibited late flowering under short-day conditions and produced significantly higher numbers of pods and seeds per plant (Cai et al., 2020a). This group was also able to demonstrate successful *base editing* in two soya bean loci GmFT2a (Glyma.16G150700) and GmFT4 (Glyma.08G363100), principally C to T mutations and C to G mutations with 6%–18% efficiency (Cai et al., 2020b).

In another study, CRISPR/Cas9-based mutation E1 genes related to flowering were reported with no off-target effects. Two novel mutations (11- and 40-bp deletions) and frameshift mutations resulted in truncated E1 proteins, and genotypes with early flowering under long-day condition were realized (Han et al., 2019).

In another study, two homoeologous loci, *GmPPD1* and *GmPPD2*, which encode the orthologs of *Arabidopsis thaliana PEAPOD* (*PPD*), were targeted. Double mutants exhibited the dome-shaped trifoliate leaves and extremely twisted pods, and produced few seeds. Continuous induction of mutations in the whole plant and advancing generations enable an efficient mutagenesis in duplicated loci in soya bean (Kanazashi et al., 2018).

Curtin and colleagues reported the comparative studies using the three different platforms (ZFN, TALEN and CRISPR/Cas9) for the generation of targeted mutations in small RNA pathway genes of soya bean. CRISPR/Cas9 reagents generated double mutants of GmDrb2a and GmDrb2b, and transgene-free null segregants were identified. Identified mutants shall be a genetic resource for further understanding the role of DRB2 function, repeat-associated small-interfering RNA (rasiRNAs) and trans-acting small-interfering RNAs (tasiRNAs) pathways and possibly, the viral replication inhibition (Curtin et al., 2018).

Con-glycinins (7S) and glycinins (11S) are the major families of storage proteins in soya bean. CRISPR-Cas8-mediated targeted mutations were reported in the three storage protein genes in soya bean hairy roots enabling soya bean breeders to engineer/develop varieties with mutations in seed storage proteins (Li et al., 2019).

In another study, CRISPR/Cas9 system was employed for mutation of four SQUAMOSA PROMOTER-BINDING PROTEIN-LIKE (SPL) transcription factor genes to improve canopy architecture. Double mutants (spl9a and spl9b) exhibited shorter plastochron length, and higher-order mutants exhibited the increased node and branch number. Notably, a feedback regulation in the pathway and delineation of active transcription factors were identified in the study (Bao et al., 2019).

Isoflavonoids derived from the phenylpropanoid pathway are concoction of secondary metabolites, distributed predominantly in leguminous plants. Isoflavone synthase (IFS) is a key enzyme in isoflavonoid biosynthesis and is a substrate competitor of other two enzymes, namely flavanone-3-hydroxylase (F3H) and flavone synthase II (FNS II). CRISPR/Cas9-mediated multiplexing was employed to simultaneously target GmF3H1, GmF3H2 and GmFNSII-1 in soya bean hairy roots and plants. Metabolome of triple-mutant leaves revealed approximately twice the leaf isoflavone content. Interestingly, these lines were also able to control specific soya bean mosaic virus (SMV) by one-third (Zheng et al., 2020).

In another study, inheritance patterns and their potential mechanisms were reported. Expected and unexpected patterns of inheritance emerged among the different lineages. Unexpected patterns comprise the following: (1) family in which transgenes and mutations were not heritable; (2) family with four unlinked transgene insertions, of which two were located at paralogous break sites; (3) family in which mutations were observed and transmitted, but without evidence of transgene integration nor transmission (Michno et al., 2020).

In another report, soya bean callus protoplast-based CRISPR/crRNA selection was identified as a potential tool to screen the efficacy of crRNAs. This finding was based on experiments whereby CRISPR/LbCpf1 RNPs were co-delivered with specific crRNA3 as complexes to edit *GlymaFAD2-1A* and *GlymaFAD2-1B* (Kim and Choi, 2020).

Most of the CRISPR/Cas9-based edits in soya bean plants are somatic and nontransmissible partly because Cas9 is driven by constitutive promoters. Zheng and co-workers designed vectors using four different egg cell-specific promoters driving Cas9 and identified a combination of AtEC1.2e1.1p:Cas9 that results in higher mutation efficiency (26.8%), besides transmission to T_2 generation (Zheng et al., 2020).

In another report, a pooled CRISPR-Cas9 platform to generate soya bean multiplex mutant populations was described. Optimized parameters in the screening protocol, starting from vector construction, sgRNA assessment, pooled transformation, sgRNA identification and gene editing verification, were reported. A total of 70 vectors targeting 102 candidate genes and their paralogs were studied in 16 batches. Population consisting of 407 primary edits was obtained with a frequency of 59.2%, of which 35.6% carry multiplex mutations. Further, double mutants with increased nodule numbers and triple-mutant lines with a decreased nodulation were also reported in the study (Bai et al., 2020).

In another report, the role of GmPRR37 in photoperiodic flowering was deciphered by knocking down the function of the gene. CRISPR/Cas9-induced Gmprr37-ZGDD mutants in soya bean exhibited early flowering under natural long-day (NLD) conditions. Further analyses indicated that GmPRR37 down-regulated the expression of the flowering-promoting FT homologues GmFT2a and GmFT5a, and up-regulated the expression of the flowering-inhibiting FT homologue GmFT1a under long-day conditions (Tables 7.1 and 7.2).

7.4 BIOINFORMATIC TOOLS FOR DESIGNING OF CRISPR REAGENTS

Designing sgRNA construct for CRISPR/Cas9 is the most important step for genome editing. Currently, various online and offline bioinformatics tools have been developed and subsequently updated for sgRNA designing and identifying possible off-targets (Table 7.3). A novel web tool for soya bean genome editing (http://stuparcrispr.cfans.umn.edu/CRISPR/) was reported for the detection of target loci for CRISPR/Cas9-mediated genome editing (Michno et al., 2015). Another commonly used online tool, CRISPR-P (http://cbi.hzau.edu.cn/crispr/), facilitates not only the designing of sgRNA for crops (available sequenced genome information) but also helps to evaluate off-targets (Liu et al., 2017). The online tool CHOPCHOP ver 3.0 (https://chopchop.cbu.uib.no) was developed by Labun and colleagues, who have upgraded algorithms since its inception for designing sgRNA, and it also assists in assessing the off-target mutation (Labun et al., 2019). The recommended strategy to design specific sgRNA for any target sequence should employ more than two tools and then validate the designed sgRNA using quick *in vitro* techniques.

TABLE 7.1
CRISPR-Based Genome Editing in Legumes

S. No.	Genotype	Explant	Genetic Transformation	Target Locus	Related Trait	Reference
Crop: Birdsfoot trefoil (*Lotus japonicus*)						
1.	MG20	Split hypocotyl	*Agrobacterium tumefaciens* (EHA105)	*Lj*SYMRK	Inactivate representative symbiotic nitrogen fixation-related genes	Wang et al. (2016)
		Hypocotyls	*Agrobacterium rhizogenes* (LBA1334)	*Lj*Lb1, *Lj*Lb2, *Lj*Lb3		
2.	Gifu B-129	Hypocotyls	*Agrobacterium rhizogenes* (ATCC15834)	CYP716A51	Triterpenoid biosynthesis pathway	Suzuki et al. (2019)
Crop: Barrel Medic (*Medicago truncatula*)						
1.	-	Cotyledonary node	*Agrobacterium rhizogenes* (ARqual)	GUS transgene	Reporter gene for transgenic development	Michno et al. (2015)
2.	R108	Leaf	*Agrobacterium tumefaciens* (AGL-1)	*Mt*PDS	Carotenoid biosynthesis gene	Meng et al. (2017)
3.	R108	Leaf	*Agrobacterium tumefaciens* (EHA105)	*MtHen1*	Small RNA processing	Curtin et al. (2018)
Crop: Alfalfa (*Medicago sativa*)						
1.	N4.4.2	-	-	SPL9	Plant development	Gao et al. (2018)
2.	-	-	-	PHO2	Phosphate hyperaccumulation	Samac et al. (2019)
Crop: Cowpea (*Vigna unguiculata*)						
1.	-	Cotyledonary nodes	*Agrobacterium rhizogenes* (K599)	*Vu*SYMRK	Inactivate representative symbiotic nitrogen fixation-related genes	Ji et al. (2019)

(Continued)

TABLE 7.1 (*Continued*)
CRISPR-Based Genome Editing in Legumes

S. No.	Genotype	Explant	Genetic Transformation	Target Locus	Related Trait	Reference
			Crop: Soya bean (*Glycine max*)			
1.	Jack, Jack-GFP	Hypocotyl	*Agrobacterium rhizogenes* (K599)	GFP	Exogenous locus encoding visible/scorable marker	Jacobs et al. (2015)
		Somatic embryos	Particle bombardment	Glyma07g14530; Glyma01g38150; Glyma11g07220; Glyma04g36150; Glyma06g18790; miR1514; miR1509	Endogenous loci encoding putative glucosyl transferase, orthologs of AtDDM1 gene, orthologs of AtMET1 and miRNA involved in a wide range of regulatory roles in plants	
2.	Zigongdongdou and Williams 82	Hypocotyl and cotyledonary node	*Agrobacterium rhizogenes* (K599)	Bar, GmFEI2, GmSHR	Signalling pathway regulating the cell wall function and root architecture	Cai et al., 2015
3.	93B86	Embryonic cultures	Particle gun bombardment and *in planta* transformation	*DD*20, *DD*43, ALS1	Defined chromosomal location in chromosome 4 and metabolic enzyme in branched-chain amino acids	Li et al. (2015)
4.	Bert	Cotyledons	*Agrobacterium rhizogenes* (ARqual)	GS1, CHI20	Nitrogen metabolism and flavonoid biosynthesis pathway	Michno et al. (2015)
5.	Williams 82	Seedling hypocotyl	*Agrobacterium rhizogenes* (K599)	Glyma06g14180, Glyma08g02290, Glyma12g37050	Uncharacterized genes	Sun et al. (2015)
		Leaf Protoplast	Polyethylene glycol (PEG)-mediated			
6.	Jack	Cotyledonary node	*Agrobacterium tumefaciens* (EHA105)	GmPDS11 GmPDS18	Carotenoid biosynthesis gene	Du et al. (2016)
		Cotyledon	*Agrobacterium rhizogenes* (K599)			
7.	Williams and Hill	Cotyledonary node	*Agrobacterium rhizogenes* (K599)	*Rj4*	Symbiotic nitrogen fixation-related genes	Tang et al. (2016)

(*Continued*)

TABLE 7.1 (*Continued*)
CRISPR-Based Genome Editing in Legumes

S. No.	Genotype	Explant	Genetic Transformation	Target Locus	Related Trait	Reference
8.	Williams 82	Protoplast	RNP-mediated genome editing (CRISPR/Cpf1)	FATTY ACID DESATURASE 2-1A (FAD2-1A, Glyma10g42470) and FAD2-1B (Glyma20g24530)	Fatty acid biosynthetic pathway	Kim et al. (2017)
9.	Jack	Cotyledonary node	*Agrobacterium tumefaciens* (EHA105)	GmFT2a	Flowering traits	Cai et al. (2018a)
10.	Jack	Cotyledonary node	*Agrobacterium tumefaciens* (EHA105)	GmFT2a, GmFT2b	Flowering traits	Cai et al. (2018b)
11.	Bert	Cotyledonary node	*Agrobacterium rhizogenes* (K599 and 18r12)	GmDrb2a, GmDrb2b, GmDicer like 3, GmHen1a	Small RNA processing	Curtin et al. (2018)
12.	Kariyutaka	Cotyledonary node	*Agrobacterium tumefaciens*	GmPPD1, GmPPD2	Cell division regulator	Kanazashi et al. (2018)
13.	Harosoy 63	Seedlings	*Agrobacterium rhizogenes* (K599)	Nine con-glycinin and glycinin genes	Seed storage protein	Li et al. (2019)
14.	Williams 82	Cotyledonary node	*Agrobacterium tumefaciens* (EHA105)	Four GmSPL9 (a-d) genes	Plant architecture	Bao et al. (2019)
15.	Jack	Protoplast	*Agrobacterium tumefaciens* (EHA105)	E1	Flowering traits	Han et al. (2019)
16.	Jack	Cotyledonary node and cotyledons	*Agrobacterium tumefaciens* (EHA105) and *Agrobacterium rhizogenes* (K599)	GmF3H1, GmF3H2 and GmFNSII-1	Isoflavone biosynthesis	Zheng et al. (2020)
17.	Jack	Cotyledonary node	*Agrobacterium tumefaciens* (EHA105)	GmFT5a	Flowering traits	Cai et al. (2020a)

(*Continued*)

TABLE 7.1 (*Continued*)
CRISPR-Based Genome Editing in Legumes

S. No.	Genotype	Explant	Genetic Transformation	Target Locus	Related Trait	Reference
18.	Bert-MN-01	Cotyledonary node	*Agrobacterium rhizogenes* (18r12, disarmed K599)	Glyma.16g090700, Glyma.18g041100, Glyma.16G209100, Glyma.09G159900	Not reported	Michno et al. (2020)
19.	Williams 82	Cotyledonary node	*Agrobacterium tumefaciens* (EHA105) and *Agrobacterium rhizogenes* (K599)	GmAGO7a, GmAGO7b	Leaf pattern	Zheng et al. (2020)
20.	Huachun 6	Hypocotyl and cotyledonary node	*Agrobacterium tumefaciens* (GV3101) and *Agrobacterium rhizogenes* (K599)	102 endogenous candidate genes and their paralogs[a]	Various traits	Bai et al. (2020)
21.	Williams 82, Kwangan and Daewon	Cotyledon or calli-derived protoplast	PEG-mediated transfection	*Glyma*FAD2-1A, *Glyma*FAD2-1B	Oil quality	Kim and Choi (2020)
22.	Jack	Cotyledonary node	*Agrobacterium tumefaciens* (EHA105)	GmFT2a, GmFT4	Flowering traits	Cai et al. (2020b)
23.	Jack and ZGDD	Cotyledonary node	*Agrobacterium tumefaciens* (EHA105)	GmPRR37-CP	Flowering traits	Wang et al. (2020)

[a] Refer to Bai et al. (2020) for list of genes.

LjSYMRK: *Lotus japonicas* SYMBIOSIS RECEPTOR KINASE; LjLb1, 2, 3: *Lotus japonicus* leghaemoglobin genes 1, 2, 3; CYP716A51: cytochrome P450 monooxygenase; MtPDS: *Medicago truncatula* phytoene desaturase; GUS: β-glucuronidase; MtHen1: *Medicago truncatula Hua enhancer* 1; SPL9: squamosa promoter-binding protein-like 9; PHO2: phosphate2; *Vu*SYMRK: *Vigna unguiculata* SYMBIOSIS RECEPTOR KINASE; Glyma07g14530: putative glucosyl transferase; GFP: green fluorescent protein; Glyma01g38150 and Glyma11g07220: orthologs of At DDM1 gene; Glyma04g36150 and Glyma06g18790: orthologs of At MET1; miR1514 and miR1509: miRNA regulating a wide range of process in plants; Bar: phosphinothricin acetyl-transferase; GmFEI2: *Glycine max* cell wall-related gene; GmSHR: *Glycine max* short root; ASL1: acetolactate synthase 1 responsible for resistance to sulfonylurea (SU); DD20 and DD43: chromosomal location in soya bean genome; GS1: glutamine synthase 1; CHI20: chalcone-flavanone isomerase; Rj4: thaumatin-like protein for nodule biosynthesis; GmFT2-a: glycine max flowering locus T2-a; GmDrb: glycine max double-stranded RNA-binding; SPL: squamosa promoter-binding promoter-like TF; E1: early flowering; GmAGO7a, GmAGO7b: glycine max regulator for leaf pattern; *Glyma*FAD2-1A, *Glyma*FAD2-1B: glycine max fatty acid desaturase 2.

TABLE 7.2
CRISPR-Induced Mutants and Their Characterization in Legumes

S. No.	Traits/Locus(i)	Mechanism and Mutation Types	Mutation Rates or Efficiency (%)	Detection of Mutation	Off-Targets	References
			Soya bean (*Glycine max*)			
1.	GFP and nine endogenous loci	NHEJ: small deletions, insertions/SNP	Up to 70%	Custom PCR amplification and sequencing	+	Jacobs et al. (2015)
2.	*bar* and two endogenous loci	NHEJ: small insertion or deletion	10.0%–93.3%	PCR/restriction enzyme (PCR/RE); T7EI assay	NR	Cai et al. (2015)
3.	Two chromosomal locations and gene *ALS1*	NHEJ: insertion/deletion HDR: insertion Targeted mutagenesis	59.0% and 76.0 %	Quantitative PCR (qPCR); custom PCR amplification/border-specific and sequencing; Southern blot hybridization	NR	Li et al. (2015)
4.	GS1 and CHI20	NHEJ: deletion	NR	CAPS-PCR, Sanger sequencing	NR	Michno et al. (2015)
5.	Three uncharacterized loci	NHEJ: insertion or deletion	3.2%–20.2%	RE-PCR assay, Sanger sequencing	+	Sun et al. (2015)
6.	GmPDS11 and GmPDS18	NHEJ: insertion or deletion	11.7%–48.1%	PCR-RE assay and PCR-T7EI/Cruiser assay	NR	Du et al. (2016)
7.	Rj4	NHEJ: insertion or deletion	NR	Custom PCR amplification and sequencing	NR	Tang et al. (2016)
8.	GmFAD2-1A & 1B	NHEJ: deletion	0.6%–11.7%	*In vitro* cleavage assay and targeted deep sequencing	-	Kim et al. (2017)
9.	GmFT2a	NHEJ: insertion or deletion	NR	Custom PCR amplification and sequencing	-	Cai et al. (2018a)
10.	GmFT2a and GmFT5a	NHEJ: large fragment deletion	12.1%–15.8%	Custom PCR amplification and sequencing	NR	Cai et al. (2018b)
11.	GmDc13a, GmDrb2a, GmDrb2b, GmDicer like 3, GmHen1a	NHEJ: deletion	~12.0% to 18.0%	PCR-digestion assay, heteroduplex and TACAS assay, whole-genome sequencing (WGS)	-	Curtin et al. (2018)
12.	GmPPD1, GmPPD2	NHEJ: insertion and deletion	Up to 33%	CAPS-PCR, Sanger sequencing	NR	Kanazashi et al. (2018)
13.	Nine conglycinin (7S) and glycinin (11S) genes	NHEJ: insertion or deletion	3.8%–43.7%	Custom PCR amplification and sequencing	NR	Li et al. (2019)

(Continued)

TABLE 7.2 (*Continued*)
CRISPR-Induced Mutants and Their Characterization in Legumes

S. No.	Traits/Locus(i)	Mechanism and Mutation Types	Mutation Rates or Efficiency (%)	Detection of Mutation	Off-Targets	References
14.	Four GmSPL9 (a-d)	NHEJ: insertion or deletion	NR	Custom PCR amplification and sequencing	NR	Bao et al. (2019)
15.	Early flowering E1 locus	NHEJ: deletion	NR	Custom PCR amplification and sequencing	-	Han et al. (2019)
16.	GmF3H1, GmF3H2 and GmFNSII-1	NHEJ: insertion and deletion	10.42%–31.25%	Custom PCR amplification and sequencing	NR	Zheng et al. (2020)
17.	GmFT5a	NHEJ: deletion	NR	Custom PCR amplification and sequencing	-	Cai et al. (2020a)
18.	Glyma.16g090700, Glyma.18g041100, Glyma.16G209100, Glyma.09G159900	NHEJ: deletion	NR	PCR amplification, Sanger sequencing, whole-genome resequencing	NR	Michno et al. (2020)
19.	GmAGO7a, GmAGO7b	NHEJ: insertion and deletion	26.8%	PCR amplification, Sanger sequencing, T7E1 assay	NR	Zheng et al. (2020)
20.	102 endogenous candidate genes and their paralogs	NHEJ: insertion and deletion	49.15%	Custom PCR amplification and sequencing	NR	Bai et al. (2020)
21.	*Glyma*FAD2-1A, *Glyma*FAD2-1B	NHEJ: insertion and deletion	10.5%–11.7%, 6.7%–9.1%	Target deep sequencing	+	Kim and Choi (2020)
22.	GmFT2a, GmFT4	Base editing (C to T; C to G)	6.0%–18.2%	Custom PCR amplification and sequencing	NR	Cai et al. (2020b)
23.	GmPRR37-CP	NHEJ: deletion	-	Custom PCR amplification and sequencing	NR	Wang et al. (2020)
			Cowpea (*Vigna unguiculata*)			
1.	VuSYMRK	NHEJ: insertion or deletion	~67.0%	Custom PCR amplification and sequencing	NR	Ji et al. (2019)

(*Continued*)

TABLE 7.2 (*Continued*)
CRISPR-Induced Mutants and Their Characterization in Legumes

S. No.	Traits/Locus(i)	Mechanism and Mutation Types	Mutation Rates or Efficiency (%)	Detection of Mutation	Off-Targets	References
			Alfalfa (*Medicago sativa*)			
1.	SPL9	NHEJ: deletion and insertion/substitution	1.4%–2.2%	Droplet digital PCR (ddPCR), custom PCR amplification and Sanger sequencing	NR	Gao et al. (2018)
2.	PHO2	NHEJ: insertion, deletion and substitution	NR	Custom PCR amplification and heteroduplex analyses/sequencing	-	Samac et al. (2019)
			Barrel medic (*Medicago truncatula*)			
1.	GUS	NHEJ: deletion	NR	CAPS-PCR analysis, Sanger sequencing	NR	Michno et al. (2015)
2.	*Mt*PDS	NHEJ: deletion	10.35%	Custom PCR amplification and sequencing	-	Meng et al. (2017)
3.	*MtHen1*	NHEJ: insertion or deletion	NR	Custom PCR amplification, TACAS and heteroduplex analyses/sequencing	-	Curtin et al. (2018)
			Birdsfoot trefoil (*Lotus japonicus*)			
1.	LjSYMRK, LjLb1-3	NHEJ: InDels and large fragment deletion	35.0%	PCR/RE assay and sequencing	NR	Wang et al. (2016)
2.	CYP716A51	NHEJ: insertion or deletion	NR	Custom PCR amplification and sequencing	NR	Suzuki et al. (2019)

+: Presence of off-target effects; –: absence of off-targets; NR: not reported; NHEJ: non-homologous end-joining; HDR: homologous-dependent recombination.

TABLE 7.3
Software for sgRNA Design and Validation

Sl.No.	Software	Link	Reference
1.	CRISPR-P 2.0	http://cbi.hzau.edu.cn/CRISPR2/	Liu et al. (2017)
2.	CRISPR-PLANT	https://www.genome.arizona.edu/crispr/	Xie et al. (2014b)
3.	CHOPCHOP 3.0	https://chopchop.cbu.uib.no	Labun et al. (2019)
4.	Benchling	https://www.benchling.com/crispr/	Benchling, San Francisco, CA
5.	sgRNAcas9	https://www.biootools.com/col.jsp?id=103/	Xie et al. (2014a)
6.	CRISPR lnc	http://www.crisprlnc.org	Chen et al. (2019)
7.	CRISPRdirect	https://crispr.dbcls.jp	Naito et al. (2015)
8.	Cas-OFFinder	http://www.rgenome.net/cas-offinder/	Bae et al. (2014)
9.	E-CRISP v 5.4	http://www.e-crisp.org/E-CRISP/designcrispr.html	-
10.	CRISPRseek	https://bio.tools/crisprseek	Zhu et al. (2014)

7.5 CONCLUSIONS

CRISPR/Cas system is currently undergoing a continuous improvement, and each step enables unprecedented ability to advance basic research and facilitates breeding in diploid and polyploid crops. This can be considered precision plant breeding, and the simplicity and robustness of CRISPR/Cas systems make genome editing a powerful tool for crop improvement. Precise knockout, knock-in, base editing and customized tinkering of gene regulation at any genomic locus (single or multiplexing strategy) are already demonstrated in legumes. However, harnessing the power of the technology requires understanding of the genetic basis of agriculturally important traits, thus enhancing the efficiency of gene targeting and prudent delivery of repurposed reagents to plant cells and subsequent plant regeneration and establishment.

Legumes are the important crops crucial for nutritional security, and hence, deployment of the genome editing tool is urgently required for its genetic improvement. Majority of the reports in legumes pertain to the proof-of-concept studies involving biosynthetic pathway genes like SNF genes, carotenoid biosynthesis, fatty acid biosynthesis and early flowering *etc* in hairy roots. Major research efforts should concentrate on biology of important traits suited to farmers and consumers, by employing and integrating the colossal genomic data generated in these crops and then by employing the powerful tool of genome editing to tailor-made these crops. Hopefully, in future besides gene editing, targeting cis-regulatory elements (CREs) or *biobrick* – which are involved in controlling gene expression – will possibly enable the turning of gene expression creating a whole range of variation in the expression of a gene (Lu et al., 2019). The potential of genome editing technology was demonstrated in soya bean, where the *Calyno oil* (modified soya bean oil) with high oleic acid was produced and likely to be marketed soon (Calyxt Inc., Minnesota, USA). Clear policy decisions on regulations of genome-edited crops and products thereof are important considerations for faster realization in market and wider public acceptability.

REFERENCES

Alok, A., D. Sandhya, P. Jogam, et al. "The rise of the CRISPR/Cpf1 system for efficient genome editing in plants." *Frontiers in Plant Science* 11 (2020): 264.

Bae, S., J. Park, J.-S. Kim. "Cas-OFFinder: A fast and versatile algorithm that searches for potential off-target sites of Cas9 RNA-guided endonucleases." *Bioinformatics* 30 (2014): 1473–1475.

Bai, M., J. Yuan, H. Kuang, et al. "Generation of a multiplex mutagenesis population *via*. pooled CRISPR-Cas9 in soybean." *Plant Biotechnology Journal* 18 (2020): 721–731.

Bao, A., H. Chen, L. Chen, et al. "CRISPR/Cas9-mediated targeted mutagenesis of GmSPL9 gene alters plant architecture in soybean." *BMC Plant Biology* 19 (2019): 131.

Bao, A., C. Zhang, Y. Huang, et al. "Genome editing technology and application in soybean improvement." *Oil Crop Science* 5 (2020): 31–40.

Bleuyard, J. Y., M. E. Gallego, C. I. White. "Recent advances in understanding of the DNA double-strand break repair machinery of plants." *DNA Repair* 5 (2006): 1–12.

Cai, Y., L. Chen, X. Liu, et al. "CRISPR/Cas9-mediated genome editing in soybean hairy roots." *PLoS One* 10, no. 8 (2015): e0136064.

Cai, Y., L. Chen, X. Liu, et al. "CRISPR/Cas9-mediated targeted mutagenesis of *Gm*FT2a delays flowering time in soybean." *Plant Biotechnology Journal* 16 (2018a): 176–185.

Cai, Y., L. Chen, S. Sun, et al. "CRISPR/Cas9-mediated deletion of large genomic fragments in soybean." *International Journal of Molecular Sciences* 19 (2018b): 3835.

Cai, Y., L. Wang, L. Chen, et al. "Mutagenesis of *GmFT2a* and *GmFT5a* mediated by CRISPR/Cas9 contributes for expanding the regional adaptability of soybean." *Plant Biotechnology Journal* 18 (2020a): 298–309.

Cai, Y., L. Chen, Y. Zhang, et al. "Target base editing in soybean using a modified CRISPR/Cas9 system." *Plant Biotechnology Journal* 18 (2020b): 1996–1998.

Cermak, T., E. L. Doyle, M. Christian, et al. "Efficient design and assembly of custom TALEN and other TAL effector-based constructs for DNA targeting." *Nucleic Acids Research* 39 (2011): e82.

Chen, K., C. Gao. "TALENs: Customizable molecular DNA scissors for genome engineering of plants." *Journal of Genetics and Genomics* 40 (2013): 271–279.

Chen, W., G. Zhang, J. Li, et al. "CRISPRlnc: A manually curated database of validated sgRNAs for lncRNAs." *Nucleic Acids Research* 47, no. D1 (2019): D63–D68.

Christenhusz, M. J. M., J. W. Byng. "The number of known plants species in the world and its annual increase." *Phytotaxa* 261 no. 3 (2016): 201–217.

Christian, M., Y. Qi, Y. Zhang, D.F. Voytas. "Targeted mutagenesis of *Arabidopsis thaliana* using engineered TAL effector nucleases." *Genes Genomes Genetics* 3 (2013): 1697–1705.

Collard, B. C. Y, D. J. Mackill. "Marker-assisted selection: An approach for precision plant breeding in the twenty-first century." *Philosophical Transactions of the Royal Society, London B: Biological Sciences* 363 (2008): 557–572.

Curtin, S. J., Y. Xiong, J. M. Michno, et al. "CRISPR/Cas9 and TALENs generate heritable mutations for genes involved in small RNA processing of *Glycine max* and *Medicago truncatula*." *Plant Biotechnology Journal* 16, no. 6 (2018): 1125–1137.

De Pater, S., L. W. Neuteboom, J. Pinas, et al. "ZFN-induced mutagenesis and gene-targeting in *Arabidopsis* through *Agrobacterium*-mediated floral dip transformation." *Plant Biotechnology Journal* 7 (2009): 821–835.

Desta, Z. A., R. Ortiz. "Genomic selection: Genome-wide prediction in plant improvement." *Trends in Plant Science* 19 (2014): 592–601.

Doudna, J. A., E. Charpentier. "The new frontier of genome engineering with CRISPR-Cas9." *Science* 346, no. 6213 (2014): 1258096.

Du, H., X. Zeng, M. Zhao, et al. "Efficient targeted mutagenesis in soybean by TALENs and CRISPR/Cas9." *Journal of Biotechnology* 217 (2016): 90–97.

Durai, S., M. Mani, K. Kandavelou, et al. "Zinc finger nucleases, custom-designed molecular scissors for genome engineering of plant and mammalian cells." *Nucleic Acids Research* 33 (2005): 5978–5990.

Gao, H., J. Smith, M. Yang, et al. "Heritable targeted mutagenesis in maize using a designed endonuclease." *Plant Journal* 61 (2010): 176–187.

Gao, R., B. A. Feyissa, M. Croft, A. Hannoufa. "Gene editing by CRISPR/Cas9 in the obligatory outcrossing *Medicago sativa*." *Planta* 247, no. 4 (2018): 1043–1050.

Geddes, B. A., I. J. Oresnik. "The mechanism of symbiotic nitrogen fixation." In *The Mechanistic Benefits of Microbial Symbionts, Advances in Environmental Microbiology*, edited by C. Hurst, pp. 69–97. Springer, Cham (2016).

Han, J., B. Guo, Y. Guo, et al. "Creation of early flowering germplasm of soybean by CRISPR/Cas9 technology." *Frontiers in Plant Science* 10 (2019): 1446.

Haun, W., A. Coffman, B. M. Clasen, et al. "Improved soybean oil quality by targeted mutagenesis of the fatty acid desaturase 2 gene family." *Plant Biotechnology Journal* 7 (2014): 934–940.

Jacobs, T. B., P. R. LaFayette, R. J. Schmitz, et al. "Targeted genome modifications in soybean with CRISPR/Cas9." *BMC Biotechnology* 15 (2015): 16.

Ji, J., C. Zhang, Z. Sun, et al. "Genome editing in cowpea *Vigna unguiculata* using CRISPR-Cas9." *International Journal of Molecular Sciences* 20 (2019): 2471.

Jinek, M., K. Chylinski, I. Fonfara, et al. "A programmable dual-RNA-guided DNA endonuclease in adaptive bacterial immunity." *Science* 337 (2012): 816–821.

Kanazashi, Y., A. Hirose, I. Takahashi, et al. "Simultaneous site-directed mutagenesis of duplicated loci in soybean using a single guide RNA." *Plant Cell Reports* 37, no. 3 (2018): 553–563.

Kim, H., J. Choi. "A robust and practical CRISPR/crRNA screening system for soybean cultivar editing using LbCpf1 ribonucleoproteins." *Plant Cell Reports* (2020). doi: 10.1007/s00299-020-02597-x.

Kim, H., S. T. Kim, J. Ryu, et al. "CRISPR/Cpf1-mediated DNA free plant genome editing." *Nature Communications* 8 (2017): 14406.

Labun, K., T. G. Montague, M. Krause, et al. "CHOPCHOP v3: Expanding the CRISPR web toolbox beyond genome editing." *Nucleic Acids Research* 47, no. W1 (2019): W171–W174.

Li, T., B. Liu, M. H. Spalding, et al. "High efficiency TALEN based gene editing produces disease resistant rice." *Nature Biotechnology* 30 (2012): 390–392.

Li, Z., Z. B. Liu, A. Xing, et al. "Cas9-guide RNA directed genome editing in soybean." *Plant Physiology* 169, no. 2 (2015): 960–970.

Li, C., V. Nguyen, J. Liu, et al. "Mutagenesis of seed storage protein genes in soybean using CRISPR/Cas9." *BMC Research Notes* 12 (2019): 176.

Liu, H., Y. Ding, Y. Zhou, et al. "CRISPR-P 2.0: An improved CRISPR/Cas9 tool for genome editing in plants." *Molecular Plant* 10, no.3 (2017): 530–532.

Liu, J., S. Gunapati, N. T. Mihelich, et al. "Genome editing in soybean with CRISPR/Cas9."In *Methods in Molecular Biology, Plant Genome Editing with CRISPR System*, edited by Y. Qi, pp. 1–16, Humana Press, New York (2019).

Lloyd, A., C. L. Plaisier, D. Carroll, et al. "Targeted mutagenesis using zinc-finger nucleases in *Arabidopsis*." *Proceedings of the National Academy of Sciences, United States of America* 102 (2005): 2232–2237.

Lu, Z., A. P. Marand, W. A. Ricci, et al. "The prevalence, evolution and chromatin signatures of plant regulatory elements." *Nature Plants* 5 (2019): 1250–1259.

Meng, Y., Y. Hou, H. Wang, et al. "Targeted mutagenesis by CRISPR/Cas9 system in the model legume *Medicago truncatula*." *Plant Cell Reports* 36, no 2 (2017): 371–374.

Michno, J. M., X. Wang, J. Liu, et al. "CRISPR/Cas mutagenesis of soybean and *Medicago truncatula* using a new web tool and modified Cas9 enzyme." *GM Crops and Food* 6 (2015): 243–252.

Michno, J. M., K. Virdi, A. O. Stec, et al. "Integration, abundance and transmission of mutations and transgenes in a series of CRISPR/Cas9 soybean lines." *BMC Biotechnology* 20 (2020): 10.

Naito, Y., K. Hino, H. Bono, et al. "CRISPRdirect: Software for designing CRISPR/Cas guide RNA with reduced off-target sites." *Bioinformatics* 31 (2015): 1120–1123.

Samac, D. A., S. S. Miller, M. R. Dornbusch, S. J. Curtin. "Genome editing in alfalfa (*Medicago sativa*) to hyperaccumulate phosphate." Retrieved from the University of Minnesota Digital Conservancy (2019).

Schrire, B. D., G. P. Lewis, M. Lavin. "Biogeography of the Leguminosae." In *Legumes of the World*, edited by G. Lewis, G. Schrire, B. Mackinder, M. Lock, pp. 21–54. Royal Botanic Gardens Kew, England (2005).

Shan, Q., Y. Wang, J. Li, et al. "Targeted genome modification of crop plants using a CRISPR-Cas system." *Nature Biotechnology* 31 (2013): 686–688.

Smith, J., S. Grizot, S. Arnould, et al. "A combinatorial approach to create artificial homing endonucleases cleaving chosen sequences." *Nucleic Acids Research* 34 (2006): e149.

Sternberg, S. H., S. Redding, M. Jinek, et al. "DNA interrogation by the CRISPR RNA-guided endonuclease Cas9." *Nature* 507, no. 6213 (2014): 1258096.

Stoddard, B. L. "Homing endonuclease from mobile group I introns: Discovery to genome engineering." *Mobile DNA* 5 (2014): 7.

Sun, N., Zhao, H. "Transcription activator-like effector nucleases (TALENs): A highly efficient and versatile tool for genome editing." *Biotechnology Bioengineering* 110 (2013): 1811–1821.

Sun, X., Z. Hu, R. Chen, et al. "Targeted mutagenesis in soybean using the CRISPR-Cas9 system." *Scientific Reports* 5 (2015): 10342.

Suzuki, H., E. O. Fukushima, Y. Shimizu, et al. "*Lotus japonicus* triterpenoid profile and characterization of the CYP716A51 and LjCYP93E1 genes involved in their biosynthesis *in planta*." *Plant and Cell Physiology* 60, no. 11 (2019): 2496–2509.

Symington, L. S., J. Gautier. "Double-strand break end resection and repair pathway choice." *Annual Review of Genetics* 45 (2011): 247–271.

Tang, F., S. Yang, J. Liu, et al. "Rj4, a gene controlling nodulation specificity in soybeans, encodes a thaumatin-like protein, but not the one previously reported." *Plant Physiology* 170 (2016): 26–32.

The Legume Phylogeny Working Group (LPWG). "A new sub-family classification of the Leguminosae based on a taxonomically comprehensive phylogeny." *Taxon* 66, no. 1 (2017): 44–77.

Wang, L., L. Wang, Q. Tan, et al. "Efficient inactivation of symbiotic nitrogen fixation related genes in *Lotus japonicus* using CRISPR-Cas9." *Frontiers in Plant Science* 7 (2016): 1333.

Wang, L., M. C. Rubio, X. Xin, et al. "CRISPR/Cas9 knockout of leghaemoglobin genes in *Lotus japonicus* uncovers their synergistic roles in symbiotic nitrogen fixation." *New Phytologist* 224, no. 2 (2019): 818–832.

Wang, L., S. Sun, T. Wu, et al. "Natural variation and CRISPR/Cas9-mediated mutation in GmPRR37 affect photoperiodic flowering and contribute to regional adaptation of soybean." *Plant Biotechnology Journal* 18 (2020): 1869–1881.

Watson, A., S. Ghosh, M. J. Williams, et al. "Speed Breeding is a powerful tool to accelerate crop research and breeding." *Nature Plants* 4 (2018): 23–29.

Wright, D. A., J. A. Townsend, R. J. Winfrey Jr, et al. "High-frequency homologous recombination in plants mediated by zinc-finger nucleases." *Plant Journal* 44 (2005): 693–705.

Xie, S., B. Shen, C. Zhang, et al. "sgRNAcas9: A software package for designing CRISPR sgRNA and evaluating potential off-target cleavage sites." *PLoS One* 9, no. 6 (2014a): e100448.

Xie, K., J. Zhang, Y. Yang. "Genome-wide prediction of highly specific guide RNA spacers for the CRISPR-Cas9 mediated genome editing in model plants and major crops." *Molecular Plant* 7, no. 5 (2014b): 923–926.

Yan, F., W. Wang, J. Zhang. "CRISPR-Cas12 and Cas13: The lesser known siblings of CRISPR-Cas9."*Cell Biology and Toxicology* 35 (2019): 489–492.

Zetsche, B., J. S. Gootenberg, O. Abudayyeh, et al. "Cpf1 is a single RNA-guided endonuclease of a class 2 CRISPR-Cas system." *Cell* 163 (2015): 759–771.

Zhang, Y., K. Massel, I. D. Godwin, et al. "Applications and potential of genome editing in crop improvement." *Genome Biology* 19 (2018): 210.

Zheng, N., T. Li, J. D. Dittman, et al. "CRISPR/Cas9-Based gene editing using egg cell-specific promoters in Arabidopsis and soybean." *Frontiers in Plant Science* 11 (2020): 800.

Zhu, L. J., B. R. Holmes, N. Aronin, et al. "CRISPRseek: A bioconductor package to identify target-specific guide RNAs for CRISPR-Cas9 genome-editing systems." *PLoS One* 9, no. 9 (2014): e108424.

8 Genome Editing in Oilseed Crops

Ritesh Kumar Raipuria
ICAR-National Institute for Plant Biotechnology

Archana Watts
ICAR-Indian Agricultural Research Institute

Nand Lal Meena
ICAR-National Bureau of Plant Genetic Resources

Anshul Watts
ICAR-National Institute for Plant Biotechnology

CONTENTS

8.1 INTRODUCTION

Both annual and perennial types of oil crops are abundant in nature. Annual oil crops are usually the oilseed crops whose seeds are used for oil extraction, whereas the seeds, fruits or mesocarp is used for oil extraction in the perennial crops. In the world, 40 different oil crops are present whose oil can be used for consumption. The soybean (*Glycine max*), rapeseed (*Brassica rapa*, *B. juncea*, *B. carinata* and *B. napus*), palm (*Elaeis guineensis*), sunflower (*Helianthus annuus*), safflower (*Carthamus tinctorius*), olive seed (*Olea europaea*), peanut (*Arachis hypogaea*), cottonseed (*Gossypium hirsutum*), jatropha (*Jatropha curcas*), tung tree (*Aleurites fordii*), jojoba (*Simmondsia chinensis*), sacha inchi (*Plukenetia volubilis*), niger seed (*Guizotia abyssinica*) and coconut are the major oilseed crops around the globe. Two types of oil are usually extracted from oilseed crops: edible and industrial oil. All or most of the above-mentioned crops are mainly used for edible oil production. Industrial oil contains some compounds that are helpful for the industry. Flax and castor bean are mainly used for industrial oil production purposes. The oilseed crops are economically important as their seeds are not only the source of vegetable oils but also used for the human consumption. After the removal of oil from these seeds, the remaining residue is used for animal feed and fodder purpose. Additionally, these oilseed crops are used as a protein supplement, to produce flour, milk, vinegar, butter, nectar, textile material, dyes, plastics, cosmetics and biofuels (Waseem et al., 2017; Villanueva-Mejia and Alvarez, 2017). The United States of America, China, India, Brazil and Argentina are the major oilseed-producing countries of the world. In 2017–2018, the production of oil crops was 592.1 million tons (FAO, 2019). At present, due to population explosion, the demand of edible oil is globally increasing every year. Therefore, for fulfilling the global oil demand, the production of oil crops needs to be increased. These oilseed crops are facing different biotic stresses like insect, pests, disease and weeds during their cultivation. Various other factors like drought susceptibility, lower yield and lack of high oil-yielding varieties/hybrids are also the major constraints in the cultivation of oilseed crops (Waseem et al., 2017). Therefore, improving production and productivity of oilseed crops is the most important requirement.

8.2 GENOME EDITING FOR IMPROVEMENT OF OILSEED CROPS

Conventionally plant breeders have used various breeding strategies and used both wild and cultivated germplasm as well as mutagenesis techniques for the improvement of various oilseed crops. Further, various biotechnological approaches such as tissue culture-based and transgenic approaches have also been used for oilseed improvement. Earlier, various chemical and physical mutagens like ethyl methanesulphonate (EMS), gamma radiations and others were used to treat the seeds, and these treated seeds were screened for the desirable traits or phenotypes. However, the drawbacks of these techniques are that they cause random mutagenesis and screening of these mutagens takes long time. Presently, genome editing techniques like zinc finger nucleases (ZFNs), transcription activator-like effector nucleases (TALENs), clustered regularly interspaced short palindromic repeats (CRISPR) and CRISPR-associated9 (Cas9) system are available and extensively being utilized for creating mutation at the specific site in the genome of various crop species. So, these genome editing technologies provide the opportunities for metabolic engineering and modification of fatty acid biosynthetic pathways. Among the available various genome editing techniques, the CRISPR/Cas9 system is robustly utilized due to its easy designing and implementation. Apart from crop improvement, genome editing can also be used as a promising tool to functionally characterize the various genes in oilseed crops. In this chapter, we have covered various genome editing techniques employed to mutate different genes in oilseed crops.

8.2.1 SOYBEAN (*GLYCINE MAX*)

Soybean (*Glycine max L.*) is one of the most important legume oilseed crops having high seed protein content and substances beneficial for human health. Soybean seed contains 35%–45% proteins, 15%–25% oil and 33% carbohydrates, and is rich in vitamin E. In addition, soybean oil is the second most produced oil worldwide after palm oil (Zulkifli et al., 2017). Different laboratories have developed edited lines using genome editing techniques such as TALEN and CRISPR/Cas9 in soybean. The detailed protocol for developing genome editing line using CRISPR/Cas9 in soybean has been given in a recently published book chapter by Bao et al. (2020).

As per the U.S. Food and Drug Administration (FDA), the consumption of oils having high monounsaturated fats (MUFAs) (oleic acid) is considered healthier due to longer shelf life and enhanced oxidative stability. In oil industries, polyunsaturated fats (PUFAs) (linoleic acid) are reduced from soybean oil by performing a partial hydrogenation, and this process is called soybean oil refining. The refined soybean oil is rich in trans-fatty acids, and its consumption raises the cholesterol and LDL (low-density lipid) levels in the blood. Hence, the consumption of trans-fatty acids is harmful to human health and can cause heart diseases (Ascherio et al., 1999). Therefore, the increment of oleic acid content in soybean seeds *via* manipulating the oil synthesis pathway is

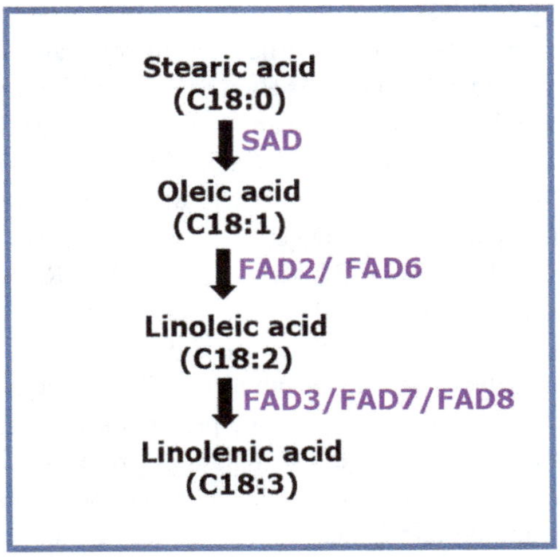

FIGURE 8.1 Conversion of saturated fatty acid to monounsaturated and monounsaturated to polyunsaturated fatty acids. Oleic acid to linoleic acid is converted by FAD2 in the ER and FAD6 in the plastid, whereas linoleic to linolenic acid is converted by FAD3 in the ER and FAD7/FAD8 in the plastid. SAD, stearoyl-acyl carrier protein desaturase; FAD, fatty acid desaturase.

a great option to improve the soybean quality. The pathway for the conversion of small-chain to longer-chain fatty acid and MUFA to PUFA is depicted in Figure 8.1. The oleic acid (MUFA) is converted to linoleic acid (PUFA) by an enzyme called fatty acid desaturase 2 (*FAD2*) (Schlueter et al., 2007). In soybean, seven paralogs of *FAD2* genes have been identified (Lakhssassi et al., 2017). In order to develop high oleic acid-containing soybean lines, TALEN was used to knock out two paralogs of *FAD2* genes (*FAD2-1A* and *FAD2-1B)* in soybean. The fatty acid profile of the T_2 seeds of homozygous edited lines (*FAD2-1A* and *FAD2-1B* both edited) showed approximately four times (80%) elevated level of oleic acid as compared to wild-type (20%). On the contrary, linoleic acid was reduced to 4%, from 50% in wild-type seeds. Subsequent screening among progenies of these edited lines for the absence of TALEN construct with incorporated mutation yielded the transgene-free high-oleic acid lines (Haun et al., 2014). In 2019, Calyxt Biotechnology Company (https://calyxt.com/) got a commercial approval for high-oleic acid oil soybean by genome editing of *FAD2-1A* and *FAD2-1B* gene in the United States. Subsequently, the same group edited fatty acid desaturase 3 (*FAD3*) gene in the same *fad2-1a/fad2-b* mutant background, and they found that oleic acid level has been increased to over 80% and linoleic and linolenic acid levels have been reduced to <3% (Demorest et al., 2016).

There have been several reports published regarding CRISPR/Cas9-mediated genome editing in soybean. Cai et al. (2015) firstly demonstrate CRISPR/Cas9-based genome editing in soybean. They selected two soybean genotypes named ‘Zigongdongdou’ (T_6 soybean transgenic line carrying bar gene) and ‘Williams 82’ (wild-type soybean). They designed one single guide RNA (sgRNA) named *bar-SP1* to mutate the *bar* transgene in Zigongdongdou genotype and three sgRNAs targeting three different sites, which are common for three endogenous genes *GmFEI1, GmFEI2* and *GmSHR* of soybean. After the hairy root transformation, DNA isolation from transformed roots, PCR and restriction enzyme assay yielded the mutations in the selected genes according to the designed sgRNA. Li et al. (2015) in the same year have also demonstrated CRISPR/Cas9 in soybean. Interestingly in this study, they have demonstrated both target mutagenesis through non-homologous end-joining (NHEJ) and target gene integration through homology-directed repair (HDR). In this study, a distal region of short arm of chromosome 4, called DD region, was selected to demonstrate the genome editing. After transformation, putative T_0 transgenic plants showed mutations for up to 76% for NHEJ and around 4% for HDR.

Further, CRISPR/Cas9 system was also used to knock out the *FAD-2* gene of soybean (Amin et al., 2019). The authors found 21% mutation efficiency in soybean transgenic plants, and entire fatty acids profiling from mutated soybean seeds showed up to 66% increased oleic acid content as compared to the control parent line. Similarly, in another study, CRISPR/Cas9-based edited homozygous soybean plants for *fatty acid desaturase 2* (*FAD2-1A* and *FAD2-1B*) were developed. These edited plants showed over 80% increment in oleic acid and a reduction of 1.3%–1.7% in linoleic acid as compared to the wild-type soybean (Do et al., 2019). Most recently, Wu et al. (2020) have also knocked out *FAD2-1A* and *GmFAD2-2A* genes using CRISPR/Cas9 strategy, and they have developed the edited lines having the oleic acid content increasing from around 20% to 72% and the linoleic acid content decreasing from around 57% to 17.27%.

The conglycinins (7S) and glycinins (11S) account for 70% of the total soybean seed proteins; hence, the quantity and quality of these seed storage proteins affect the quality of soya products. Recently, an attempt was made to mutate these seed storage proteins using CRISPR/Cas9 technology. In this study, *Agrobacterium rhizogenes*-mediated hairy root transformation was attempted, and nine different seed storage protein genes were targeted. Out of these nine genes, mutations were recorded only in three seed storage protein-encoding genes with an editing efficiency ranging from 3.8% to 43.7% (Li et al., 2018). This study clearly demonstrated the use of CRISPR/Cas9 for mutating seed storage proteins, and thus, this strategy can be used for targeted mutagenesis of seed storage proteins in other crops as well.

Isoflavonoids are the secondary metabolites, which are present 100 times more in soybean than in other legumes (Yu et al., 2003). They provide resistance to many diseases in different plants (Kant et al., 2019). Recently, an attempt was made to increase isoflavonoids content and check their effect on soybean mosaic virus (SMV) infection. The simultaneous knockout of three genes, namely two paralogs of flavanone-3-hydroxylase gene (*GmF3H1* and *GmF3H2*) and one flavone synthase gene (*GmFNSII-1*), which encodes for key enzymes of isoflavonoid pathway, was done. The analysis of T_3 triple-mutant (mutated for three *GmF3H1*, *GmF3H2* and *GmFNSII-1* genes) homozygous plants showed a significant increase in isoflavone content (3094.32 μg/g) as compared to wild-type (1600.26 μg/g). Additionally, these plants showed a significant one-third reduction after 3 weeks of infection in the leaf (Zhang et al., 2019).

Photoperiod is a major factor governing the flowering and plant's adaptation to different climates. Flowering locus T (FT) is an important gene that regulates flowering and induces floral initiation (Kong et al., 2010). Knockout of FT is known to cause a delay in flowering, whereas the overexpression of FT causes early flowering in the model plant *Arabidopsis thaliana* (Koornneef et al., 1991; Kobayashi et al., 1999). In soybean, out of ten homologs of FT genes, *GmFT2a* (*Glyma16g26660*) and *GmFT5a* (*Glyma16g04830*) are known to activate flowering (Kong et al., 2010; Nan et al., 2014). The usage of CRISPR/Cas9 strategy to knock out *GmFT2a* gene obtained the edited soybean lines showing a delayed flowering under both long- and short-day periods against the wild-type soybean (Cai et al., 2018). Further, the same group of researchers used CRISPR/Cas9 to knock out both *GmFT2a* and *GmFT5a* genes. They found that these two genes collectively regulate the flowering time in soybean. *GmFT2a* is more prominent than *GmFT5a* under short-day condition, while *GmFT5a* is more prominent than *GmFT2a* under long-day condition. Double mutants showed a delayed flowering by around 32 days in soybean under short-day condition (Cai et al., 2020). These kinds of studies can be beneficial for regional adaptability of soybean in future. Similarly, E1 locus encodes a transcription factor, which is also known to regulate flowering time (Xia et al., 2012). The CRISPR/Cas9-based knockout of E1 gene (*Glyma.06G207800*) in soybean induced 20-day early flowering as compared to wild-type under long-day condition (Han et al., 2019). Such genes can be used for the development of early-maturing soybean lines.

In plants, SQUAMOSA PROMOTER BINDING PROTEIN-LIKE (SPL) transcription factor family is well known to control the transition from juvenile to adult phase, to regulate the plastochron length, leaf size, shoot branching, plant height and panicle length (Luo et al., 2012; Lu et al., 2013). In soybean, CRISPR/Cas9-based editing of four genes of SPL9 family named *GmSPL9a*,

GmSPL9b, GmSPL9c and *GmSPL9d* was attempted by targeting *via* four different sgRNAs represented as SP1(for *GmSPL9a* and *GmSPL9b*), SP2 (for *GmSPL9a* and *GmSPL9b*), SP3 (for *GmSPL9c* and *GmSPL9d*) and SP4 (only for *GmSPL9d*). Analysis of different transgenic events, like T_2 homozygous double mutant for *GmSPL9a* and *GmSPL9b* genes, showed a shorter plastochron length. Similarly, T_4 quadruple mutant showed an increased nod and branch number as compared to non-edited plants (Bao et al., 2019). Similarly, to check the function of MYB transcription factor named late elongated hypocotyl (*LHY*), it was mutated using CRISPR/Cas9 system. The homozygous transgene-free T_2 quadruple-mutant plants for four *GmLHY* genes (*GmLHY1a, GmLHY1b, GmLHY2a* and *GmLHY2b*) showed shorter internodes and shorter plant height, including the reduced expression level of gibberellic acid (GA3) biosynthesis pathway genes as compared to control plants (Chang et al., 2019).

Apart from the crop improvement, genome editing technologies can also be used for the functional validation of genes. For example, fast neutron irradiation (FN)-mutant line *R59C46* of soybean showed smaller and thinner trichomes along with smaller cell nuclei as compared to its parental line M92–220. Further, a detailed analysis showed that in these mutant lines, deletions in chromosome six including six putative genes are responsible for this phenotype. Of these six genes, one gene *Glyma.06g145800* whose ortholog showed constitutive expression of PR gene 5 (CPR5) in *Arabidopsis* was found to be responsible for altered trichome phenotype (Kirik et al., 2002). To validate this, CRISPR/Cas9-based knockout of *GmCPR5* (*Glyma.06g145800*) gene yielded the similar shorter trichome phenotype in T_2 transgenic plants similar to the mutant line R59C46. Interestingly, the plants harboring homozygous in-frame mutation or heterozygous in-frame/frameshift mutation exhibited an intermediate trichome size. Therefore, they have demonstrated that along with a complete gene knockout, CRISPR/Cas9 can also be useful for the creation of novel allelic series for a gene (Campbell et al., 2019).

8.2.2 Rapeseed

In the *Brassica* genus, four species *B. rapa*, *B. juncea*, *B. carinata* and *B. napus* are cultivated around the world for oilseed production (Gupta, 2015). Among these species till date, genome editing has been demonstrated in *B. napus* only. The seeds in this family are usually protected in seed pod called siliques. When these seeds get matured and dry, the pod gets open for seed dispersal *via* shattering. The seed pod has three distinct regions: (1) valves, (2) the replum and (3) the valve margins. The valve of the seed pod wall surrounds the developing seeds that are attached to replum. This replum forms a middle ridge that connects the pod to the plant. The valve margins are very short boundaries between the valve and the replum, which are essential for seed dispersal or pod shattering (Liljegren et al., 2004). One of the major problems associated with the rapeseed production is siliques shattering, which can contribute up to 25% post-harvest yield loss (Price et al., 1996). In the model plant *Arabidopsis thaliana*, four transcription factors named SHATTERPROOF1 (*SHP1*), SHATTERPROOF2 (*SHP2*), INDEHISCENT (*IND*) and ALCATRAZ (*ALC*) are known to express in the valve margin for the formation of dehiscence zone (Ballester and Ferrandiz, 2017). The pods of individual knockout *Arabidopsis*-mutant lines for *shp1 shp2* and *ind* genes were found to be fully indehiscent due to the absence of lignified and separation layers, whereas *alc*-knockout mutant lines showed defects in the separation layer (Rajni and Sundaresan, 2001; Liljegren et al., 2004). To induce pod shattering resistance and to lower the seed loss during threshing, CRISPR/Cas9-mediated mutation in *B. napus* was attempted. In this report, simultaneous knockout of two copies of *ALC* gene (*BnaA.ALC.a* and *BnaC.ALC.a*) with a single common sgRNA was done. The double *ALC*-mutant T_2 plants did not show any phenotypic variation in pod as compared to wild-type plants (Braatz et al., 2017). Similarly, in another report, CRISPR/Cas9-mediated loss-of-function mutant was created individually for *IND* and *ALC* genes in *B. napus.* Morphological analysis of siliques revealed that homozygous T_2 mutant lines of *BnALC* gene (two copies) have the well-defined dehiscence zones similar to wild-type, whereas siliques of mutant line of *BnIND* gene (two copies) were devoid of defined dehiscence zones, which makes them more resistant to pod shattering (Zhai et al., 2019). These results showed that *IND* is more promising candidate for inducing shatter resistance rather than *ALC* gene in *B. napus*. The *JAG* gene encodes for a C_2H_2 zinc finger transcription factor, which acts as a mediator among genes involved in the regulation of identity of tissues and lateral organs (Ding et al., 2015). To know the effect of this gene upon pod shattering resistance in *B. napus*, CRISPR/Cas9-based multiplex genome editing of five homologs of *JAGGED* (*JAG, BnJAG.A02, BnJAG.A07, BnJAG.A08, BnJAG.C02, BnJAG.C06*) genes was carried out, which drastically affected the development of lateral organ to form the pod shape and size. However, the *BnJAG.A08*-knockout mutant showed significant changes in pod dehiscence zone with larger replum region conferring increased pod shattering resistance over the wild-type plants (Zaman et al., 2019).

In *A. thaliana* and *B. napus*, bilocular siliques are found, while in some accessions of *B. rapa* and *B. juncea*, multilocular siliques are found (Katiyar et al., 1998). Such multilocular types of siliques are very advantageous as they can contain a greater number of seeds and can provide better shattering resistance. Mutation in CLAVATA3 (CLV3) gene and retrotransposon insertion in CLAVATA1 (CLV1) gene cause multilocular siliques in *B. rapa* and *B. juncea* (Fan et al., 2014; Xu et al., 2017). Therefore, to obtain multilocular siliques phenotype in *B. napus*, Yang et al. (2018) attempted to knock out three CLAVATA genes: *CLAVATA1*, *CLAVATA2* and *CLAVATA3*. Two paralogs of each of *CLAVATA1*, *CLAVATA2* and *CLAVATA3* genes were targeted. Multilocular siliques were obtained by knockout of all copies of each CLAVATA (*BnCLV*) gene. However, among all the three CLAVATA-knockout genes, double-mutant line of *BnCLV3* was very efficient and yielded more leaves, multilocular siliques with significantly increased seed number per silique and seed weight compared to control lines (Yang et al., 2018).

Improving seed oil quality has always been one of the major objectives of Brassica breeding. Towards achieving this target, knockout of fatty acid desaturase 2 (*FAD2*) gene was attempted. It is reported that *B. napus* genome contains four *FAD2* paralogs (*BnaA.FAD2.a, BnaA.FAD2, BnaC.FAD2.a* and *BnaC.FAD2.b*) (Yang et al., 2012). CRISPR/Cas9-based editing of *BnaA.FAD2.a* and *BnaA.FAD2.b* genes in *B. napus* significantly enhanced the oleic acid content (73%–80%) and lowered the linoleic acid content (16%–9%) as compared to control lines without affecting the plant growth and other traits (Okuzaki et al., 2018). Further, through genome editing, not only knockout of the genes, but gene activation was also attempted. The reduction in saturated fatty acid (palmitic acid) content from canola oil can further improve its quality. The fatty acid chain elongation is catalysed by the enzyme β-ketoacyl-ACP synthase II (*KASII*), which converts palmitoyl-ACP to stearoyl-ACP before its desaturation to oleoyl-ACP (Browse and Summerville, 1991). There are reports available in different species that alteration of regulation of *KASII* activity can manipulate the palmitic acid content with fatty acid profile (Rubin-wilson et al., 2001; Pidkowich et al., 2007). Therefore, Gupta et al. (2012) have fused zinc finger transcription with VP16 activation domain and transformed in *B. napus* with the purpose of increasing the expression of *KAS II* gene. This approach was used to improve the fatty acid profile of canola oil. The artificial ZF-TFs fused with VP16 activation domain were designed to bind the 5' UTR regulatory region common in the two paralogs of *KASII* gene. The seeds of T_2 transgenic plants carrying ZFP-TFs showed the increased *KASII* transcript level with the significantly decreased palmitic acid content as compared to null siblings. Through this, it was demonstrated that fatty acid content can be modulated by the activation of *KAS II* gene through ZF-VP16 activation domain fusion strategy (Gupta et al., 2012).

The lysophosphatidic acid acyltransferase (*LPAT*) gene is crucial for the biosynthesis of triacylglycerols (TAG) and membrane lipid (Baud and Lepiniec, 2010). The mutant lines for *BnLPAT2* (with seven homologs) and *BnLPAT5* (with four homologs) developed using CRISPR/Cas9 showed enlarged oil bodies but the number of oil bodies and seed size was decreased. The size of oil bodies was increased due to coalescence between two neighbouring oil bodies. However, the seed oil content of the *BnLAPT2*/*BnLAPT5* double-mutant lines was declined by 39% as compared to WT seeds. This study has demonstrated that *BnLAPT2* and *BnLAPT5* genes play a key role in regulating the size and morphology of oil bodies in *B. napus*. Change in the total oil content shows that *BnLPAT2/BnLPAT5* plays an important role in oil biosynthesis pathway (Zhang et al., 2019).

Development of biotic stress resistance is another major objective in Brassica breeding. Among several pathogens, *Sclerotinia sclerotiorum* is one of the necrotrophic fungi that causes sclerotinia stem rot disease in different Brassica species, including *B. napus*, which ultimately reduces the yield ranging from 5% to 100% (Saharan and Mehta, 2008). The WRKY transcription factors are known to consist of WRKY domain and to be involved mainly in the plant immunity (Jiang et al., 2017b). Previously, it has been reported that *WRKY11* and *WRKY70* are involved in the pathogen response in *Arabidopsis* and *B. napus* (Shim et al., 2013; Wang et al., 2014). To check the response of these transcription factors upon *S. sclerotiorum* infection, loss-of-function mutants and overexpression lines were created (Sun et al., 2018). The genome of *B. napus* contains two paralogs of *BnWRKY11* and four paralogs of *BnWRKY70*, which were knocked out using common sgRNA. *BnWRKY11* mutants did not show any *Sclerotinia* resistance, similar to the wild-type plants. In contrast, the *BnWRKY70*-overexpressing lines showed a reduction in resistance. Therefore, *BnWRKY70* gene might play a crucial role in *Sclerotinia* resistance in *B. napus*. The WRKY transcription factors are known to play a role in SA and JA-induced disease resistance to pathogen. CRISPR/Cas9-mediated knockout mutants for WRKY transcription factors (*BnWRKY70*) showed an enhanced resistance to *S. sclerotiorum* when compared with the wild-type plant (Sun et al., 2018).

To increase the seed oil content, efforts were also made by preventing the synthesized lipids from decomposing during seed maturation (Kanai et al., 2019; Ding et al., 2019). In a recent report, CRISPR/Cas9-mediated simultaneous knockout of SEED FATTY ACID REDUCER (*BnSFAR4 and BnSFAR5,* encode for GDSL lipase) gene family members yielded a range of enhanced seed oil-containing mutant lines having no pleiotropic effects on seed vigour, germination and oil mobilization (Karunarathna et al., 2020).

Weeds are the major threat to crop production, and to manage them, herbicides are sprayed to the crops. The gene acetolactate synthase (*BnALS*) encodes for an enzyme responsible for the synthesis of the branched-chain amino acids such as leucine, isoleucine and valine, which are the target sites of several important herbicides (Powles and Yu, 2010). In *B. napus*, five paralogs of *BnALS* genes are found, and out of which, two *ALS1* and *ALS3* are constitutively expressed. Recently, to create the herbicide-resistant *B. napus* line, cytosine base editing strategy was used. Amino acid at the position 197 (P197) is conserved, which can be mutated using a base editor to develop the herbicide-resistant *B. napus*. Therefore, in this paper, the authors have converted C to T (amino acid P197) of both *BnALS1* and *BnALS3* genes. They found two C-T substitutions in some edited lines and three C-T substitutions in some lines resulting in the replacement of proline with other amino acid. The homozygous mutant line for both the genes showed the survival at the 30 mg ai/L dosage of acetolactate synthase inhibitors such as tribenuron-methyl herbicide as compared to wild-type (Wu et al., 2020).

Yellow-seeded *B. napus* is a breeding priority as yellow-seeded lines have high oil and low fibre content as compared to brown-seeded lines (Rehman and McVetty, 2011). In *Arabidopsis*, several genes related to the proanthocyanidin pathway have been identified, which play a key role in regulating seed colour (Appelhagen et al., 2014). In *B. napus*, TT2 (*BnaA.TT2*, *BnaA08g29930D* and *BnaC.TT2*, *BnaC08g07960D*) genes were knocked out using CRISPR/Cas9, which resulted in the creation of yellow-seeded lines. These *TT2*-knockout mutants have reduced flavonoid accumulation and improved fatty acid composition (Xie et al., 2020).

Phytic acid (PA) is an important constituent of seed, which serves as a major reservoir of phosphorous. PA is negatively charged compound capable of chelating essential minerals. Therefore, high-PA content could hamper the seed meal quality. Additionally, high-PA content is undesirable as it cannot be metabolized by monogastric animals (Campbell et al., 2015). So far, several mutant lines have been identified for low-PA content in different crop species (Sparvoli and Cominelli, 2015). PA is synthesized predominantly in seeds *via* a lipid-independent pathway (Raboy, 2009). In *B. napus*, EMS-treated population of 7680 plants was screened for identifying knockout mutants of five key genes of PA biosynthesis, and a double mutant of phosphoglyceric acid kinase (*Bn.2-PGK2*) showed up to 25.4% PA reduction as compared to non-mutated plants (Sashidhar et al., 2020a). Recently, the same group used CRISPR/Cas9 system to knock out the two paralogs of inositol tetrakisphosphate kinase gene (*BnITPK1* and *BnITPK4*). The analysis of T4 homozygous non-transgenic triple-mutant lines for *BnITPK1* showed the reduced (27.2%–35.3%) seed PA content as compared to control plants (Sashidhar et al., 2020b).

8.2.3 Camelina Sativa

Camelina sativa is the hexaploid oilseed crop of *Brassicaceae* family. It has a short growing season and bears good productivity in limited rainfall and low soil fertility (Iskandarov et al., 2014). To improve the oil quality of Camelina seeds, CRISPR/Cas9 system has been used. Like soybean and *B. napus*, *FAD2* genes have also been targeted for improving the oil quality in *C. sativa*. CRISPR/Cas9 system-based complete knockout of *FAD2* gene in *Arabidopsis* and *Camelina* increased the oleic acid content from 16% to >50%, decreased the linolenic acid content from 16% to <4% and decreased the linoleic acid content (from 35% to 10%) (Jiang et al., 2017a). Similarly, Celine-Morineau et al. (2017) have also knocked out the *FAD2* genes of *C. sativa*, which also resulted in the enhanced oleic acid and decreased linoleic and linolenic acid contents. In another report, CRISPR/Cas9-based complete inactivation of diacylglycerol acyl transferase (*CsDGAT1*) and phospholipid diacylglycerol acyltransferase (*CsPDAT1*) yielded wrinkled seeds, reduced oil content and altered fatty acid profile (Aznar-Moreno and Durrett, 2017).

The nutritional quality of seed meal is mostly dependent on the composition of specific abundant proteins in the seeds, which also determine their economic value (Wanasundara, 2011). In the Brassicaceae family, the cruciferins (11S) and napins (2S) are the majorly found classes of seed storage proteins, and more than 60% seed proteins are cruciferins in *Camelina sativa* (Reggiani and Russo, 2015). At the time of seed maturation, the available carbon and nitrogen sources are directed for the formation of these cruciferins and napins (Herman et al., 2014). Therefore, cruciferins could be a potential target to improve the seed meal quality in Brassicaceae. In a recent report, to improve the *Camelina* seed meal quality, CRISPR/Cas9-mediated knockout lines for genes encoding the seed storage protein *CRUCIFERIN C* (*CsCRUC*) were generated. These mutant lines did not alter the total seed protein and oil content but significantly altered the amino acid composition (increased alanine, cysteine and proline) and fatty acid profile (Lyzenga et al., 2019).

8.2.4 Sunflower (*Helianthus annuus*)

Sunflower (*Helianthus annuus L.*) is an important oilseed crop cultivated around the world mainly in the temperate and semi-dry regions. Sunflower seeds are mainly used for the extraction of oil, which is mainly composed of proteins (around 20%) and fats (≥50%). The 90% of the sunflower oil consists of unsaturated fatty acids, for example oleic acid (16%–19%), linoleic acid (69%–72%) and the rest of the 10% comprising palmitic acid, steric acid, myristic acid etc. (Davey and Jan, 2010). Till date, sunflower breeding has been done to achieve high oil yield varieties, resistant varieties to various biotic stresses like root parasite *Orobanche* resistance, tolerant varieties to drought, herbicide-resistant hybrids, confectionary hybrids, ornamental hybrids, high-protein and low-fat hybrids (Jocic et al., 2015; Dimitrijevic and Horn 2018). In the past, several mutant lines have been developed by EMS treatment and gamma radiations, and screened for the traits or phenotypes important for breeding purposes, for example flowering time, oil content, fatty acid biosynthesis etc. (Dimitrijevic and Horn, 2018). Sunflower is known to be recalcitrant for transformation and regeneration protocols. Therefore, for the successful implementation of genome editing techniques in sunflower, a proper transformation and regeneration system is required. Root parasite broomrape is one of the serious problems in sunflower cultivation. Using genome editing techniques by knockout of the strigolactone biosynthetic pathways, CCD8 broomrape-resistant genotypes can be developed in a short span of time as compared to conventional breeding (Cvejic et al., 2020).

8.2.5 Oil Palm (*Elaeis guineensis*)

Oil palm (*Elaeis guineensis*) are perennial trees belonging to the Arecaceae family. These trees are the most efficient source of palm vegetable oil and palm kernel oil. The oil palm trees have the bunch of large red spherical fruits, and their pulp and kernels are the source of oil extraction. It is grown in more than 16 million hectares around the world, which makes it world's major cultivated crop. The oil palm is a cross-pollinated tree species, and their natural seed germination rate is very poor. Additionally, long life spans and absence of axillary shoot, the breeding programmes and vegetative propagation are not suitable for oil palm cultivation. Therefore, genetic transformation, somatic embryogenesis and recent genome editing techniques offer a great opportunity for the genetic improvement of oil palm. Oil palm is very difficult to get transformed by available approaches like Agrobacterium-mediated transformation, particle bombardment, PEG-mediated or DNA microinjection (Masani et al., 2018). Though, several attempts have been made in the past for the transformation of oil palm (Parveez et al., 2000; Masli et al., 2009; Masani et al., 2014; Hiei et al., 2014). The genome sequences of African oil palm (*Elaeis guineensis*) and South American oil palm (*Elaeis oleifera*) were published by Singh et al. (2013), and PalmXplore – the oil palm gene database (http://palmxplore.mpob.gov.my.) (Sanusi et al., 2018) – is available, which can be used for genomic-assisted breeding as well as genome editing purpose, if any method can be optimized to generate a stable transgenic oil palm.

Till date, several important genes have been identified and characterized in the oil palm (Masura et al., 2017). In oil palm, to demonstrate the genome editing, *Agrobacterium* [carrying the CRISPR/Cas9 construct for two genes isoflavone reductase (*IFR*) and metallothionein-like protein (*MT*)]-mediated transformation was attempted in oil palm calli (Budiani et al., 2018). After transformation, PCR and Sanger sequencing confirmed the editing in the selected genes.

8.2.6 Peanut (*Arachis hypogaea L.*)

Peanut/groundnut (2n=4×=40, AABB) is an important self-pollinated allotetraploid legume family member and oilseed crop. It is mainly grown in rainfed and semi-arid tropic regions around the world. Peanut is cultivated for its edible seeds and oil extraction. The peanut seed contains approximately 22%–30% protein and 35%–60% oil; hence, it is highly nutritive, as out of 80% of fatty acids, more than 45% oleic acid present is a desirable and healthy parameter for cardiovascular health (Chen et al., 2015). Peanut productivity is severely affected by various biotic stresses (insects, pests, diseases) and abiotic stresses (drought, salinity, heat etc.). Different breeding programmes have been run around the world for the improvement of groundnut to obtain the high pod yield, high oil content and high-oleic acid, to get early maturity, high shelling percentage and resistance to various biotic and abiotic stresses (Desmae et al., 2019). In one of these breeding programmes, EMS treatment-based high-oleate spontaneous mutant line 'F435' containing 80% oleic acid and 2% linoleic acid (identified in 1987) has been extensively used as one of the parents to develop high-oleate cultivars (Chu et al., 2009).

Transgenic approaches can be very effective to combat with the various abiotic and biotic stresses. Peanut can be transformed *via Agrobacterium*-mediated transformation approach or particle bombardment, and stable transgenic plants have been developed for different abiotic stresses (Bhatnagar-Mathur et al., 2014; Sarkar et al., 2016; Patel et al., 2017). As reported in other crops, in peanut the transformation efficiency is genotype dependent. As per Krishna et al. (2015), out of 37 peanut genotypes, four of them (YueYou116, Toalson, Tatu and Luhua9) are unable to be transformed due to their recalcitrant nature. There are only few reports of utilization of genome editing technology for the improvement of peanut till date. For instance, towards improving oil quality,

TALEN system was used to mutate *FAD2* genes (*AhFAD2A* and *AhFAD2B*), which successfully enhanced the oleic acid 0.5–2-fold as compared to control plants (Wen et al., 2018). Similarly, the first report of successful demonstration of CRISPR/Cas9 system-based genome editing in peanut was reported by Yuan et al. (2019). They also targeted the *AhFAD2A* and *AhFAD2B* genes and confirmed the editing after protoplast transfection and hairy root transformation.

8.2.7 Olive Seed (*Olea europaea*)

The olive (*Olea europaea L.*) is a perennial tree species, which belongs to *Oleaceae* family. Earlier, olive trees were used to be cultivated mainly in Mediterranean Basin but now they are also cultivated in America and Australia. In the current scenario, the olive trees are one of the most important oil tree species. Olives are the fruit of olive trees, which are extremely bitter in the taste; therefore, they are used for the production of olive oil. The natural fruit juice of olive is called 'virgin olive oil', which bears great nutritional qualities and promising health benefits. In addition, leaf extract and olive oil also bear the antioxidant, antibacterial, antifungal, antidiabetic, analgesic and anti-inflammatory properties (Munoz-Merida et al., 2013). Today, olive oil is considered to be the best oil for human consumption due to its high oxidative stability and balanced fatty acid composition (Olmo-Garcia et al., 2018). Unlike other oilseed crops, olive oil consists of 55%–83% oleic acid and 3.5%–21% linoleic acid having high levels of MUFAs naturally (https://www.oliveoilsource.com/), making it superior than other available edible oils. Olive trees have also been reported to be drought- and salinity-tolerant along with cold sensitivity (Gucci and Tattini, 1997). In the last few years, olive oil production is also affected by a fungal pathogen *Verticillium dahliae Kleb* causing Verticillium wilt disease, which reduced the olive yield from 56% to 84% in Spain (Keykhasaber et al., 2017; Narvaez et al., 2020). Like other tree species, to reach the flowering stage, olive has to cross a long juvenile phase because of that a long time is required to screen and develop new varieties for disease resistance *via* breeding. Therefore, genome editing is one of the good options for the improvement of olive oil. *Agrobacterium*-mediated genetic transformation strategy by using somatic embryos as explant is well documented in olive (Jafarzadeh-Bajestani et al., 2011; Palomo-Rios et al., 2017; Narvaez et al., 2020). Therefore, genome editing technologies can be potentially useful to shorten the juvenile period, in order to increase the oil content in olive seeds.

8.2.8 Safflower (*Carthamus tinctorius*)

Safflower (*Carthamus tinctorius L.*) is also known false saffron belonging to Asteraceae family. It is a xerophilous species mainly cultivated in arid and semi-arid regions of the world. In addition, safflower has been used for various medicinal purposes: for instance, purgative, analgesic and anti-pyretic properties that are used to treat patients with poisoning, can manage menstrual cramps and are used to treat whooping cough and chronic bronchitis; the water extracts act as anticoagulant, vasodilator, immunosuppressive, anticancer agent and many more. The dried and ground petals of safflower are used to extract a dye called carthamin, which is used to colour the fabrics (Asgarpanah and Kazemivash, 2013; Delshad et al., 2018). The safflower seed is mainly composed of the hull and the kernel, and 95% of the oil is found in the kernel. Safflower oil bears the high nutritional quality, which is almost similar to olive oil. The most of the available commercial safflower oil consists of 72%–78% linoleic acid and 15%–20% oleic acid (Hall, 2016). Alteration in the fatty acid profile to obtain high-oleic acid oilseed varieties is always a preferable task for breeders and oil industries. Genome editing techniques along with breeding programmes can potentially accelerate the safflower improvement. Several attempts have been made for the genetic transformation of safflower. In *in vitro* conditions, safflower shoots are reported to be sensitive to hyperhydration and poor formation of roots, which leads to the death of plant when transferred to soil (Shilpa et al., 2010). To overcome this problem, after *an Agrobacterium*-mediated transformation and obtaining transgenic shoots, an effective grafting method has been developed, from which near 50% transgenic shoots were developed into mature plants bearing T_1 seeds (Belide et al., 2011). Similarly, *Agrobacterium*-mediated floral dip transformation method has also been optimized to get transgenic seeds (Tu et al., 2019). Though, the transformation method is well optimized in safflower but the attempts have not been made to conduct the genome editing experiments till now. Table 8.1 summarizes the list of various genes targeted through genome editing.

TABLE 8.1
List of Different Targeted Genes of Various Oilseed Crops *via* Genome Editing Techniques

Oilseed Crop	Technique	Targeted Gene	Purpose	References
Glycine max (*L.*)	TALENs	Fatty acid desaturase 2 (*FAD2*) *FAD2-1A* and *FAD2A*	To improve oil quality (high-oleic acid and low linoleic acid)	Haun et al. (2014)
Glycine max (*L.*)	TALENs	*FAD2-1A*, *FAD2-1B*, *FAD3A*	To improve oil quality (high-oleic acid and low linoleic acid)	Demorest et al. (2016)
Glycine max (*L.*)	CRISPR/Cas9	bar transgene, *GmFEI2* and *GmSHR*	To demonstrate genome editing	Cai et al. (2015)
Glycine max (*L.*)	CRISPR/Cas9	Flower locus T (*GmFT2a*)	To delay flowering	Cai et al. (2018)
Glycine max (*L.*)	CRISPR/Cas9	Microsomal omega-6 desaturase *FAD2-1* and *FAD2-2*	To improve oil quality (high-oleic acid and low linoleic acid)	Amin et al. (2019)
Glycine max (*L.*)	CRISPR/Cas9	Fatty acid desaturase 2 *(FAD2-1A and FAD2-1B)*	To improve oil quality (high-oleic acid and low linoleic acid)	Do et al. (2019)
Glycine max (*L.*)	CRISPR/Cas9	Flavanone-3-hydroxylase (*GmF3H1*, *GmF3H2*) and flavone synthase (*GmFNSII-1*)	To increase isoflavonoid content and soybean mosaic virus (SMV) resistance	Zhang et al. (2019)
Glycine max (*L.*)	CRISPR/Cas9	*E1* gene (unique transcription factor)	To induce early flowering	Han et al. (2019)
Glycine max (*L.*)	CRISPR/Cas9	Squamosal promoter binding protein-like (*SPL*) transcription factors	To check the knockout effect of these transcription factors	Bao et al. (2019)
Glycine max (*L.*)	CRISPR/Cas9	Late elongated hypocotyl (*LHY*): a MYB transcription factor	To check the knockout effect of these transcription factors	Chang et al. (2019)
Brassica napus L.	CRISPR/Cas9	Alcatraz (*ALC*)	To induce shatter resistance	Braatz et al. (2017)
Brassica napus L.	CRISPR/Cas9	Indehiscent (*IND*) and Alcatraz (*ALC*)	To induce shatter resistance	Zhai et al. (2019)
Brassica napus L.	CRISPR/Cas9	Jagged (*BnJAG.A02*, *BnJAG.A07*, *BnJAG.A08*, *BnJAG.C02*, *BnJAG.C06*)	To alter the pod shape	Zaman et al. (2019)
Brassica napus L.	CRISPR/Cas9	CLAVATA3	To obtain multilocular siliques	Yang et al. (2018)
Brassica napus L.	CRISPR/Cas9	Fatty acid desaturase (*FAD2* and *FAD3*)	To improve oil quality (high-oleic acid and low linoleic acid)	Okuzaki et al. (2018)

(*Continued*)

TABLE 8.1 (*Continued*)
List of Different Targeted Genes of Various Oilseed Crops *via* Genome Editing Techniques

Oilseed Crop	Technique	Targeted Gene	Purpose	References
Brassica napus L.	CRISPR/Cas9	Lysophosphatidic acid acyltransferase (*BnLPAT2* (with 7 homologs) and *BnLPAT5* (with 4 homologs))	To increase oil content	Zhang et al. (2019)
Brassica napus L.	CRISPR/Cas9	WRKY transcription factors (*BnWRKY70*))	To provide resistance for *S. sclerotiorum*	Sun et al. (2018)
Camelina sativa (L.) Crantz	CRISPR/Cas9	Diacylglycerol acyltransferase (*CsDGAT1*) and phospholipid diacylglycerol acyltransferase (*CsPDAT1*)	To improve oil quality and yield	Aznar-Moreno and Durrett (2017)
Camelina sativa (L.) Crantz	CRISPR/Cas9	Fatty acid desaturase 2 (*FAD2*)	To improve oil quality (high-oleic acid and low linoleic acid)	Jiang et al. (2017a) and Celine-Morineau et al. (2017)
Camelina sativa (L.) Crantz	CRISPR/Cas9	Cruciferin C (*CsCRUC*)	To improve seed meal quality	Lyzenga et al. (2019)
Arachis hypogaea	TALENs	Fatty acid desaturase 2 (FAD2A and FAD2B)	To improve oil quality (high-oleic acid and low linoleic acid)	Wen et al. (2018)

8.3 NEW OPPORTUNITIES AND CHALLENGES OF GENOME EDITING IN OILSEED CROPS

Genome editing provides new opportunities for crop improvement, which were not possible earlier. It has been shown that by the use of genome editing, crop diversity can be increased and wild species can be modified to use them as a cultivated species (Chen et al., 2019). It has been demonstrated in case of the wild tomato and *Physalis pruinosa*, which is a wild relative of tomato, where CRISPR/Cas9 technologies have been used for their improvement (Li et al., 2018; Zsogon et al., 2018; Lemmon et al., 2018). However, in oilseed, there are certain wild species that have some potential characteristics to use them as a cultivated species. One example is pennycress (*Thlaspi arvense*), which belongs to the Brassicaceae family. It is having a short life cycle, cold tolerance and high productivity, which make it a new alternative oilseed crop. Therefore, using genome editing technology, certain genes related to seed dormancy, oil quality, oil yield and glucosinolate content can be modified. If these characteristics can be modified, then it provides a new crop in the list of already-available oilseed crop (Sedbrook et al., 2014; Chen et al., 2019). Decreasing the antinutritional compounds such as glucosinolate and erucic acid in rapeseed is another possibility, which can be achieved through genome editing technologies much easier than through conventional approaches (Rani et al., 2016). Glucosinolate is transported to the seed from the maternal tissue. Glucosinolate transporter genes are playing a key role in transporting the glucosinolate to the seeds. Nour-Eldin et al. (2017) have mutated the *B. rapa* and *B. juncea* GTR genes and found that this strategy can be efficient in lowering the glucosinolate content in the seeds. This can also be done through the use of targeting GTR genes using the genome editing strategy. Additionally, fatty acid elongase (FAE) genes can be knocked out by using genome editing, and the amount of erucic acid can be reduced.

Apart from various opportunities provided by genome editing, there are still some technical challenges for the use of these technologies in the improvement of oilseed crops. The response of the crop for transformation and regeneration after transformation is the major constraint to utilize these modern genome editing techniques in any crop. The major issue is the availability of an efficient transformation and *in vitro* regeneration (from explant to intact plant) method in all the cultivars of a species. Next, the off-target effect in the genome is also a major issue especially with CRISPR/Cas9 system, which needs to be properly studied for the employment of these techniques. Although genome editing requires transformation and development of transgenic line, in further generation transgene-free genome editing can be possible through crossing and removal of editing constructs. However, in some countries especially in the European Union, any kind of change through genome editing is regulated as transgenic (Callaway, 2018). Otherwise, these genome editing techniques have the potential to revolutionize the crop improvement programmes. There is still not a suitable method available for developing edited lines in many oilseed crops.

8.4 CONCLUSION AND FUTURE PROSPECTS

Oilseed crops are cultivated around the globe not only for vegetable oil production but also for the production of biofuels, various industrial products and animal feed. Vegetable oils and biofuels demand is continuously increasing at the global level, and therefore, alternative strategies are required for increasing the production of oilseed crops. Improving oil yield and oil quality is difficult to achieve through conventional breeding programmes as there is a lacuna of availability of diverse germplasm. In contrast, by engineering the biochemical pathway of oil synthesis, both the oil quality and quantity can be improved by using the modern genome editing techniques. Genetic diversity is the key for the improvement of any crop variety. Through genome editing, negative regulators or repressors of a gene can be knocked out, which can improve the oil yield and oil quality. Additionally, through genome editing, genes can be activated, gene regulation can be modulated and specific amino acid can be changed through which gene diversity can be created. In addition, by using CRISPR/Cas9 system, several genes can be knocked out in a single attempt. Hence, the usage of multiple sgRNAs to mutate different rate-limiting biochemical pathway genes can potentially enhance the oil quality and quantity simultaneously. However, till date, successful genome editing has been established in few oilseed crops. The major factor essential for the employment of these techniques is the response of crop under *in vitro* conditions for regeneration after transformation. As per the above discussion, many oilseed crops don't have a suitable transformation and regeneration system. Therefore, a stable method of transformation is the utmost requirement in any plant species for genome editing. Finally, more efforts should be given to the optimization of transformation and regeneration methods so that these genome editing techniques can be easily utilized in all oilseed crops. However, till date, genome editing has been successfully demonstrated in soybean, rapeseed and Camelina.

REFERENCES

Amin, N. A., N. Ahmad, N. Wu, et al. "CRISPR-Cas9 mediated targeted disruption of FAD2–2 microsomal omega-6 desaturase in soybean (*Glycine max L*)." *BMC Biotechnology* 19, no. 1 (2019): 9.

Appelhagen, I., K. Thiedig, N. Nordholt, et al. "Update on transparent testa mutants from *Arabidopsis thaliana*: Characterization of new alleles from an isogenic collection." *Planta* 240 (2014): 955–970.

Ascherio, A., M. B. Katan, P. L. Zock, et al. "Trans fatty acids and coronary heart disease." *New England Journal of Medicine* 340 (1999): 1994–1998.

Asgarpanah, J. and N. Kazemivash. "Phytochemistry, pharmacology and medicinal properties of *Carthamus tinctorius L*." *Chinese Journal of Integrative Medicine* 19, no. 2 (2013): 153–159.

Aznar-Moreno, J. A. and T. P. Durrett. "Simultaneous targeting of multiple gene homeologs to alter seed oil production in *Camelina sativa*." *Plant Cell Physiology* 58, no. 7 (2017): 1260–1267.

Ballester, P. and C. Ferrandiz. "Shattering fruits: Variations on a dehiscent theme." *Current Opinion in Plant Biology* 35 (2017): 68–75.

Bao, A., H. Chen, L. Chen, et al. "CRISPR/Cas9-mediated targeted mutagenesis of GmSPL9 genes alters plant architecture in soybean." *BMC Plant Biology* 19 (2019): 131.

Bao, A., L. S. P. Tran and D. Cao. "CRISPR/Cas9-based gene editing in Soybean". In *Legume Genomics: Methods in Molecular Biology*, edited by M. Jain and R. Garg, pp. 349–364, vol. 2107. Humana: New York (2020).

Baud, S. and L. Lepiniec. "Physiological and developmental regulation of seed oil production." *Progress in Lipid Research* 49 (2010): 235–249.

Belide, S., L. Hac, S. P. Singh, et al. "*Agrobacterium*-mediated transformation of safflower and the efficient recovery of transgenic plants via grafting." *Plant Methods* 7 (2011): 12.

Bhatnagar-Mathur, P., J. S. Rao, V. Vadez, et al. "Transgenic peanut overexpressing the DREB1A transcription factor has higher yields under drought stress." *Molecular Breeding* 33 (2014): 327–340.

Braatz, J., H. J. Harloff, M. Mascher, et al. "CRISPR-Cas9 targeted mutagenesis leads to simultaneous modification of different homoeologous gene copies in polyploid oilseed rape (*Brassica napus*)." *Plant Physiology* 174 (2017): 935–942.

Browse, J. and C. Summerville. "Glycerolipid metabolism, biochemistry and regulation." *Annual Review of Plant Physiology and Plant Molecular Biology* 42 (1991): 467–506.

Budiani, A., R. A. Putranto, I. Riyadi, et al. "Transformation of oil palm calli using CRISPR/Cas9 system: Toward genome editing of oil palm." *Earth and Environmental Science* 183 (2018): 012003.

Cai, Y., L. Chen, X. Liu, et al. "CRISPR/Cas9-mediated genome editing in soybean hairy roots." *PLoS One* 10 (2015): e0136064. doi: 10.1371/journal.pone.0136064.

Cai, Y., L. Chen, X. Liu, et al. "CRISPR/Cas9-mediated targeted mutagenesis of GmFT2a delays flowering time in soya bean." *Plant Biotechnology Journal* 16 (2018): 176–185.

Cai, Y., L. Wang, L. Chen, et al. "Mutagenesis of GmFT2a and GmFT5a mediated by CRISPR/Cas9 contributes for expanding the regional adaptability of soybean." *Plant Biotechnology Journal* 18 (2020): 298–309.

Callaway, E. "CRISPR plants now subject to tough GM laws in European Union". *Nature* 560 (2018): 16.

Campbell, L., C. B. Rempel and J. P. Wanasundara. "Canola/rapeseed protein: Future opportunities and directions." – workshop proceedings of IRC." *Plants* 5, no. 2 (2015): e17.

Campbell, B. Z., J. B. Hoyle, B. Bucciarelli, et al. "Functional analysis and development of a CRISPR/Cas9 allelic series for a CPR5 ortholog necessary for proper growth of soybean trichomes". *Scientific Reports* 9 (2019): 14757.

Celine-Morineau, C., Y. Bellec, F. Tellier, et al. "Selective gene dosage by CRISPR-Cas9 genome editing in hexaploid *Camelina sativa*." *Plant Biotechnology Journal* 15 (2017): 729–739.

Chang, Q., L. Dong, T. Su, et al. "CRISPR/Cas9-mediated targeted mutagenesis of GmLHY genes alters plant height and internode length in soybean." *BMC Plant Biology* 19 (2019): 562.

Chen, S., Y. Lei, X. Xu, et al. "The Peanut (*Arachis hypogaea L.*) gene AhLPAT2 increases the lipid content of transgenic *Arabidopsis* seeds." *PLoS One* 10, no. 8 (2015): e0136170.

Chen, K., Y. Wang, R. Zhang, et al. "CRISPR/Cas genome editing and precision plant breeding in agriculture." *Annual Review of Plant Biology* 70 (2019): 667–697.

Chu, Y., C. C. Holbrook and P. Ozias-Akins. "Two alleles of ahFAD2B control the high oleic acid trait in cultivated peanut." *Crop Science* 49 (2009): 2029–2036.

Cvejic, S., A. Radanovic, B. Dedic, et al. "Genetic and genomic tools in Sunflower breeding for broomrape resistance." *Genes* 11, no. 2 (2020): 152.

Davey, M. R. and M. Jan. "Sunflower (*Helianthus annuus L.*): Genetic improvement using conventional and in vitro technologies." *Journal of Crop Improvement* 24 (2010): 349–391.

Delshad, E., M. Yousefi, P. Sasannezhad, et al. "Medical uses of *Carthamus tinctorius* L. (Safflower): A comprehensive review from traditional medicine to modern medicine." *Electronic Physician* 10, no. 4 (2018): 6672–6681.

Demorest, Z. L., A. Coffman, N. J. Baltes, et al. "Direct stacking of sequence- specific nuclease- induced mutations to produce high oleic and low linolenic soybean oil." *BMC Plant Biology* 16 (2016): 225.

Desmae, H., P. Janila, P. Okori, et al. "Genetics, genomics and breeding of groundnut (*Arachis hypogaea L.*)." *Plant Breeding* 138 (2019): 425–444.

Dimitrijevic, A. and R. Horn. "Sunflower hybrid breeding: From markers to genomic selection." *Frontiers in Plant Sciences* 8 (2018): 2238.

Ding, L., S. Yan, L. Jiang, et al. "HANABA TARANU (HAN) bridges meristem and organ primordia boundaries through PINHEAD, JAGGED, BLADE-ON-PETIOLE2 and CYTOKININ OXIDASE 3 during flower development in *Arabidopsis*." *PLoS Genetics* 11, no. 9 (2015): e1005479.

Ding, L. N., X. J. Guo, M. Li, et al. "Improving seed germination and oil contents by regulating the GDSL transcriptional level in *Brassica napus*." *Plant Cell Reports* 38 (2019): 243–253.

Do, P. T., C. X. Nguyen, H. T. Bui, et al. "Demonstration of highly efficient dual gRNA CRISPR/Cas9 editing of the homeologous GmFAD2–1A and GmFAD2–1B genes to yield a high oleic, low linoleic and α-linolenic acid phenotype in soybean." *BMC Plant Biology* 19 (2019): 311.

Fan, C., Y. Wu, Q. Yang, et al. "A novel single-nucleotide mutation in a CLAVATA3 gene homolog controls a multilocular silique trait in *Brassica rapa L.*" *Molecular Plant* 7 (2014): 1788–1792.

FAO. Food outlook: Biannual report on global food markets, November 2019, Rome.

Gucci, R. and M. Tattini. "Salinity tolerance in Olive." In *Horticultural Reviews*, edited by J. Janick, pp. 177–214. John Wiley & Sons, Inc.: Oxford, UK (1997).

Gupta, S. K. "Brassicas" In *Breeding Oilseed Crops for Sustainable Production Opportunities and Constraints*, edited by S. K. Gupta, pp. 33–54. Elsevier Science Direct : San Diego, CA (2015).

Gupta, M., R. C. DeKelver, A. Palta, et al. "Transcriptional activation of *Brassica napus* b- ketoacyl-ACP synthase II with an engineered zinc finger protein transcription factor." *Plant Biotechnology Journal* 10 (2012): 783–791.

Hall, C. "Overview of the oilseed Safflower (*Carthamus tinctorius L.*)." In *Reference Module in Food Science*. Elsevier: Amsterdam (2016). doi: 10.1016/B978-0-08-100596-5.00030-5.

Han, J., B. Guo, Y. Guo, et al. "Creation of early flowering germplasm of soybean by CRISPR/Cas9 technology." *Frontiers in Plant Sciences* 10 (2019): 1446.

Haun, W., A. Coffman, M. Benjamin, et al. "Improved soybean oil quality by targeted mutagenesis of the fatty acid desaturase 2 gene family." *Plant Biotechnology Journal* 12 (2014): 934–940. doi: 10.1111/pbi.12201.

Herman, E. M. "Soybean seed proteome rebalancing." *Frontiers in Plant Sciences* 5 (2014): 437.

Hiei, Y., Y. Ishida and T. Komari. "Progress of cereal transformation technology mediated by *Agrobacterium tumefaciens*." *Frontiers in Plant Science* 5 (2014): 5628.

Iskandarov, U., H. J. Kim and E. B. Cahoon. "Camelina: An emerging oilseed platform for improved biofuels and bio-based materials." In *Plants and Bio Energy*, edited by N. Carpita, M. McCann and M. S. Buckeridge, pp. 131–140. Springer: New York (2014).

Jafarzadeh-Bajestani, M., M. Khodai-Kalaki, N. Motamed, et al. "Genetic transformation of olive somatic embryos through *Agrobacterium tumefaciens* and regeneration of transgenic plants." *African Journal of Biotechnology* 10, no. 28 (2011): 5468–5475.

Jiang, W. Z., I. M. Henry, P. G. Lynagh, et al. "Significant enhancement of fatty acid composition in seeds of the allohexaploid, *Camelina sativa*, using CRISPR/Cas9 gene editing." *Plant Biotechnology* 15 (2017a): 648–657. doi: 10.1111/pbi.1266.

Jiang, J. J., S. H. Ma, N. H. Ye, et al. "WRKY transcription factors in plant responses to stresses." *Journal of Integrative Plant Biology* 59 (2017b): 86–101.

Jocic, S., D. Miladinovic and Y. Kaya. "Breeding and genetics of sunflower". In *Sunflower: Chemistry, Production, Processing, and Utilization*, edited by E. Martinez-Force, N. T. Dunford and J. J. Salas, pp. 1–26. AOCS: Urbana, IL (2015).

Kanai, M., T. Yamada, M. Hayashi, et al. Soybean (*Glycine max. L.*) triacylglycerol lipase GmSDP1 regulates the quality and quantity of seed oil." *Scientific Reports* 9 (2019): 8924.

Kant, R., K. Tyagi, S. Ghosh, et al. "Host alternative NADH: Ubiquinone oxidoreductase serves as a susceptibility factor to promote pathogenesis of *Rhizoctonia solani* in plants." *Phytopathology* 109 (2019): 1741–1750.

Karunarathna, N. L., H. Wang, H. J. Harloff, et al. "Elevating seed oil content in a polypoid crop by induced mutations in seed fatty acid reducer genes." *Plant Biotechnology Journal* (2020). doi: 10.1111/pbi.13381.

Katiyar, R. K., R. Chamola and V. L. Chopra. "Tetralocular mustard, Brassica juncea: New promising variability through interspecific hybridization." *Plant Breeding* 117 (1998): 398–399.

Keykhasaber, M., B. P. H. J. Thomma and J. A. Hiemstra. "Verticillium wilt caused by *Verticillium dahliae* in woody plants with emphasis on olive and shade trees." *European Journal for Plant Pathology* 150 (2017): 21–37.

Kirik, V., D. Bouyer, U. Schobinger, et al. "CPR5 is involved in cell proliferation and cell death control and encodes a novel transmembrane protein." *Current Biology* 11 (2002): 1891–1895.

Kobayashi, Y., H. Kaya, K. Goto, et al. "A pair of related genes with antagonistic roles in mediating flowering signals." *Science* 286 (1999): 1960–1962.

Kong, F., B. Liu, Z. Xia, et al. "Two coordinately regulated homologs of flowering locus T are involved in the control of photoperiodic flowering in soybean." *Plant Physiology* 154 (2010): 1220–1231.

Koornneef, M., C. J. Hanhart and J. H. Vander-veen. "A genetic and physiological analysis of late flowering mutants in *Arabidopsis thaliana*." *Molecular Genetics and Genomics* 229 (1991): 57–66.

Krishna, G., B. K. Singh, E. K. Kim, et al. "Progress in genetic engineering of peanut (*Arachis hypogaea L.*): A review." *Plant Biotechnology Journal* 13 (2015): 147–162.

Lakhssassi, N., Z. Zhou, S. Liu, et al. "Characterization of the FAD2 gene family in soybean reveals the limitations of gel-based TILLING in genes with high copy number." *Frontiers in Plant Science* 8 (2017): 324. doi: 10.3389/fpls.2017.00324.

Lemmon, Z. H., N. T. Reem, J. Dalrymple, et al. "Rapid improvement of domestication traits in an orphan crop by genome editing." *Nature Plants* 4 (2018): 766–770.

Li, Z., Z. B. Liu, A. Xing, et al. "Cas9-Guide RNA directed genome editing in soybean." *Plant Physiology* 169 (2015): 960–970.

Li, T., X. Yang, Y. Yu, et al. "Domestication of wild tomato is accelerated by genome editing." *Nature Biotechnology* 36 (2018): 1160–1163.

Liljegren, S. J., A. H. K. Roeder, S. A. Kempin, et al. "Control of fruit patterning in *Arabidopsis* by indehiscent." *Cell* 116 (2004): 843–853.

Lu, Z. F., H. Yu, G. S. Xiong, et al. "Genome-wide binding analysis of the transcription activator Ideal Plant Architecturel reveals a complex network regulating rice plant architecture." *Plant Cell* 25 (2013): 3743–3759.

Luo, L., W. Q. Li, K. Miura, et al. "Control of tiller growth of rice by OsSPL14 and strigolactones, which work in two independent pathways." *Plant Cell Physiology* 53 (2012): 1793–1801.

Lyzenga, W. J., M. Harrington, D. Bekkaoui, et al. "CRISPR/Cas9 editing of three CRUCIFERIN C homoeologues alters the seed protein profile in *Camelina sativa*." *BMC Plant Biology* 19 (2019): 292.

Masani, M. Y. A., G. Noll, G. K. A. Parveez, et al. "Efficient transformation of oil palm protoplasts by PEG-mediated transfection and DNA microinjection." *PLoS One* 9, no. 5 (2014): e96831.

Masani, M. Y. A., A. M. D. Izawati, O. A. Rasid, et al. "Biotechnology of oil palm: Current status of oil palm genetic transformation." *Biocatalysis and Agricultural Biotechnology* 15 (2018): 335–347. doi: 10.1016/j.bcab.2018.07.008.

Masli, D. I. A., G. K. A. Parveez and A. M. M. Yunus. "Transformation of oil palm using *Agrobacterium tumefaciens*." *Journal of Oil Palm Research* 21 (2009): 643–652.

Masura, S. S., N. I. Tahir, O. A. Rasid, et al. "Post-genomic technologies for the advancement of oil palm research." *Oil Palm Research* 29 (2017): 469–486.

Munoz-Merida, A., J. J. Gonzalez-Plaza, A. Canada, et al. "De-novo assembly and functional annotation of the Olive (*Olea europaea*) transcriptome." *DNA Research* 20 (2013): 93–108.

Nan, H., D. Cao, D. Zhang, et al. "GmFT2a and GmFT5a redundantly and differentially regulate flowering through interaction with and up regulation of the bZIP transcription factor GmFDL19 in soybean." *PLoS One* 9 (2014): e97669.

Narvaez, I., P. C. Pliego, E. Palomo-Rios, et al. "Heterologous expression of the AtNPR1 gene in olive and its effects on fungal tolerance." *Frontiers in Plant Science* 11 (2020): 308.

Nour-Eldin, H. H., S. R. Madsen, S. Engelen, et al. "Reduction of anti-nutritional glucosinolates in *Brassica* oilseeds by mutation of genes encoding transporters." *Nature Biotechnology* 35 (2017): 377–382.

Okuzaki, A., T. Ogawa, C. Koizuka, et al. "CRISPR/Cas9-mediated genome editing of the fatty acid desaturase 2 gene in *Brassica napus*." *Plant Physiology and Biochemistry* 131 (2018): 63–69. doi: 10.1016/j.plaphy.2018.04.025.

Olmo-Garcia, L., N. Kessler, H. Neuweger, et al. "Unravelling the distribution of secondary metabolites in *Olea europaea L.*: Exhaustive characterization of eight olive-tree derived matrices by complementary platforms (LC-ESI/APCI-MS and GC-APCI-MS)." *Molecules* 23 (2018): 2419.

Palomo-Rios, E., S. Cerezo, J. S. Mercado, et al. "Generation and selection of transgenic olive plants." *Bio-Protocol* 7, no. 22 (2017): e2611. doi: 10.21769/BioProtoc.2611.

Parveez, G. K. A., O. Rasid, A. Zainal, et al. "Transgenic oil palm: Production and projection." *Biochemical Society Transactions* 28 (2000): 969–972.

Patel, K. G., R. Thankappan, G. P. Mishra, et al. "Transgenic Peanut (*Arachis hypogaea L.*) overexpressing mtld gene showed improved photosynthetic, physio-biochemical, and yield-parameters under soil-moisture deficit stress in lysimeter system." *Frontiers in Plant Sciences* 8 (2017): 1881.

Pidkowich, M. S., H. T. Nguyen, I. Heilmann, et al. "Modulating seed b-ketoacyl-acyl carrier protein synthase II level converts the composition of a temperate seed oil to that of a palm-like tropical oil." *Proceedings of the National Academy of Sciences of the United States of America* 104 (2007): 4742–4744.

Powles, S. B. and Q. Yu. "Evolution in action: plants resistant to herbicides". *Annual Review in Plant Biology* 61 (2010): 317–347.

Price, J. S., R. N. Hobson, M. A. Neale, et al. "Seed losses in commercial harvesting of oilseed rape". *Journal of Agricultural Engineering Research* 65 (1996): 183–191.

Raboy, V. "Approaches and challenges to engineering seed phytate and total phosphorus." *Plant Sciences* 177 (2009): 281–296.

Rajni, S. and V. Sundaresan. "The *Arabidopsis* myc/bHLH gene ALCATRAZ enables cell separation in fruit dehiscence." *Current Biology* 11 (2001): 1914–1922.

Rani, R., P. Yadav, K. M. Barbadikar, et al. "CRISPR/Cas9: A promising way to exploit genetic variation in plants." *Biotechnology Letters* 38, no. 12 (2016): 1991–2006.

Reggiani, R. and R. Russo. "Seed protein in *Camelina sativa (L.)* Crantz var. Calena". *International Journal of Plant Soil Science* 8, no. 2 (2015): 1–6.

Rehman, M. and P. B. E. McVetty. "A review of *Brassica* seed colour". *Canadian Journal of Plant Sciences* 91 (2011): 437–446.

Rubin-Wilson, B. C., S. A. Young and O. Folkerts. "Nucleotide sequences of maize and soybean b-ketoacyl-acyl carrier protein synthase II and their use in content of oil." *US Patent* 6 (2001): 323–395.

Saharan, G. S. and N. Mehta. *"Sclerotinia Diseases of Crop Plants: Biology, Ecology and Disease Management."* Springer: Heidelberg, Germany (2008).

Sanusi, N. S. N. M., R. Rosli, M. A. Halim, et al. "PalmXplore: Oil palm gene database." *Database* 2018 (2018): bay0951.

Sarkar, T., R., Thankappan, A. Kumar, et al. "Stress inducible expression of AtDREB1A transcription factor in transgenic peanut (*Arachis hypogaea L.*) conferred tolerance to soil- moisture deficit stress." *Frontiers in Plant Sciences* 7 (2016): 935.

Sashidhar, N., H. J. Harloff and C. Jung. "Identification of phytic acid mutants in oilseed rape (*Brassica napus*) by large-scale screening of mutant populations through amplicon sequencing." *New Phytologist* 225 (2020a): 2022–2034.

Sashidhar, N., H. J. Harloff, L. Potgieter et al. "Gene editing of three BnITPK genes in tetraploid oilseed rape leads to significant reduction of phytic acid in seeds." *Plant Biotechnology Journal* 18 (2020b): 1–10.

Schlueter, J., I. Vasylenko-Sanders, S. Deshpande, et al. "The FAD2 gene family of soybean: Insights into the structural and functional divergence of a paleopolyploid genome." *Crop Science* 47 (2007): 14–26.

Sedbrook, J. C., W. B. Phippen and M. D. Marks. "New approaches to facilitate rapid domestication of a wild plant to an oilseed crop: Example pennycress (*Thlaspi arvense L.*)." *Plant Science* 227 (2014): 122–132.

Shilpa, K. S., V. D. Kumar and M. Sujatha. "*Agrobacterium* mediated genetic transformation of safflower (*Carthamus tinctorius L.*)." *Plant Cell Tissue and Organ Culture* 103 (2010): 387–401.

Shim, J. S., C. Jung, S. Lee, et al. "AtMYB44 regulates WRKY70 expression and modulates antagonistic interaction between salicylic acid and jasmonic acid signaling." *The Plant Journal* 73, no. 3 (2013): 483–495.

Singh, R., M. Ong-Abdullah, E. T. Low, et al. "Oil palm genome sequence reveals divergence of interfertile species in old and new worlds." *Nature* 500 (2013): 335–339.

Sparvoli, F. and E. Cominelli. "Seed biofortification and phytic acid reduction: A conflict of interest for the plant?" *Plants* 4 (2015): 728–755.

Sun, Q., L. Lin, D. Liu, et al. "CRISPR/Cas9-mediated multiplex genome editing of the BnWRKY11 and BnWRKY70 genes in *Brassica napus L.*". *International Journal of Molecular Sciences* 19 (2018): 2716.

Tu, Y., B. He, S. Gao, et al. "CtACO1 overexpression resulted in the alteration of the flavonoids profile of Safflower." *Molecules* 24 (2019): 1128.

Villanueva-Mejia, D. and J. C. Alvarez. "Genetic improvement of oilseed crops using modern biotechnology." In *Advances in Seed Biology*, edited by J. C. Jimenez-Lopez, pp. 295–317. In Tech: London, UK (2017).

Wanasundara, J. P. "Proteins of Brassicaceae oilseeds and their potential as a plant protein source." *Critical Reviews in Food Science and Nutrition* 51 (2011): 635–677.

Wang, Z., H. D. Fang, Y. Chen, et al. "Overexpression of BnWRKY33 in oilseed rape enhances resistance to *Sclerotinia sclerotiorum*." *Molecular Plant Pathology* 15 (2014): 677–689.

Waseem, S., S. R. Imadi, A. Gul, et al. "Oilseed crops: Present scenario and future prospects." In *Oilseed Crops: Yield and Adaptations under Environmental Stress*, edited by P. Ahmad, pp. 1–18. John Wiley and Sons Ltd: Hoboken, NJ (2017).

Wen, S., H. Liu, X. Li, et al. "TALEN-mediated targeted mutagenesis of fatty acid desaturase 2 (FAD2) in peanut (*Arachis hypogaea L.*) promotes the accumulation of oleic acid." *Plant Molecular Biology* 97 (2018): 177–185.

Wu, N., Q. Lu, P. Wang, et al. "Construction and analysis of GmFAD2-1A and GmFAD2-2A Soybean fatty acid desaturase mutants based on CRISPR/Cas9 Technology." *International Journal of Molecular Sciences* 21, no. 3 (2020): 1104. doi: 10.3390/ijms21031104.

Xia, Z., S. Watanabe, T. Yamada, et al. "Positional cloning and characterization reveal the molecular basis for soybean maturity locus E1 that regulates photoperiodic flowering." *Proceedings of the National Academy of Sciences of the United States of America* 109, no. 32 (2012): e2155.

Xie, T., X. Chen, T. Guo, et al. "Targeted knockout of BnTT2 homologues for yellow-seeded *Brassica napus* with reduced flavonoids and improved fatty acid composition." *Journal of Agricultural and Food Chemistry* 68, no. 20 (2020): 5676–5690.

Xu, P., S. Cao, K. Hu, et al. "Trilocular phenotype in *Brassica juncea L.* resulted from interruption of CLAVATA1 gene homologue (BjMc1) transcription." *Scientific Reports* 7 (2017): 3498.

Yang, Q., C. Fan, Z. Gua, et al. "Identification of FAD2 and FAD3 genes in *Brassica napus* genome and development of allele-specific markers for high oleic and low linoleic acid contents." *Theoretical and Applied Genetics* 125 (2012): 715–729.

Yang, Y., Z. Kaiyu, L. Huailin, et al. "Precise editing of CLAVATA genes in *Brassica napus L.* regulates multilocular silique development." *Plant Biotechnology Journal* 16 (2018): 1322–1335. doi: 10.1111/pbi.12872.

Yu, O., J. Shi, A. O. Hession, et al. "Metabolic engineering to increase isoflavone biosynthesis in soybean seed." *Phytochemistry* 63 (2003): 753–763.

Yuan, M., J. Zhu, L. Gong, et al. "Mutagenesis of FAD2 genes in peanut with CRISPR/Cas9 based gene editing." *BMC Biotechnology* 19 (2019): 24.

Zaman, Q. U., W. Chu, M. Hao, et al. "CRISPR/Cas9-mediated multiplex genome editing of JAGGED gene in *Brassica napus L.*" *Biomolecules* 9 (2019): 725.

Zhai, Y., S. Cai, L. Hu, et al. "CRISPR/Cas9 mediated genome editing reveals differences in the contribution of INDEHISCENT homologues to pod shatter resistance in *Brassica napus L.*" *Theoretical and Applied Genetics* 132 (2019): 2111–2123. doi: 10.1007/s00122-019-03341-0.

Zhang, P., H. Du, J. Wang, et al. "Multiplex CRISPR/Cas9-mediated metabolic engineering increases soya bean isoflavone content and resistance to soya bean mosaic virus." *Plant Biotechnology Journal* 18 (2019): 1384–1395. doi: 10.1111/pbi.13302.

Zsogon, A., T. Cermak, E. R. Naves, et al. "De novo domestication of wild tomato using genome editing." *Nature Biotechnology* 36 (2018): 1211–1216.

Zulkifli, Y., A. Norziha, M. H. Naqiuddin, et al. "Designing the oil palm of the future." *Journal of Oil Palm Research* 29 (2017): 440–455.

9 Towards 'CRISPrY' Vegetables: Status of Genome Editing in Vegetable Crops

Amit Pareek and Amolkumar U. Solanke
ICAR-National Institute for Plant Biotechnology

Vikas Mangal
ICAR-Central Research Institute for Jute and Allied Fibres

Gitanjali Jiwani and Mohd Tasleem
ICAR-National Institute for Plant Biotechnology

Om Prakash Gupta
ICAR-Indian Institute of Wheat and Barley Research

Suhas Gorakh Karkute
ICAR-Indian Institute of Vegetable Research
ICAR-National Institute for Plant Biotechnology

CONTENTS

9.1 INTRODUCTION

Horticultural crops, which include fruits and vegetables, are very important for survival and longevity of humans. In addition, they are rich source of nutrition and thus play a vital role in food and nutritional security. At the same time, horticultural crops are important for farmers' livelihood security as they fetch higher price in the market. There is an increasing need for higher food production, food safety and security to feed the increasing world population. However, production constraints like adverse environmental factors, including both biotic and abiotic stresses, pose a major threat to achieve food security. Traditional breeding techniques have made a great impact on introducing new varieties, yet there is an urgent need to develop new varieties resistant or tolerant to these stresses for increasing vegetable crop production. Genome editing has proven to be an efficient and less time-consuming technology for improving the vegetable crop breeding scenario (Erpen-Dalla Corte et al. 2019).

Genome editing technology proposes to introduce monogenic and polygenic traits in an efficient manner, which is less time-consuming, unlike traditional breeding techniques. Metabolic engineering research can be carried out by CRISPR (clustered regularly interspaced short palindromic repeats)/Cas9 (CRISPR-associated) technique to improve and enhance the nutritional value of vegetable crops wherein the entire metabolic pathway genes can be introduced at the same time (Karkute et al. 2017). CRISPR/Cas9 technique has been applied and established for many vegetable crops. Increasing knowledge on this technique shall help in the development of non-transgenic plants and hence bring about a revolution in the generation of improved horticultural crops (Erpen-Dalla Corte et al. 2019). The genes targeted for genome editing in vegetable crops have been of varied functions, but the prime focus was on genes controlling the development followed by genes driving metabolism and then stress response. Out of all the vegetable crops, tomato has been the most worked-upon crop with 42% genome editing studies conducted on it. Potato has also been in limelight with 13% genome editing studies done on it. Vegetable crops have received major consideration as 72% genome editing studies have focused on them, while few floral and medicinal crops have also been targeted (Xu, Hua, and Lang 2019).

Genome editing technology is based on the utilization of chimeric nucleases, which consist of sequence-specific DNA binding domain attached to a non-specific DNA cleavage component. This technique can manipulate almost any gene in various numbers of organisms and different cell types. These nucleases are capable of bringing about a specific genetic change or modification by introducing a double-stranded break (DSB). Further, these DSBs induce DNA repair mechanism, which includes non-homologous end-joining (NHEJ) repair and homology-directed repair (HDR) (Gaj, Gersbach, and Barbas 2013). To validate the function of a gene, a general approach followed by scientist was to either knock down or knock out the gene or overexpress a gene by increasing its activity. Ectopic expression of gene was another way of checking the gene function. In order to test these attributes of a gene, a transgenic approach was followed along with a homologous recombination (HR). Further, viral vectors were also used to modify the genome of any organism. All these techniques are very lengthy and labour-intensive. Hence, researchers came up with more direct and effective techniques, which include ZFNs (zinc finger nucleases), TALENs (transcription activator-like effector nucleases) and CRISPR /Cas (Nemudryi et al. 2014). ZFNs are site-specific endonucleases wherein two different zinc finger proteins are linked to cleavage domain of FokI endonuclease (Kim, Cha, and Chandrasegaran 1996). The ZFNs have a modular structure due to the presence of 'zinc finger' domains, which help in the recognition of one triplet of nucleotides. This method was the criteria to edit cultured cells which included model and non-model plants. ZFNs have changed the face of genome editing by displaying the ability of manipulating genomic sites of interest and opened avenues for both basic and applied research. TALENs are chimeric nucleases which involve TALE proteins that recognize and activate specific plant promoters *via* a set of tandem repeats (Kamburova 2017). The most recent, less time-taking and effective out of all the genome editing techniques is the CRISPR/Cas. This system involves a RNA-driven nuclease called Cas9, which introduces a DSB in the genome. These DSBs are being repaired by the repair machinery of cell and bring about a successful genome modification (Kumar and Jain 2015).

CRISPR-Cas9 system has been developed from type II CRISPR-Cas systems, which equip bacteria with adaptive immunity against viruses and plasmids. Cas9, which is CRISPR-associated protein, is an endonuclease that utilizes an association of guide sequence and RNA duplex in order to make base pairing with the target DNA sequence (tracrRNA : crRNA). Following this, the Cas9 is able to introduce the site-specific DSBs in the DNA sequence. A single guide RNA (sgRNA) consists of dual component: tracrRNA and crRNA. The sgRNA consists of two important features: first is a sequence at 5′ end which decides the target site in DNA on the basis of Watson-Crick base pairing. Second feature is a duplex RNA structure at the 3′ end, which interacts with Cas9. This two-component system involves changes in guide RNA sequence that directs Cas9 to aim the specific desired DNA sequence and bring about successful genome editing for any trait (Doudna and Charpentier 2014). Post-introduction of sgRNA-Cas9 system into the cell, the sgRNA directs the Cas9 to cleave DNA 3 bp upstream to PAM (protospacer-adjacent motif) site. Two DNA repair mechanisms get activated after the double-stranded DNA break: first is the NHEJ and second is the HDR. In case of NHEJ, religation of the broken DNA molecule occurs in the absence of a homologous template, which generally leads to the introduction of insertion, deletion or substitution at the site of DSB. However in case of HDR, the presence of a donor sequence results in the introduction of novel alleles or correction of mutations; also the introduction of a new sequence of interest can occur (Zhou et al. 2020). CRISPR/Cas system has been classified into two predominant classes. Class 1 consists of multi-subunit effectors, while Class 2 contains single protein effectors. These classes are further subdivided into five types. Class 2 system has been utilized for genome editing purpose up until now (Lander 2016). Class 2 CRISPR/Cas system consists of type II, type V and type VI. Types II and V are being used for DNA editing; however, type VI is used for RNA targeting. The type V contains various subtypes that have been characterized. These subtypes include Cas12a, Cas12b, Cas12e and Cas14. All of these subtypes have a different architecture. Cas12a needs only a single crRNA but the others need an additional tracrRNA along with crRNA. Cas14 has both ssDNA and dsDNA cleavage activities. Cas13 is a RNA nuclease, which belongs to type VI (Moon et al. 2019, Knott and Doudna 2018). CRISPR/Cas9 system has been successfully implemented for the improvement of fruit crops in respect of disease susceptibility, fruit quality, flower architecture and improved fruit yield. The CRISPR/Cas9 technique has proved to be very impactful in the improvement of vegetable crops (Zhou et al. 2020).

9.2 CRISPR/Cas SYSTEM IN VEGETABLE CROPS

As reported with various crops, vegetable crops are also susceptible to biotic stresses as well as environmental stresses that bring down their production as well as quality. To overcome this situation, resistant and tolerant cultivars are required. In addition to this, other aspects like architecture, metabolite profiles and shelf life are needed to improve. The role of CRISPR/Cas approach of genome editing has been proved in many vegetable crops to achieve the desired goals (Figure 9.1).

In the beginning, experiments of genome editing using CRISPR/Cas9 were carried out to demonstrate its functionality in various crops, enhancing the efficiency and also for more advanced applications. In one such experiment in tomato, two genes *phytoene desaturase* (*SlPDS*) and *phytochrome interacting factor* (*SlPIF4*) were targeted to create mutations. The gene *SlPDS* encodes a key enzyme of carotenoid biosynthesis pathway, while *SlPIF4*, a homologous gene of Arabidopsis *PIF4*, is a transcription factor that interacts with a phytochrome photoreceptor. Knockout of *PDS* gene resulted in albino phenotype and was transmitted to the next generation without any modifications. The mutation and inheritance patterns were calculated in the T_0 as well as sequential generations by sequencing technique and genotyping. Lacking off-target events suggested high specificity of CRISPR/Cas9 system in tomato, and it can be efficiently used for the basic-level research and genetic refinement in tomato plants (Pan et al. 2016). Similar demonstrations of genome editing by CRISPR/Cas9 are reported in other crops like Chinese kale (*Brassica oleracea* var. *alboglabra*) by targeting *BaPDS1* and *BaPDS2* genes (Sun et al. 2018), chicory by targeting *CiPDS* gene (Bernard et al. 2019) and *Cucumis melo* by targeting *CmPDS* gene (Hoogvhorst et al. 2019).

Delivery of sufficient donor DNA repair templates (DRTs), which are essential for repairing DSB, is the major obstacle in achieving gene targeting. To overcome this situation, Geminivirus replicons (GVRs) have been used for delivering sufficient DRTs into tomato plant for CRISPR/Cas9-mediated HDR. GVR was used to produce heritable changes in the genome of tomato plants at much higher frequencies than traditional methods through *Agrobacterium*-mediated transformation. A strong promoter cauliflower mosaic virus 35S was attached upstream of the *anthocyanin 1 (ANT1)* gene, which controls anthocyanin biosynthesis in plant tissues. A precise insertion was achieved and carried to the next generation without any modifications. Both the pigmentation and

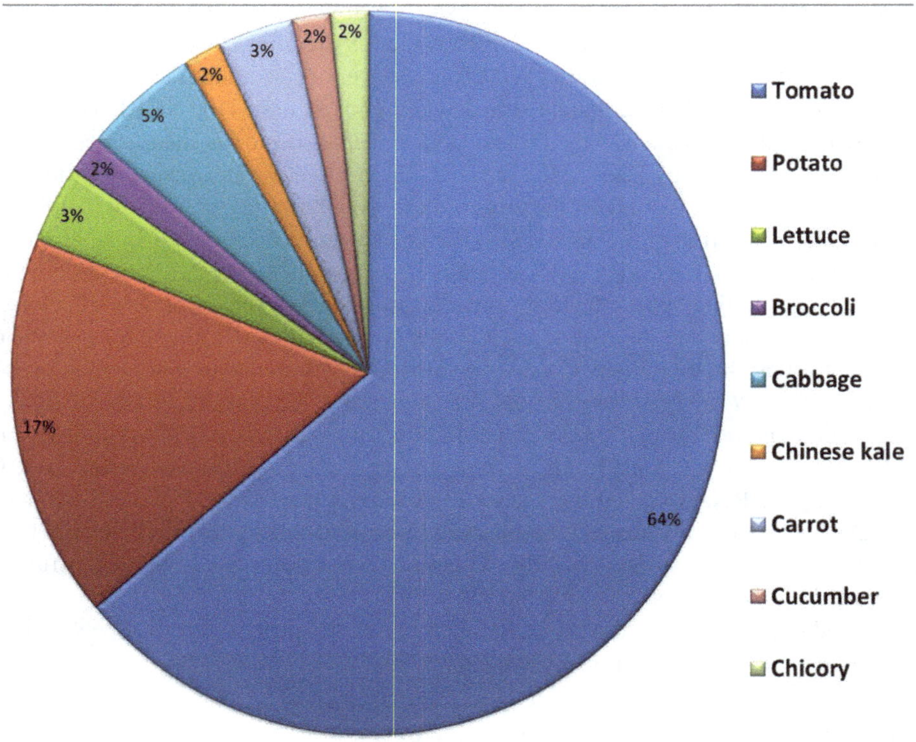

FIGURE 9.1 Graphical representation of CRISPR/Cas-based edited vegetable crops covered in the present literature.

kanamycin resistance marker enabled the identification and analysis of HDR events (Cermak et al. 2015, Li and Xia 2020).

The other similar experiment was carried out by combining CRISPR/Cas9 with bean yellow dwarf virus rolling circle replicon, which resulted in targeted mutagenesis and gene replacement by targeting *phytoene synthase 1* (*PSY1*) and *carotenoid isomerase* (*CRTISO*) genes of the carotenoid biosynthesis pathway in tomato. Gemini viral replicon amplification provides a huge sum of donor template for repairing DSB in the selected gene, through HR. This study showed that coherent gene targeting can be managed without using any selection or reporters with a customizable construct that is all-around compatible to other targets in tomato (Dahan-Meir et al. 2018).

Further, to cope with the naturally occurring uncontrolled crossing over, which lacks preciseness, Hayut et al. (2017) showed the targeted crossing over in between two heterozygous tomato plants individually expressing gRNA and Cas9. This study also provides evidence of precise reshuffling of specific alleles between homologous chromosomes in somatic tissues. This experimental study gained much attention for reshuffling of inaccessible parts present on the chromosome by the targeted crossing-over mechanism.

9.2.1 Abiotic Stress Tolerance

Tomato is an economic vegetable crop grown throughout the world as well as a well-known model plant, in which extensive genetic studies have been done. One of the harshest environmental factors is a drought, which limits plant growth and development. Several studies using CRISPR/Cas9 have unearthed the mechanism of stress tolerance in plants that can be utilized to develop stress tolerance in them.

Mitogen-activated protein kinases (MAPKs) play an important role in stress tolerance. As reported in a previous study, overexpression of tomato (*Solanum pimpinellifolium*) *SpMAPK3* gene in Arabidopsis showed higher seed germination under polyethylene glycol (PEG) treatment (Li et al. 2014). Following this, in another study, *Solanum lycopersicum*, *slmapk3* mutants were generated by introducing frameshift mutation using CRISPR/Cas9 system. The resulting knockout lines showed a reduced drought tolerance reflected by increased H_2O_2, malondialdehyde (MDA) content and ion leakage, suggesting a severe oxidative damage under drought stress condition. Furthermore, lower enzymatic activities of SOD, POD, APX and CAT and lower expression of genes encoding stress-responsive transcription factors in mutants compared to wild type (WT) revealed the involvement of antioxidant enzyme system mediated by *SlMAPK3* to provide drought tolerance. *SlMAPK3* might therefore work as a positive regulator in drought stress response, which expands our knowledge about the involvement of *SlMAPK3* gene in abiotic stress tolerance (Wang et al. 2017). Instead of above, *slmapk3* mutants displayed a contrasting feature of tolerance against heat stress compared to WT plants, suggesting the role of *SlMAPK3* as a negative regulator in thermotolerance. Additionally, transcript levels of various *HSPs* (like *SlHSP70*, *SlHSP90*, *SlHSP100*) and *HSFs* (like *SlHSFA1a*, *SlHSFA2* and *SlHSFA3*) were significantly higher in *slmapk3* mutants than in WT plants. A reduction in ROS accumulation and an elevation in antioxidant enzyme activity lead to heat tolerance. This study showed the specific role of *SlMAPK3* as a negative regulator in heat stress (Yu et al. 2019).

Tomato is a chilling-sensitive crop; its fruits are prone to chilling stress, which causes damage in a great manner. The presence of complete functional CBF cold-response pathways directed the researcher to find out their regulatory mechanism in cold tolerance. To find out the role of *SlCBF1* (C-repeat/dehydration-responsive element binding factors) in the cold-response pathway, *SLCBF1* mutants were generated by Li et al. using CRISPR/Cas9 system. *SLCBF1* mutants exhibited a reduced chilling tolerance with higher electrolyte leakage in comparison with WT plants. This study showed that *SlCBF1* protects plants from chilling injury (Li et al. 2018e) (Table 9.1).

A master regulator *non-expressor of pathogenesis-related gene 1*(*NPR1*), a receptor of SA (salicylic acid) important in defence against pathogens, was modified using CRISPR/Cas9 genome editing, and *slnpr1* mutants were generated. These mutants showed that high sensitivity against drought condition with lower membrane stability index (MSI), increased MDA content and hydrogen peroxide (H_2O_2) levels resulted in severe oxidative damage. Antioxidant enzyme activity when compared to WT showed the suppression of antioxidant enzyme system under stress conditions. Drought sensitivity also showed the reduced expression of key genes, including *SlGST*, *SlDREB* and *SlDHN* (Li et al. 2019).

Phosphorus is a major part of molecular constituents, which is important for the proper cell structure and functioning of plants. Due to its deficiency, plants show an abnormal growth, which

TABLE 9.1
Application of CRISPR-Based Genome Editing Approach in Plants for Abiotic Stress Tolerance

Crop	Targeted Gene	Trait/Function	References
Tomato	*SlMAPK3*	Drought tolerance	Wang et al. (2017)
	SlMAPK3	Heat	Yu et al. (2019)
	SlNPR1	Drought tolerance	Li et al. (2019)
	SlCBF1	Chilling tolerance	Li et al. (2018)
	SlycPHO1;1	Phosphate	Zhao, You, and Lei (2019)
Potato	*StPHO1*	Phosphate	Zhou et al. (2017)
Lettuce	*LsNCED4*	Thermo-inhibition of seed germination	Bertier et al. (2018)

leads to changes at physiological, biochemical and molecular levels. Phosphate transporters are important for P_i acquisition and translocation. *SlPHO1;1* deletion mutants generated by CRISPR/Cas9-based genome editing technique displayed the decreased shoot fresh weight (SFW) and increased root fresh weight (RFW) phenotypes of P_i starvation. More anthocyanin and soluble P_i contents were observed in the root, while less amount in the shoot was observed. Collectively, the above-mentioned results showed the importance of SlPHO1; 1 in P_i transport at the seedling stage in tomato (Zhao, You, and Lei 2019).

A study showed that the enhanced expression of *StMYB44* gene in potato led to slash down the phosphorous accumulation in the shoot organ. Transcriptomic study revealed low abundance of *Solanum tuberosum PHOSPHATE1* (*StPHO1*), a P_i transport-related gene, in transgenic lines. However, knockout of *StMYB44* by a CRISPR/Cas9 system resulted in no visible phenotype without any difference in phosphate content and increased transcription of *StPHO1*gene as compared to WT. It has been shown that *StMYB44* interacts with *AtWRKY6*, a known transcription factor involved in the regulation of *PHO1* expression. This study indicates that *StMYB44* might play a negative role as a component of the regulatory network controlling *StPHO1* transcription (Zhou et al. 2017).

Lettuce is a representative member of the most eminent family of flowering plants. The *9-cis-EPOXYCAROTENOID DIOXYGENASE4* (*NCED4*) gene is a key player in ABA biosynthesis pathway, which governs thermo-inhibition of seed germination. Different constructs, each comprising Cas9 and a different gRNA aiming distinct sites in *LsNCED4*, were made and transformed into lettuce (*Lactuca sativa*) cvs. Salinas and Cobham Green. Knockouts showed a considerable gain in the highest temperature for the maximum seed germination. Phenotypic and genotypic data of knockouts showed the possibility of *NCED4* as a selectable marker without any pleiotropic effects. *NCED4* can be used in a co-editing manner to maximize the frequency of edited events by seed germination at the maximum temperature (Bertier et al. 2018).

9.2.2 Biotic Stress Tolerance

Different types of molecular approaches have been used for improving resistance against disease-causing microbes, mainly fungus, virus and bacteria, which are responsible for critical losses of vegetable crops production worldwide. Food demand of the growing population is a major challenge in the present situation. Disease-resistant vegetable crops are in demand to overcome this situation, which will help in ensuring food security worldwide.

CRISPR/Cas system-mediated disease resistance is an emerging field of discovery in vegetable crops and getting attention of researcher day by day. CRISPR/Cas9 system was employed to produce the disease-resistant tomato by the inactivation of *SlDMR6-1* (downy mildew resistance 6-1) gene orthologue. Small deletions in the targeted genes resulted in frameshift mutation, and the mutants showed a broad-spectrum disease resistance against a variety of pathogens, including *P. syringae*, *P. capsici* and *Xanthomonas* spp. without any growth and developmental defects (Paula de Toledo Thomazella et al. 2016).

CRISPR/Cas9 system has been used to achieve fungal resistance in crops by targeting the susceptibility genes, e.g. *mildew resistance locus O* (*Mlo*), a well-known S gene locus. Powdery mildew disease caused by ascomycete fungus *Oidium neolycopersici* is the most commonly found and most devastating disease in tomato as all cultivars are prone to this disease. A powdery mildew-resistant non-transgenic tomato variety Tomelo has been developed by knockdown of tomato *mildew resistant locus O* (*SlMlo*) using CRISPR/Cas9 technology. Unlike traditional approaches, it leads to the disease-resistant variety in <1 year using locally adapted varieties of tomato (Nekrasov et al. 2017, Das, Sharma, and Prasad 2018, Koseoglou 2017). Another S gene *powdery mildew resistance 4* (*PMR4*), other than MLO associated with powdery mildew disease, has been reported in *Arabidopsis*, which encodes callose synthase. CRISPR/Cas9-mediated knockout of this gene showed resistance against pathogen *O. neolycopersici* (Jaganathan et al. 2018, Koseoglou 2017). Another tomato fungal pathosystem is *Fusarium oxysporum*, which is the cause of the most devastating disease in tomato. CRISPR/Cas9-mediated knockout event of *Solyc08g075770 gene* has proved its role in tolerance against *Fusarium* wilt by showing disease susceptibility like reduced mycorrhizal colonization (rmc) mutants. The rmc mutant is known for lacking skill to maintain mycorrhiza and for high susceptibility to *Fusarium* wilt as compared to isogenic line 76R (Prihatna, Barbetti, and Barker 2018).

Tomato yellow leaf curl virus (TYLCV) is a monopartite virus of the genus Begomovirus of the Geminiviridae family, and it can cause 100% crop loss. CRISPR/Cas9 systems have been used to provide immunity against TYLCV by using sgRNA targeting the coding sequences of coat protein (CP) and replicase (Rep), which resulted in lower accumulation of the viral DNA in the transgenic plants. This study provides an effective platform to engineer virus resistance in plants (Tashkandi et al. 2018) (Table 9.2).

Due to the life style variations, effective defence to combat biotrophic pathogens concurrently results in an increased susceptibility against necrotrophs. *Pseudomonas syringae* (*Pto*) DC3000 is the main causative agent of bacterial speck disease in tomato plants by producing coronatine (COR) toxin, which induces stomatal opening. The AtJAZ2 (jasmonate ZIM-domain-2) acts as a co-receptor for COR toxin. An orthologue of *AtJAZ2*, *SlJAZ2*, was mutated using CRISPR/Cas9 system to produce the dominant JAZ2 repressor lacking C-terminal Jas domain, which resulted in the prevention of COR and prompted stomatal opening and resistance against microbe *Pseudomonas syringae* DC3000. Mutants also showed an unaltered resistance against necrotrophs, which solved the problem of defence trade-off toward necrotrophs and biotrophs by fruitful uncoupling of relation between the jasmonic acid (JA) and SA hormonal-mediated defence, respectively (Ortigosa et al. 2019, Koseoglou 2017).

Occurrence of the two-layered inducible defence system in plants, including the first Pathogen associated molecular pattern (PAMP)-triggered immunity (PTI) and second NOD-like receptor (NLR)-triggered immunity (NTI), provides a broad-spectrum resistance against pathogenic microbes. Wall-associated kinase is greatly known for its involvement in plant immunity against *Pseudomonas syringae* pv. Tomato (Pst) but the molecular mechanism is yet to be elucidated. Recently, two homozygous *SlWak1* tomato mutant lines were produced by using CRISPR/Cas9 tool to identify the role of *SlWak1* in defence against Pst. Flagellin-derived peptide (flg22 & flgII-28)-triggered PTI was compromised in the apoplast, while they remain unaffected at the leaf surface in these mutants. In detail, analysis of these *Slwak1* mutants exhibited the importance of Wak1 protein as a positive regulator in later stages of flagellin-derived peptide (flg22 & flgII-28)-mediated PTI response in the apoplast and in association with PRR (Fls2 and Fls3) to trigger immune signalling (Zhang et al. 2020).

TABLE 9.2
Application of CRISPR-Based Genome Editing Approach in Plants for Biotic Stress Tolerance

Crop	Targeted Gene	Targeted Trait	References
Tomato	*SlDMR6*	Broad-spectrum disease resistance	Paula de Toledo Thomazella et al. (2016)
	SlMlo1	Powdery mildew resistance	Nekrasov et al. (2017)
	SlPMR4	Powdery mildew resistance	Koseoglou (2017)
	Solyc08g075770	Fusarium wilt	Prihatna, Barbetti, and Barker (2018)
	REP,CP	Tomato yellow leaf curl virus	Tashkandi et al. (2018)
	SlJAZ2	Bacterial speck disease	Ortigosa et al. (2019)
	SlWak1	Immunity against various pathogens	Zhang et al. (2020)
Potato	Coilin gene	Enhanced resistance to biotic agents	Makhotenko et al. (2019)
	Potato virus Y (PVY)	Resistance to viral diseases	Zhan et al. (2019)
	StALS1	Herbicide resistance	Butler et al. (2015, 2016)
Cucumber	*eIF4E*	Virus resistance	Chandrasekaran et al. (2016)

An alternative approach to achieve transgene-free genome edits is CRISPR/Cas9 RNP, which is assembled from Cas9 and sgRNA that target coilin gene to engineer virus-resistant potato plants. Coilin is the structural component of Cajal bodies involved in RNA metabolism and assembly of various ribonucleoprotein complexes (RNPs). The edited plants generated by the nanoparticle-mediated delivery of CRISPR complex showed resistance to potato virus Y (Makhotenko et al. 2019). Zhan and colleagues also generated potato virus Y (PvY)-resistant potatoes by using Class 2 type VI CRISPR/LshCas13a, which provides immunity against RNA phages. Transgenic potato showed the suppressed PvY accumulation and resistance against PvY (Zhan et al. 2019).

The *eukaryotic translation initiation factor* (*eIF*) genes were reported as recessive resistance genes in various host plants. Integration of these genes, e.g. inactive *eIF4E*, needs an extensive back-crossing, which may take many years to achieve resistance, while CRISPR/Cas9 method can accelerate the resistance development in a great manner. By using this technique, virus-resistant cucumber plants were obtained through the disruption of the recessive *eIF4E* (*eukaryotic translation initiation factor 4E*) gene. The mutant plants conferred immunity to cucumber vein yellowing virus (Ipomovirus), poty viruses, papaya ring spot mosaic virus-W (PRSMV) and Zucchini yellow mosaic virus (ZYMV) (Chandrasekaran et al. 2016).

Potato has been subjected to CRISPR/Cas9-mediated genome editing of *StIAA2* gene in a double haploid cultivar (Wang et al. 2015) and *StALS1* gene in both diploid and tetraploid plants (Butler et al. 2015). Mutation rates were high as 83% and 60% of regenerated lines, respectively. Herbicide-resistant trait in vegetable crops would be beneficial in weed management. Following earlier work, Butler et al. (2016) purposefully tested CRISPR/Cas9 and other sequence-specific nucleases (SSNs) in association with GVR, along with different repair templates, to achieve point mutation within *ALS1* locus in potato by HDR. Regenerated transformed lines showed variable degrees of herbicide resistance, as a result of multiallelic mutagenesis.

9.2.3 Quality Trait Improvement

Among vegetable crops, tomato is the most studied with CRISPR/Cas9 system, due to easiness of agrobacterium transformation and economic perspective. A transient expression study was performed in hairy roots of tomato by introducing CRISPR/Cas9 and sgRNA-associated complex aiming the *SHORT-ROOT* (*SHR*) gene. A range of mutations in the coding region of *SlSHR* gene was observed, along with the reduced size of meristem in transformants (Ron et al. 2014). Thus, transformants produced by hairy root transformation limit their use in research by means of inability to carry mutation over a next generation. To combat this drawback, in another study, CRISPR/Cas9 editing system was used in tomato for the stable gene integration where sgRNAs and Cas9 target different sites in the II exon of *ARGONAUTE7* (*SlAGO7*), which results in a major deletion. The mutant's phenotypic characteristic is the first leaf with leaflets but no petioles, and later, without laminae (Brooks et al. 2014, Koltun et al. 2018, Yamamoto et al. 2018).

Tomato is an ideal plant for studying climacteric fleshy fruit ripening. A ripening process is controlled by various numbers of transcription factors, including ethylene hormone and their target downstream effector genes. As reported earlier, transcription factors APETALA2a (AP2a), NON-RIPENING (NOR) and FRUITFULL (FUL1/TDR4 and FUL2/MBP7) were identified as master regulators in tomato fruit ripening process. To re-evaluate the functional study of these transcription factors, CRISPR/Cas9-mediated system was used to produce knockout plants for the above transcription factor-encoding genes. Phenotypic characteristics of *ap2a* mutants were same as reported in earlier studies, while null mutant *nor* showed a milder phenotype in comparison with spontaneously generated *nor* mutant, which is caused by a dominant-negative allele. Individual and combined null alleles of *FUL1* and *FUL2* showed partial redundant functions of both genes in fruit ripening, including an extra role of *FUL2* in early fruit development (Wang et al. 2019a) (Table 9.3).

Regulation of ripening in fleshy fruits (tomato) is the most important concern. The tomato *RIN* gene also known as master regulator gene, encoding MADS-box transcription factor, plays an

TABLE 9.3
Application of CRISPR-Based Genome Editing Approach in Plants for Improvement of Quality Traits

Crop	Target	Trait/Function	References
Tomato (*Solanum lycopersicum*)	ARGONAUTE 7 (*SlAGO7*)	Leaf morphology	Brooks et al. (2014)
	Ripening inhibitor (*RIN*)	Fruit ripening	Ito et al. (2015)
	SlORRM4	Fruit ripening	Yang et al. (2017)
	lncRNA1459	Fruit ripening	Li et al. (2018a)
	Pectate lyase (PL), polygalacturonase 2a (PG2a) and β-galactanase (TBG4)	Fruit shelf life	Wang et al. (2019a)
	Phytoene synthase 1 (*Psy1*)	Fruit colour	Hayut et al. (2017)
	Alcobaca (*ALC*)	Fruit shelf life	Yu et al. (2017)
	Phytoene desaturase (*SlPDS*), phytochrome interacting factor (*SlPIF4*)	Carotenoid biosynthesis	Pan et al. (2016)
	Phytoene synthase 1 (*Psy1*) and β-carotene hydroxylase 2 (*CrtR-b2*)	Carotenoid biosynthesis	D'Ambrosio, Stigliani, and Giorio (2018)
	Anthocyanin 1 (*ANT1*)	Anthocyanin biosynthesis	Cermak et al. (2015)
	Carotenoid isomerase (*CRTISO*) and phytoene synthase 1 (*Psy1*)	Carotenoid biosynthesis	Dahan-Meir et al. (2018)
	SlIAA9	Parthenocarpy	Ueta et al. (2017)
	AGAMOUS-LIKE 6 (*SlAGL6*)	Parthenocarpy	Klap et al. (2017)
	Glutamate decarboxylase (*GAD*)	γ-Aminobutyric acid biosynthesis	Nonaka et al. (2017)
	Stay green 1 (*SGR1*), β-lycopene cyclase (*Blc*), lycopene ε-cyclase (*LCY-E*), lycopene β-cyclase 1(*LCY-B1*), lycopene β-cyclase 2 (*LCY-B2*)	Lycopene content	Li et al. (2018c)
	GABA-TP1, *GABA-TP2*, *GABA-TP3*, *CAT9* and *SSADH*	γ-Aminobutyric acid metabolism	Li et al. (2018b)

(Continued)

TABLE 9.3 (*Continued*)
Application of CRISPR-Based Genome Editing Approach in Plants for Improvement of Quality Traits

Crop	Target	Trait/Function	References
Potato (*Solanum tuberosum*)	Steroid16 α-hydroxylase (*St16DOX*)	Steroidal glycoalkaloid (SGA) metabolism	Nakayasu et al. (2018)
	StIAA2	Aux/IAA protein	Wang et al. (2015)
	Granule-bound starch synthase (*GBSS*)	Starch quality	Andersson et al. (2017)
	Starch-branching enzyme *SBE1*, *SBE2*	Starch quality	Tuncel et al. (2019)
	Granule-bound starch synthase (*GBSS*)	Starch quality	Kusano et al. (2018)
Lactuca sativa (Lettuce)	*BRASSINOSTEROID-INSENSITIVE* 2 (*LsBIN2*)	Impaired brassinosteroid response	Woo et al. (2015)
Broccoli (*B. oleracea* var. *alboglabra* X *B. oleracea* var. *italica)*	*BolC.GA4.a*	Gibberellin biosynthesis	Lawrenson et al. (2015)
Cabbage (*B. oleracea* var. *capitata*)	Phytoene desaturase gene (*BoPDS*), the S-receptor kinase gene (*BoSRK*) and male sterility-associated gene (*BoMS1*)	Carotenoid biosynthesis, self-incompatibility, male sterility	Ma et al. (2019)
	PDS and *FRIGIDA*	Carotenoid biosynthesis and vernalization	Murovec et al. (2018)
	Centromere-specific histone H3 (*CENH3*)	Haploid lines induction	Stajič et al. (2019)
Chinese kale (*Brassica oleracea* var. *alboglabra*)	*BaPDS1* and *BaPDS2*	Anthocyanin biosynthesis	Sun et al. (2018)
Carrot (*Dacus carota*)	Flavanone-3-hydroxylase (*F3H*)	Anthocyanin biosynthesis	Klimek-Chodacka et al. (2018)
	DcPDS, *DcMYB 113*- like	Anthocyanin biosynthesis	Xu, Feng, and Xiong (2019)
Chicory (*Cichorium intybus L.*)	*CiPDS*	Albino phenotype	Bernard et al. (2019)

important role in fruit ripening. Three patterns of sgRNAs were made to knock out the *RIN* gene by insertion or deletion mutation. The *rin* mutant phenotype in T_0 lines suggested the presence of mutation in homozygous alleles. The RIN protein-defective mutants showed that lower red colour pigmentation in fruits resulted in incomplete fruit ripening. The *rin* mutants generated by this method might be used as a resource for breeding programmes, with the extended shelf life (Ito et al. 2015).

Analysis of the expression profiling of RNA editing factors exhibited the identification of 11 putative factors related to fruit ripening in tomato plants. In a previous study, a rapid virus-induced gene silencing (VIGS) assay showed that the organelle RNA recognition motif-containing protein (SlORRM4) affects tomato fruit ripening. Mutation in *SlORRM4* using a CRISPR/Cas9 strategy showed the lowered respiratory rate and lower ethylene production, resulting in a delayed fruit ripening in tomato. In addition, fruit ripening-associated gene and mitochondrial function-related gene expression changed significantly in mutated plants (Yang et al. 2017). In association with emergence of high-throughput sequencing, the importance of long non-coding RNAs has been proved in diverse biological processes like fruit ripening. To explore the importance of lncRNA1459 in fruit ripening, knockout tomato mutants were developed by CRISPR/Cas9 system. Delayed ripening process as compared to WT fruits was observed in lncRNA1459 mutants. Molecular characterization of mutants showed the downregulation of genes related to ethylene and carotenoid biosynthesis as compared to WT fruits. In addition, significant changes were observed in the expression of various ripening-related genes in mutant plants. However, some clarification about the mechanism and targeted genes of lncRNA1459 is yet to be discovered (Li et al. 2018a).

Shelf life of tomato is a key parameter related to cell wall. Another study reported the production of CRISPR-based mutants of different genes, which encode the pectin-degrading enzymes pectate lyase (PL), polygalacturonase 2a (PG2a) and β-galactanase (TBG4). The comparison of physio-chemical properties among different knockouts showed that only PL mutation resulted in firmer fruits. Localization and distribution of pectin and solubility of the pericarp cells were also checked in mutant fruits. This study showed that PL, PG2a and TBG4 act separately on cell wall domains, and significance of pectin is emulated by its higher accumulation in different mutants (Wang et al. 2019b). Post-harvest losses are a matter of concern in the production chain of fleshy vegetables, which results in lower profits. CRISPR/Cas9 system was used to generate tomato fruits having a longer shelf life by the replacement of *ALC* (alcobaca) gene with recessive alc in the presence of homologous repair template. Subsequently, alc homozygous breeding elites were obtained with the characteristic of long storage. It shows that the gene replacement *via* HDR is a worthwhile method to produce breeding elites (Yu et al. 2017, Erpen-Dalla Corte et al. 2019).

Carotenoids are secondary metabolites synthesized in plastids, which are an essential component of the diet of superior animals because of their unique role in major physiological processes. Nutrients like vitamins, carotenoids, etc. in food to enhance their nutritional value are very high in demand by consumers. Therefore, the nutrient content becomes a priority in breeding programmes of vegetable and fruit crops. The *beta-carotene hydroxylase 2* (*CrtR-b2*) and *phytoene synthase* 1 (*Psy1*) were selected due to their association with simply detectable colour change of the specific organ. A number of transformed plants (~69%) displayed the loss-of-function characteristic results of mutation in both alleles of a single locus. Though, 15% plants showed a normal phenotype, which contained mutation in one allele only. This study showed that CRISPR-based genome editing method can be used to create beneficial mutations in tomato plants that lead to implementation in breeding programmes (D'Ambrosio, Stigliani, and Giorio 2018).

In earlier time, crossing or backcross breeding was frequently practised to produce various kinds of mutations for metabolic engineering research. Nevertheless, these conventional approaches are time-consuming so at the present time, metabolic engineering research is more promising to be accomplished at the molecular level (Lu, Tang, and Li 2016, Li et al. 2018b). In recent times, the CRISPR/Cas systems have turned into a robust method to employ a site-directed mutagenesis in a range of plants, including various vegetable crops. γ-Aminobutyric acid (GABA) gains much attention as it has health-promoting hypotensive effects resulting in low blood pressure. Tomato contains higher levels of GABA in comparison with other major crops. A key enzyme of GABA biosynthesis pathway, named as glutamate decarboxylase (GAD), has a C-terminal autoinhibitory domain to regulate the enzymatic function, and deletion of this domain causes the increased GAD activity. Five *GAD* genes (*SlGAD1–5*) have been reported in tomato till now. Deletion of autoinhibitory domain of *SlGAD2* and *SlGAD3* by using CRISPR/Cas9 technology resulted in the increased GABA content. It was the first attempt which elaborated the usefulness of CRISPR/Cas9 system to improve GABA content in tomato fruits (Nonaka et al. 2017). Furthermore, a multiplex pYLCRISPR/Cas9 system was used to target five key genes of GABA pathway. Regenerated plants represented single to quadruple mutants as an outcome of these events. The GABA accumulation was much high in the leaves of quadruple mutants in comparison with WT tomato plants (Li et al. 2018b).

Lycopene, a bioactive component, is well known for use in the treatment of chronic diseases and in lowering the possibility of cardiovascular disease. A two-way scheme, namely induction of lycopene biosynthesis and inhibition of the conversion process of lycopene to α- and β-carotene, was employed to enhance lycopene accumulation. Five genes were targeted with a site-specific genome editing technology, which resulted in about 5.1-fold increased lycopene content in tomato fruits. This study showed the usefulness of CRISPR/Cas9 system for lycopene content enhancement in tomato fruits, including benefits of rare off-target mutations in progeny and stable transmission to the next generation (Li et al. 2018c). Overall results showed that multiplex CRISPR/Cas9 system can create multiple knockout mutations, which might be helpful in metabolic engineering research.

Potato (*Solanum tuberosum*) is also an important food crop around the world. Almost all the tissues of potato contain toxic substances steroidal glycoalkaloids (SGAs) like α-solanine and α-chaconine. SGAs provide a bitter taste to humans and show toxicity against different kinds of organisms like bacteria, insects, fungi and humans, and therefore, its reduction in tubers is the necessity for breeding purpose in potato. To achieve this objective, CRISPR/Cas9 technique was used to knock out the *St16DOX* gene, which encodes a steroid 16α-hydroxylase in SGA biosynthesis pathway. This resulted in the total elimination of the SGA accumulation (Nakayasu et al. 2018).

Starch in potato tubers was the target of various kinds of genetic alteration, for diverse purposes. CRISPR/Cas9 technique has been employed to generate starches in tuber with distinct properties. Starch-branching enzymes *SBE1*, *SBE2* or both were targeted, and various lines were raised. Some of those lines had mutations in all homeoalleles, and other than those had mutation in some of the homeoalleles of *SBE* genes. Some mutants showed abnormalities in starch granules with reductions in SBE1/ SBE2 protein as was expected from genotype, while a large reduction in the activity of both *SBE* genes maintained an extreme level of starch phenotype in tubers. Lowering of short amylopectin chains and increased long chains, including a large reduction in branching of starch, were observed. An immense reduction in the expression of both *SBEs* resulted in a reduced amylopectin branching in the course of granule growth, while alone a reduction of *SBE2* affects the numbers of starch granule initiations. Overall, this study showed the efficiency of *SBE* genes to generate starch with new valuable properties (Tuncel et al. 2019).

Another study showed a transient application of CRISPR/Cas9 system in the protoplasts of tetraploid potato along with mutations in all four alleles of *GBSS* gene through a single delivery, however with less percentage of (up to 2%) regenerated lines. The *GBSS* gene encoding granule-bound starch synthase was targeted under various experimental conditions. A PCR-based high-resolution fragment analysis method (HRFA) was used for the identification of four allele-mutated lines. Complete inhibition of GBSS enzyme activity was confirmed in mutated lines by phenotypic studies of starch (Andersson et al. 2017).

There are various kinds of obstructions in genome editing process of potato plants because of their complex heterozygosity and tetraploid nature. In this case, it is very difficult to obtain homozygous mutants carrying mutations in all the targeted genes (Das Dangol et al. 2019). Kusano et al. (2018) developed a more robust genome editing system with the use of a portion of the *OsMac3* mRNA 5′ UTR region named as translational enhancer dMac3. This enhancer was attached at the 5′-end of Cas9 for the enrichment of its expression level with multiple gRNAs. The efficiency of the targeted mutagenesis of *granule-bound starch synthase I* (*GBSSI*) gene was high using dMac3-Cas9 gene. Cleaved amplified polymorphic sequence (CAPS) analysis showed 25% frequency of transformants having mutation in four alleles. Regenerated plants also showed the low-level amylose starch in tubers (Kusano et al. 2018).

Multiplexed CRISPR/Cas9 systems with two sgRNA were used for targeting *flavanone-3-hydroxylase* (*F3H*) gene, which blocks anthocyanin biosynthesis in callus of purple-coloured carrot, thus resulting in the production of discoloured carrot calli (Klimek-Chodacka et al. 2018). CRISPR/Cas9 system was also employed to knock out *DcPDS, a gene for chlorophyll biosynthesis* in orange carrots, and the *DcMYB113-like* gene, a gene for anthocyanin biosynthesis in purple carrots. Knockout of *DcPDS* and *DcMYB113*-like gene showed albino and depigmented carrot plantlets, respectively (Xu, Feng, and Xiong 2019).

9.2.4 Improvement of Yield Traits

Crop yield depends on several factors, and thus, it is a complex trait to target. However, the ultimate goal of the breeding programmes is to enhance the yield, and therefore, extreme efforts have been done to improve yield traits by using modern tools, including genome editing. This section focuses on the improvement of all the yield-related traits by genome editing approaches.

Plant architecture, including inflorescence, and flowering architecture are very grievous for fruit production. In accordance with the evolution of plants, flowering is affected by the modulation of day length, which limits its geographical range of production. Day length-sensitive flowering in tomato plants is triggered by *SELF-PRUNING 5G* (*SP5G*), a flowering repressor gene. Expression analysis of *SP5G* gene showed a variation among wild species and cultivated tomato by the presence of variation in cis-regulatory elements (CREs). CRISPR/Cas9-induced mutations in *SlSP5G* gene resulted in a quick flowering and enhancement of compact conclusive growth, followed by an increased flower production outcome into an early yield of tomato fruits (Soyk et al. 2017b, Schlegel 2017). Formation of multiflowered inflorescences is dependent on a specific event of meristem development mediated by the *TERMINATING FLOWER* (*TMF*) gene. TMF protein works jointly with the homologs of *A. thaliana* BLADE-ON-PETIOLE (BOP) transcriptional cofactors, as a transcriptional complex. CRISPR/Cas9 system was used for silencing of tomato *blade-on-petiole* (*BOP*) gene to test the functional similarity with its homolog in *A. thaliana*, which affects inflorescence architecture. A new flowering module showed an interaction of SlBOP-TMF, which ensures faster flowering in *CRISPR-bop1/2/3* triple mutant than in the WT, with simplified inflorescences (Xu et al. 2016) (Table 9.4).

Shoot meristems contain stem cells, which replenished in a continuous manner by a feedback loop with the involvement of *WUSCHEL* (*WUS*) gene and the *CLAVATA* (*CLV*) gene signalling pathway. CRISPR/Cas-mediated mutation in genes of the CLV signalling pathway (*SlCLV1*, SlCLV2 and SlCLV3) or encoding an arabinosyl transferase (*SlRRA3a*) exhibited inflorescence branching as well as extra flowers and fruit organs (Xu et al. 2015). A group of scientists precisely edited the promoter sequence of *CLAVATA3* (*SlCLV3*) gene in tomato by producing number of mutations in CREs. A wide range of new alleles for major phenotypic traits in tomato (plant architecture, flowering time, fruit shape, fruit size and inflorescence branching) were generated to assess the association with CREs. It shows the utility of the CRISPR/Cas9 system in the production of novel variations in plants (Rodriguez-Leal et al. 2017).

Inflorescence architecture with surpassed flower production and productivity is very common to various domesticated crops. In case of tomato inflorescences, breeders avoid excessive branching because of low fertility. Three members of the MADS-box transcription factors, JOINTLESS-2(J2), ENHANCER-OF-JOINTLESS2 (EJ2) and LONG INFLORESCENCE (LIN), are involved in meristem maturation and inflorescence architecture. Among three genes, the *J2* gene controls the presence of the joint on the fruit pedicel and *EJ2* controls the formation of large calyx. Their concurrent presence in market varieties produced excessive branching along with a reduction in

TABLE 9.4
Application of CRISPR-Based Genome Editing Approach in Plants for Improvement of Yield Traits

Crop	Target	Trait/Function	References
Tomato (*Solanum lycopersicum*)	Self-pruning *5G* (*SlSP5G*)	Day length-sensitive flowering	Soyk et al. (2017b)
	Blade-on-petiole (*SlBOP*)	Inflorescence architecture	Xu et al. (2016)
	SlCLV1, *SlCLV2*, *SlCLV3*	Inflorescence architecture	Xu et al. (2015)
	SlWUS, *SlCLV3*	Plant architecture, flowering time, fruit shape, fruit size and inflorescence branching	Rodriguez-Leal et al. (2017)
	Methyltransferase 1 (*SlMET1*)	DNA methylation	Yang et al. (2019)
Tomato (*Solanum pimpinellifolium*)	JOINTLESS-2 (J2), ENHANCER-OF-JOINTLESS2 (EJ2), LONG INFLORESCENCE (LIN)	Inflorescence architecture	Soyk et al. (2017a)
	SP, *MULT*, *FAS*, *CycB*, *OVATE* and *FW2.2*	Fruit size, number and lycopene content	Zsogon et al. (2018)
	SP, *SP-5G*, *CLV3* and *WUS*, *GGP1*	Plant architecture, flowering, fruit production, ascorbic acid synthesis	Li et al. (2018d)
Potato (*Solanum tuberosum*)	*StIAA2*	Aux/IAA protein	Wang et al. (2015)
Cabbage (*B. oleracea* var. *capitata*)	Phytoene desaturase gene (*BoPDS*), the S-receptor kinase gene (*BoSRK*) and male sterility-associated gene (*BoMS1*)	Carotenoid biosynthesis, self-incompatibility, male sterility	Ma et al. (2019)
	Centromere-specific histone H3 (*CENH3*)	Haploid lines induction	Stajič et al. (2019)

fruit set due to epistasis, which causes flower dropping. To overcome this negative epistasis without introducing branching suppressors, the use of gene editing technology is a convenient way for fine-tuning of gene dosage and expression with higher productivity (Soyk et al. 2017a).

Plant hormones have an essential role in growth and development and thus in crop yield. Parthenocarpy is regulated by the cross-talk of phytohormones as AUXIN-INDUCED (Aux/IAA) and AUXIN-RESPONSE FACTORS (ARFs) play a key role in the auxin signalling pathway. In this pathway, the auxin receptor TIR1 promotes a reduction in Aux/IAAs (a negative regulator) level to activate the ARFs by inhibiting their interaction with Aux/IAAs. This editing system was used for the production of parthenocarpic plants *via* mutation of *SlIAA9* gene in tomato with up to 100% mutation rate. Knockout plants exhibited simple leaves rather than WT compound leaves, and seedless fruits (Ueta et al. 2017).

Microsporogenesis process is moderately affected by temperature fluctuations, a major constrain for sexual reproduction in tomato, which limits the year-around cropping. In consequence, breeding for parthenocarpy is considered an utmost goal for maintaining sustainable agriculture during adverse conditions. Screening of EMS-mutagenized tomato population resulted in a mutant, which showed high-quality parthenocarpic fruits under heat stress conditions. NGS and marker-assisted mapping followed by CRISPR/Cas9-mediated knockout assured that a mutation in *SlAGAMOUS-LIKE 6* (*SlAGL6*) gene sequence is liable for parthenocarpic phenotype. In comparison with other recessive monogenic mutants in tomato, homeotic changes were not observed in mutants. Parthenocarpic fruits had the normal weight and shape, unaffected viability of pollens, without compromising the sexual reproduction capacity (Klap et al. 2017, Erpen-Dalla Corte et al. 2019).

In one of the most interesting attempts of crop breeding in tomato plants, six important genes of key traits, associated with domestication of tomato plants, have been modified by multiplex CRISPR/Cas9 genome editing strategy, and the traits are general plant growth habit (*SELFPRUNING*), fruit size (*FASCIATED*), fruit shape (*OVATE*) (fruit number (*MULTIFLORA*), fruit weight (*FW 2.2*) and nutritional quality (*LYCOPENE β CYCLASE*). Altered morphology, nutritional value along with the size, shape and number of the fruits were observed. Notably, lycopene content is enhanced by 500% as compared to extensively cultivated *S. lycopersicum*. Editing of these trait-related loci enabled de novo domestication of wild *Solanum pimpinellifolium* (Zsogon et al. 2018). Inbreeding often comes with fitness penalties and genetic diversity losses during crop improvement. Genome editing can speed up the domestication of crops. Advantageous traits were introduced into stress-tolerant wild tomato accessions by targeting the *cis*-regulatory regions to upstream of open reading frames and coding regions of genes related to morphological trait, flowering and fruit production, and ascorbic acid synthesis. In another similar experiment, four genes *SP* (*SELF PRUNNING*), *SP5G* (*SELF PRUNNING 5-G*), *CLV3* (*CLAVATA3*) and *WUS* (*WUSCHEL*) were targeted using six sgRNA-containing constructs, to improve the desired traits (Li et al. 2018d).

A negative regulator of the brassinosteroid (BR) signalling pathway, homolog of *A. thaliana BIN2* (*BRASSINOSTEROID-INSENSITIVE 2*) gene, has been knocked out in lettuce. PEG-treated protoplasts were used for the transformation of a mixture of a sgRNA and Cas9. Forty percent of calli containing biallelic mutations with no off-target mutations were obtained. Plants regenerated from the mutant calli were able to carry forward the mutations to the next generation (Woo et al. 2015).

A doubled haploid genotype of broccoli derived from a cross between *B. oleracea* ssp. *alboglabra* and *B. oleracea* ssp. *italica* was used for targeting multicopy genes *BolC.GA4.a* (*gibberellin 3-beta-dioxygenase 1*) by CRISPR/Cas9 system. Indel mutations were introduced at two sites of the *BolC.GA4.a* gene. Regenerated plants displayed several targeted mutations in the gene with dwarf phenotype and alterations in pod valve margins (Lawrenson et al. 2015).

The self-incompatibility characteristics and prolonged vernalization for flowering of cabbage (*Brassica oleracea* var. *capitata*) are the main constrains in the development of new germplasm *via* conventional physical or chemical mutagenesis approaches. A study was employed in cabbage to create inheritable mutagenesis by using CRISPR/Cas9 system-based endogenous tRNA processing. Three genes, namely the male sterility-associated gene (*BoMS 1*), phytoene desaturase (*BoPDS*) gene and S-receptor kinase (*BoSRK*) gene, were chosen for multisite and multiple gene mutations. Mutation of *BoSRK3* gene vanquished self-incompatibility effectively, while the *BoMS1* gene mutation resulted in a complete male-sterile plant, having high cross-compatibility with its non-mutant isoline at the flowering stage. Interestingly, multiple-site mutation efficiency was detected, which overall showed great efficiency of the CRISPR/Cas9 system, coupled with an endogenous tRNA-processing system (Ma et al. 2019). Another study showed the use of RNPs for targeted CRISPR/Cas9 genome editing in cabbage protoplast using two endogenous genes: vernalization-determinant FRIGIDA (*FRI*) and phytoene desaturase (*PDS*). High mutation rate up to 24.51% directed a way to regenerate precisely mutated cabbage plants (Murovec et al. 2018). Cabbage has been targeted at four different genomic sites: two model genes (*FRI*, *PDS*) and the remaining two from trait-specific centromere-specific histone H3 (*CENH3*) gene, and mutants were observed to have mutation at all the target sites with frequencies up to 14.42% (Stajič et al. 2019).

Genome editing is also playing a vital role in the functional characterization of important genes. DNA methylation plays an important role during the plant development process. A study was conducted in tomato to identify the critical function of *methyltransferase 1* (*SlMET1*) in plant development using CRISPR/Cas9-mediated editing. *slmet1* mutants were generated with a frameshift mutation, which showed severe developmental defects in leaves and inflorescence, and exhibited parthenocarpy. Genome-wide investigations signify the role of *SlMET1* for maintaining a normal DNA methylome and transcriptome of tomato (Yang et al. 2019).

9.3 CONCLUSION

Over various conventional techniques, a CRISPR/Cas system of genome editing has revolutionized research through targeted and precise changes at the genome level. A great number of applications of this system in vegetable crops affirms its multifaceted nature in functional genomics. In combination with other breeding techniques, it may speed up the strengthening of genetic resources, which results in improved vegetable crops. In vegetables, the majority of modifications have been done in tomato and potato plants. Lack of enough genomic resources and efficient delivery methods in a large number of vegetables limits the use of CRISPR/Cas in these crops. CRISPR/Cas targets can be coding sequence (results in null alleles), regulatory elements of promoter region (results in novel alleles) and upstream ORFs to modify at the protein level. This editing system has been applied in vegetable crops for functional studies of genes as well as for the improvement of important traits such as abiotic and biotic stress tolerance, quality and yield, which shows the relevance of the technique. This editing system has been successfully used to modify the plant architecture along with inflorescence patterns. Improved tolerance to stress is highly fascinating since unfavourable conditions greatly affect vegetable crops. To cope with this situation, highly materialized cultivation system is required to grow vegetables, which make higher cost. To overcome this situation, knowledge on vegetable crops will guide us to modify the endogenous genetic traits along with the addition of new beneficial traits, which leads to obtain 'environmentally fitted' plants that will help to achieve food and nutrition security for growing human population.

REFERENCES

Andersson, M., H. Turesson, A. Nicolia, A. S. Falt, M. Samuelsson, and P. Hofvander. 2017. "Efficient targeted multiallelic mutagenesis in tetraploid potato (*Solanum tuberosum*) by transient CRISPR-Cas9 expression in protoplasts." *Plant Cell Rep* 36 (1):117–128.

Bernard, G., D. Gagneul, H. Alves Dos Santos, A. Etienne, J. L. Hilbert, and C. Rambaud. 2019. "Efficient Genome Editing Using CRISPR/Cas9 Technology in Chicory." *Int J Mol Sci* 20 (5):1155.

Bertier, L. D., M. Ron, H. Huo, K. J. Bradford, A. B. Britt, and R. W. Michelmore. 2018. "High-resolution analysis of the efficiency, heritability, and editing outcomes of CRISPR/Cas9-induced modifications of NCED4 in lettuce (Lactuca sativa)." *G3 (Bethesda)* 8 (5):1513–1521.

Brooks, C., V. Nekrasov, Z. B. Lippman, and J. Van Eck. 2014. "Efficient gene editing in tomato in the first generation using the clustered regularly interspaced short palindromic repeats/CRISPR-associated9 system." *Plant Physiol* 166 (3):1292–1297.

Butler, N. M., P. A. Atkin, D. F. Voytas, D. S Douches. 2015. "Generation and inheritance of targeted mutations in potato (*Solanum tuberosum* L.) using the CRISPR/Cas system." *PLoS ONE* 10(12):e0144591.

Cermak, T., N. J. Baltes, R. Cegan, Y. Zhang, and D. F. Voytas. 2015. "High-frequency, precise modification of the tomato genome." *Genome Biol* 16:232.

Chandrasekaran, J., M. Brumin, D. Wolf, D. Leibman, C. Klap, M. Pearlsman, A. Sherman, T. Arazi, and A. Gal-On. 2016. "Development of broad virus resistance in non-transgenic cucumber using CRISPR/Cas9 technology." *Mol Plant Pathol* 17 (7):1140–1153.

D'Ambrosio, C., A. L. Stigliani, and G. Giorio. 2018. "CRISPR/Cas9 editing of carotenoid genes in tomato." *Transgenic Res* 27 (4):367–378.

Dahan-Meir, T., S. Filler-Hayut, C. Melamed-Bessudo, S. Bocobza, H. Czosnek, A. Aharoni, and A. A. Levy. 2018. "Efficient in planta gene targeting in tomato using geminiviral replicons and the CRISPR/Cas9 system." *Plant J* 95 (1):5–16.

Das Dangol, S., A. Barakate, J. Stephens, M. E. Çalıskan, and A. Bakhsh. 2019. "Genome editing of potato using CRISPR technologies: Current development and future prospective." *Plant Cell Tissue Organ Cult (PCTOC)* 139 (2):403–416.

Das, A., N. Sharma, and M. Prasad. 2018. "CRISPR/Cas9: A novel weapon in the arsenal to combat plant diseases." *Front Plant Sci* 9:2008.

Doudna, J. A., and E. Charpentier. 2014. "Genome editing: The new frontier of genome engineering with CRISPR-Cas9." *Science* 346 (6213):1258096.

Erpen-Dalla Corte, L., L. M. Mahmoud, T. S. Moraes, Z. Mou, J. W. Grosser, and M. Dutt. 2019. "Development of improved fruit, vegetable, and ornamental crops using the CRISPR/Cas9 genome editing technique." *Plants (Basel)* 8 (12):601.

Filler Hayut, S., C. Melamed Bessudo, A.A. and Levy. 2017. "Targeted recombination between homologous chromosomes for precise breeding in tomato." *Nat Commun* 8:15605.

Gaj, T., C. A. Gersbach, and C. F. Barbas, 3rd. 2013. "ZFN, TALEN, and CRISPR/Cas-based methods for genome engineering." *Trends Biotechnol* 31 (7):397–405.

Hooghvorst, I. 2019. "1,2 Camilo López-Cristoffanini,1 and Salvador Nogués1Efficient knockout of phytoene desaturase gene using CRISPR/Cas9 in melon." *Sci Rep* 9:17077.

Ito, Y., A. Nishizawa-Yokoi, M. Endo, M. Mikami, and S. Toki. 2015. "CRISPR/Cas9-mediated mutagenesis of the RIN locus that regulates tomato fruit ripening." *Biochem Biophys Res Commun* 467 (1):76–82.

Jaganathan, D., K. Ramasamy, G. Sellamuthu, S. Jayabalan, and G. Venkataraman. 2018. "CRISPR for crop improvement: An update review." *Front Plant Sci* 9:985.

Kamburova, V. S. 2017. "Genome editing in plants: An overview of tools and applications." *Int J Agron* 2017:15.

Karkute, S. G., A. K. Singh, O. P. Gupta, P. M. Singh, and B. Singh. 2017. "CRISPR/Cas9 mediated genome engineering for improvement of horticultural crops." *Front Plant Sci* 8:1635.

Kim, Y. G., J. Cha, and S. Chandrasegaran. 1996. "Hybrid restriction enzymes: Zinc finger fusions to FokI cleavage domain." *Proc Natl Acad Sci U S A* 93 (3):1156–1160.

Klap, C., E. Yeshayahou, A. M. Bolger, T. Arazi, S. K. Gupta, S. Shabtai, B. Usadel, Y. Salts, and R. Barg. 2017. "Tomato facultative parthenocarpy results from SlAGAMOUS-LIKE 6 loss of function." *Plant Biotechnol J* 15 (5):634–647.

Klimek-Chodacka, M., T. Oleszkiewicz, L. G. Lowder, Y. Qi, and R. Baranski. 2018. "Efficient CRISPR/Cas9-based genome editing in carrot cells." *Plant Cell Rep* 37 (4):575–586.

Knott, G. J., and J. A. Doudna. 2018. "CRISPR-Cas guides the future of genetic engineering." *Science* 361 (6405):866–869.

Koltun, A., L. E.-D. Corte, L. M. Mertz-Henning, and L. S. A. Gonçalves. 2018. "Genetic improvement of horticultural crops mediated by CRISPR/Cas: A new horizon of possibilities." *Hortic Bras* 36:290–298.

Koseoglou, E. 2017. "The study of SlPMR4 CRISPR/Cas9-mediated tomato allelic series for resistance against powdery mildew." Wageningen University and Research, Wageningen.

Kumar, V., and M. Jain. 2015. "The CRISPR-Cas system for plant genome editing: Advances and opportunities." *J Exp Bot* 66 (1):47–57.

Kusano, H., M. Ohnuma, H. Mutsuro-Aoki, T. Asahi, D. Ichinosawa, H. Onodera, K. Asano, T. Noda, T. Horie, K. Fukumoto, M. Kihira, H. Teramura, K. Yazaki, N. Umemoto, T. Muranaka, and H. Shimada. 2018. "Establishment of a modified CRISPR/Cas9 system with increased mutagenesis frequency using the translational enhancer dMac3 and multiple guide RNAs in potato." *Sci Rep* 8 (1):13753.

Lander, E. S. 2016. "The heroes of CRISPR." *Cell* 164 (1–2):18–28.

Lawrenson, T., O. Shorinola, N. Stacey, C. Li, L. Ostergaard, N. Patron, C. Uauy, and W. Harwood. 2015. "Induction of targeted, heritable mutations in barley and Brassica oleracea using RNA-guided Cas9 nuclease." *Genome Biol* 16:258.

Li, S., and L. Xia. 2020. "Precise gene replacement in plants through CRISPR/Cas genome editing technology: Current status and future perspectives." *aBIOTECH* 1 (1):58–73.

Li, C., P. P. Chang, K. M. Ghebremariam, L. Qin, and Y. Liang. 2014. "Overexpression of tomato SpMPK3 gene in Arabidopsis enhances the osmotic tolerance." *Biochem Biophys Res Commun* 443 (2):357–362.

Li, R., D. Fu, B. Zhu, Y. Luo, and H. Zhu. 2018a. "CRISPR/Cas9-mediated mutagenesis of lncRNA1459 alters tomato fruit ripening." *Plant J* 94 (3):513–524.

Li, R., X. Li, D. Fu, B. Zhu, H. Tian, Y. Luo, and H. Zhu. 2018b. "Multiplexed CRISPR/Cas9-mediated metabolic engineering of gamma-aminobutyric acid levels in *Solanum lycopersicum*." *Plant Biotechnol J* 16 (2):415–427.

Li, X., Y. Wang, S. Chen, H. Tian, D. Fu, B. Zhu, Y. Luo, and H. Zhu. 2018c. "Lycopene is enriched in tomato fruit by CRISPR/Cas9-mediated multiplex genome editing." *Front Plant Sci* 9:559.

Li, T., X. Yang, Y. Yu, X. Si, X. Zhai, H. Zhang, W. Dong, C. Gao, and C. Xu. 2018d. "Domestication of wild tomato is accelerated by genome editing." *Nat Biotechnol* 36:1160–1163.

Li, R., L. Zhang, L. Wang, L. Chen, R. Zhao, J. Sheng, and L. Shen. 2018e. "Reduction of tomato-plant chilling tolerance by CRISPR-Cas9-mediated SlCBF1 mutagenesis." *J Agric Food Chem* 66 (34):9042–9051.

Li, R., C. Liu, R. Zhao, L. Wang, L. Chen, W. Yu, S. Zhang, J. Sheng, and L. Shen. 2019. "CRISPR/Cas9-mediated SlNPR1 mutagenesis reduces tomato plant drought tolerance." *BMC Plant Biol* 19 (1):38.

Lu, X., K. Tang, and P. Li. 2016. "Plant metabolic engineering strategies for the production of pharmaceutical terpenoids." *Front Plant Sci* 7:1647.

Ma, C., M. Liu, Q. Li, J. Si, X. Ren, and H. Song. 2019. "Efficient BoPDS gene editing in cabbage by the CRISPR/Cas9 system." *Hortic Plant J* 5:164–169.

Makhotenko, A. V., A. V. Khromov, E. A. Snigir, S. S. Makarova, V. V. Makarov, T. P. Suprunova, N. O. Kalinina, and M. E. Taliansky. 2019. "Functional analysis of coilin in virus resistance and stress tolerance of potato *Solanum tuberosum* using CRISPR-Cas9 editing." *Dokl Biochem Biophys* 484 (1):88–91.

Moon, S. B., D. Y. Kim, J. H. Ko, and Y. S. Kim. 2019. "Recent advances in the CRISPR genome editing tool set." *Exp Mol Med* 51 (11):1–11.

Murovec, J., K. Gucek, B. Bohanec, M. Avbelj, and R. Jerala. 2018. "DNA-free genome editing of Brassica oleracea and B. rapa protoplasts using CRISPR-Cas9 ribonucleoprotein complexes." *Front Plant Sci* 9:1594.

Nakayasu, M., R. Akiyama, H. J. Lee, K. Osakabe, Y. Osakabe, B. Watanabe, Y. Sugimoto, N. Umemoto, K. Saito, T. Muranaka, and M. Mizutani. 2018. "Generation of alpha-solanine-free hairy roots of potato by CRISPR/Cas9 mediated genome editing of the St16DOX gene." *Plant Physiol Biochem* 131:70–77.

Nekrasov, V., C. Wang, J. Win, C. Lanz, D. Weigel, and S. Kamoun. 2017. "Rapid generation of a transgene-free powdery mildew resistant tomato by genome deletion." *Sci Rep* 7 (1):482.

Nemudryi, A. A., K. R. Valetdinova, S. P. Medvedev, and S. M. Zakian. 2014. "TALEN and CRISPR/Cas genome editing systems: Tools of discovery." *Acta Nat* 6 (3):19–40.

Nonaka, S., C. Arai, M. Takayama, C. Matsukura, and H. Ezura. 2017. "Efficient increase of -aminobutyric acid (GABA) content in tomato fruits by targeted mutagenesis." *Sci Rep* 7 (1):7057.

Ortigosa, A., S. Gimenez-Ibanez, N. Leonhardt, and R. Solano. 2019. "Design of a bacterial speck resistant tomato by CRISPR/Cas9-mediated editing of SlJAZ2." *Plant Biotechnol J* 17 (3):665–673.

Pan, C., L. Ye, L. Qin, X. Liu, Y. He, J. Wang, L. Chen, and G. Lu. 2016. "CRISPR/Cas9-mediated efficient and heritable targeted mutagenesis in tomato plants in the first and later generations." *Sci Rep* 6:24765.

Paula de Toledo Thomazella, D., Q. Brail, D. Dahlbeck, and B. Staskawicz. 2016. "CRISPR-Cas9 mediated mutagenesis of a DMR6 ortholog in tomato confers broad-spectrum disease resistance." bioRxiv:064824.

Prihatna, C., M. J. Barbetti, and S. J. Barker. 2018. "A novel tomato fusarium wilt tolerance gene." *Front Microbiol* 9:1226.

Rodriguez-Leal, D., Z. H. Lemmon, J. Man, M. E. Bartlett, and Z. B. Lippman. 2017. "Engineering quantitative trait variation for crop improvement by genome editing." *Cell* 171 (2):470–480 e8.

Ron, M., K. Kajala, G. Pauluzzi, D. Wang, M. A. Reynoso, K. Zumstein, J. Garcha, S. Winte, H. Masson, S. Inagaki, F. Federici, N. Sinha, R. B. Deal, J. Bailey-Serres, and S. M. Brady. 2014. "Hairy root transformation using agrobacterium rhizogenes as a tool for exploring cell type-specific gene expression and function using tomato as a model." *Plant Physiol* 166 (2):455–469.

Schlegel, R. H. J. 2017. History of Plant Breeding. CRC Press: Boca Raton, FL.

Soyk, S., Z. H. Lemmon, M. Oved, J. Fisher, K. L. Liberatore, S. J. Park, A. Goren, K. Jiang, A. Ramos, E. van der Knaap, J. Van Eck, D. Zamir, Y. Eshed, and Z. B. Lippman. 2017a. "Bypassing negative epistasis on yield in tomato imposed by a domestication gene." *Cell* 169 (6):1142–1155 e12.

Soyk, S., N. A. Muller, S. J. Park, I. Schmalenbach, K. Jiang, R. Hayama, L. Zhang, J. Van Eck, J. M. Jimenez-Gomez, and Z. B. Lippman. 2017b. "Variation in the flowering gene SELF PRUNING 5G promotes day-neutrality and early yield in tomato." *Nat Genet* 49 (1):162–168.

Stajič, E., A. Kiełkowska, J. Murovec, and B. Bohanec. 2019. "Deep sequencing analysis of CRISPR/Cas9 induced mutations by two delivery methods in target model genes and the CENH3 region of red cabbage (Brassica oleracea var. capitata f. rubra)." *Plant Cell Tissue Organ Cult (PCTOC)* 139 (2):227–235.

Sun, B., A. Zheng, M. Jiang, et al. 2018. "CRISPR/Cas9-mediated mutagenesis of homologous genes in Chinese kale." *Sci Rep* 8:16786. https://doi.org/10.1038/s41598-018-34884-9

Tashkandi, M., Z. Ali, F. Aljedaani, A. Shami, and M. M. Mahfouz. 2018. "Engineering resistance against tomato yellow leaf curl virus via the CRISPR/Cas9 system in tomato." *Plant Signal Behav* 13 (10):e1525996.

Tuncel, A., K. R. Corbin, J. Ahn-Jarvis, S. Harris, E. Hawkins, M. A. Smedley, W. Harwood, F. J. Warren, N. J. Patron, and A. M. Smith. 2019. "Cas9-mediated mutagenesis of potato starch-branching enzymes generates a range of tuber starch phenotypes." *Plant Biotechnol J* 17 (12):2259–2271.

Ueta, R., C. Abe, T. Watanabe, S. S. Sugano, R. Ishihara, H. Ezura, Y. Osakabe, and K. Osakabe. 2017. "Rapid breeding of parthenocarpic tomato plants using CRISPR/Cas9." *Sci Rep* 7 (1):507.

Wang, L., L. Chen, R. Li, R. Zhao, M. Yang, J. Sheng, and L. Shen. 2017. "Reduced drought tolerance by CRISPR/Cas9-mediated SlMAPK3 mutagenesis in tomato plants." *J Agric Food Chem* 65 (39):8674–8682.

Wang, R., Ecdr Tavano, M. Lammers, A. P. Martinelli, G. C. Angenent, and R. A. de Maagd. 2019a. "Re-evaluation of transcription factor function in tomato fruit development and ripening with CRISPR/Cas9-mutagenesis." *Sci Rep* 9 (1):1696.

Wang, D., N. H. Samsulrizal, C. Yan, N. S. Allcock, J. Craigon, B. Blanco-Ulate, I. Ortega-Salazar, S. E. Marcus, H. M. Bagheri, L. Perez Fons, P. D. Fraser, T. Foster, R. Fray, J. P. Knox, and G. B. Seymour. 2019b. "Characterization of CRISPR mutants targeting genes modulating pectin degradation in ripening tomato." *Plant Physiol* 179 (2):544–557.

Wang, S., S. Zhang, W. Wang, X. Xiong, F. Meng, X. Cui. 2015. "Efficient targeted mutagenesis in potato by the CRISPR/Cas9 system." *Plant Cell Rep* 34:1473–1476.

Woo, J. W., J. Kim, S. I. Kwon, C. Corvalan, S. W. Cho, H. Kim, S. G. Kim, S. T. Kim, S. Choe, and J. S. Kim. 2015. "DNA-free genome editing in plants with preassembled CRISPR-Cas9 ribonucleoproteins." *Nat Biotechnol* 33 (11):1162–1164.

Xu, C., K. L. Liberatore, C. A. MacAlister, Z. Huang, Y. H. Chu, K. Jiang, C. Brooks, M. Ogawa-Ohnishi, G. Xiong, M. Pauly, J. Van Eck, Y. Matsubayashi, E. van der Knaap, and Z. B. Lippman. 2015. "A cascade of arabinosyltransferases controls shoot meristem size in tomato." *Nat Genet* 47 (7):784–792.

Xu, C., S. J. Park, J. Van Eck, and Z. B. Lippman. 2016. "Control of inflorescence architecture in tomato by BTB/POZ transcriptional regulators." *Genes Dev* 30 (18):2048–2061.

Xu, Z. S., K. Feng, and A. S. Xiong. 2019. "CRISPR/Cas9-mediated multiply targeted mutagenesis in orange and purple carrot plants." *Mol Biotechnol* 61 (3):191–199.

Xu, J., K. Hua, and Z. Lang. 2019. "Genome editing for horticultural crop improvement." *Hortic Res* 6:113.

Yamamoto, T., S. Kashojiya, S. Kamimura, T. Kameyama, T. Ariizumi, H. Ezura, and K. Miura. 2018. "Application and development of genome editing technologies to the Solanaceae plants." *Plant Physiol Biochem* 131:37–46.

Yang, Y., G. Zhu, R. Li, S. Yan, D. Fu, B. Zhu, H. Tian, Y. Luo, and H. Zhu. 2017. "The RNA editing factor SlORRM4 is required for normal fruit ripening in tomato." *Plant Physiol* 175 (4):1690–1702.

Yang, Y., K. Tang, T. U. Datsenka, W. Liu, S. Lv, Z. Lang, X. Wang, J. Gao, W. Wang, W. Nie, Z. Chu, H. Zhang, A. K. Handa, and J. K. Zhu. 2019. "Critical function of DNA methyltransferase 1 in tomato development and regulation of the DNA methylome and transcriptome." *J Integr Plant Biol* 61 (12):1224–1242.

Yu, Q. H., B. Wang, N. Li, Y. Tang, S. Yang, T. Yang, J. Xu, C. Guo, P. Yan, Q. Wang, and P. Asmutola. 2017. "CRISPR/Cas9-induced targeted mutagenesis and gene replacement to generate long-shelf life tomato lines." *Sci Rep* 7 (1):11874.

Yu, W., L. Wang, R. Zhao, J. Sheng, S. Zhang, R. Li, and L. Shen. 2019. "Knockout of SlMAPK3 enhances tolerance to heat stress involving ROS homeostasis in tomato plants." *BMC Plant Biol* 19 (1):354.

Zhan, X., F. Zhang, Z. Zhong, R. Chen, Y. Wang, L. Chang, R. Bock, B. Nie, and J. Zhang. 2019. "Generation of virus-resistant potato plants by RNA genome targeting." *Plant Biotechnol J* 17 (9):1814–1822.

Zhang, N., M. A Pombo, H. G Rosli, and G. B. Martin. 2020. "Tomato wall-associated kinase SlWak1 acts in an Fls2- and Fls3-dependent manner to promote apoplastic immune responses to Pseudomonas syringae." bioRxiv:2020.01.27.921460.

Zhao, P., Q. You, and M. Lei. 2019. "A CRISPR/Cas9 deletion into the phosphate transporter SlPHO1;1 reveals its role in phosphate nutrition of tomato seedlings." *Physiol Plant* 167 (4):556–563.

Zhou, X., M. Zha, J. Huang, L. Li, M. Imran, and C. Zhang. 2017. "StMYB44 negatively regulates phosphate transport by suppressing expression of PHOSPHATE1 in potato." *J Exp Bot* 68 (5):1265–1281.

Zhou, J., D. Li, G. Wang, F. Wang, M. Kunjal, D. Joldersma, and Z. Liu. 2020. "Application and future perspective of CRISPR/Cas9 genome editing in fruit crops." *J Integr Plant Biol* 62 (3):269–286.

Zsogon, A., T. Cermak, E. R. Naves, M. M. Notini, K. H. Edel, S. Weinl, L. Freschi, D. F. Voytas, J. Kudla, and L. E. P. Peres. 2018. "De novo domestication of wild tomato using genome editing." *Nat Biotechnol.* 36:1211–1216

10 Genome Editing in Fruit Tree Crops: Prospects and Challenges

Israr Ahmad and Anshul Watts
ICAR-Central Institute for Subtropical Horticulture
ICAR-National Institute of Plant Biotechnology

CONTENTS

10.1 INTRODUCTION

Traditionally, DNA-level genetic modifications have resulted in the advancement of superior crop varieties. The present cultivated plants are the result of natural genetic modifications like mutations at random positions in the wild relatives of these plants. Induced modifications using mutagenesis approaches, i.e. chemical agents or radiation, have helped in cultivars improvement (Dhaliwal et al., 2015; Sikora et al., 2011). These approaches cause randomly, many double-stranded breaks (DSBs) in the genome of plants followed by the repair of DNA. These natural DNA repair mechanisms, evolved over the period of time, are very important for the survival of plant species. Results of induced mutagenesis approaches are not predictable, involve the selection of several plants and are hard to get a required trait/phenotype (Oladosu et al., 2016; Pacher & Puchta, 2017). Genomic modification by using nucleases will accurately replace, insert or delete sequences at a specific location in the plant genome, having a desirable novel trait/phenotype. Genome modification requires synthetic nucleases for the commencement of DSBs at a precise location in the plant genome and cell DNA repair machinery for the modifications. The ability to create a specific break in dsDNA and its repair machinery makes the basic principle for gene editing (Puchta, 2004; Voytas, 2013). For genome editing, zinc finger nucleases (ZFNs) and transcription activator-like effector nucleases (TALENs) were primarily used. ZFNs are engineered DNA-binding proteins having zinc finger motifs and FokI endonuclease domain that facilitate the targeted genome editing by generating DSBs in DNA at required locations (Weinthal et al., 2010). Transcription activator-like effector (TALE) proteins are sequence-programmable transcription factors (TFs) with DNA-binding domains derived from plant bacterial pathogens. TALE proteins were released into the host plant cell at the time of bacterial infection. TALENs are preferred over ZFNs, as these are more specific and monomers don't have the cross-reactivity problem (Christian et al., 2010; Li et al., 2010). Each technique has its own advantages and drawbacks but both are laborious and costly, which limits its wide applications (Gaj et al., 2013; Wright et al., 2014). CRISPR/Cas9 mechanism is a component of adaptive defence mechanism in bacteria (Jiang et al., 2013; Song et al., 2016). Small non-coding RNAs are the major components of this system, for guiding a complex of proteins to make DSBs at a specific DNA sequence. The mutant generation ability among CRISPR/Cas9, TALENs and ZFNs is correlated (Gaj et al., 2013). In the past few years, CRISPR/Cas9 has extensively received researcher's interest, and many examples of gene mutation and transcriptional regulation control were available for various plants (Khatodia et al., 2016; Liu et al., 2017). CRISPR/Cas9 is cheaper and more accurate than TALEN and ZFN techniques of editing DNA.

10.2 GENOME EDITING IN FRUIT TREE CROPS

In 2013, the first report on plant genome editing was published using CRISPR/Cas9 system. Since then, this system has been applied in various horticulture fruit crops, i.e. banana, kiwifruit, cacao, grape, pear etc. In these studies, phytoene desaturase (PDS) gene encoding carotenoid biosynthesis pathway enzyme was mutated. The mutation of PDS gene using CRISPR/Cas9 system inhibits the carotenoid and chlorophyll production, giving albino phenotype (Qin et al., 2007). In banana, transgenic Cas9 expression using sgRNA (19 bp) targeting the PDS genes gives a complete albino and variegated phenotype (Kaur et al., 2018). Using polycistronic gRNAs, a higher editing efficiency is reported in banana (Naim et al., 2018). CRISPR/Cas9-mediated mutations in the PDS gene confer albino phenotype in pear, apple (Charrier et al., 2019), grapes (Nakajima et al., 2017), kiwifruit (Wang et al., 2018b) and kumquat (Zhu et al., 2019).

In several perennial fruit crops, long juvenile phase (3–10 years) results in extensive off-flowering time. Such a long juvenile phase is one of the major limitations in conventional breeding approach for new cultivars development. Juvenility is generally related to high concentration of terminal flowering protein (TFL) (Pillitteri et al., 2004). TFL is a negative regulator, inhibiting the expression of several proteins like APETALA1 (AP1), LEAFY (LFY) and FLOWERING LOCUS T (FT) that induce flowering (Pillitteri et al., 2004; Liljegren et al., 1999). Using CRISPR/Cas9 system, Charrier et al. (2019) edited apple and pear TFL1 through two different sgRNAs (sgRNA1 and sgRNA1). In apple and pear, 93% and 9% TFL1.1 gene transgenic lines show early flowering, respectively. Mismatch between the target gene and sgRNA1 in pear explains the lower percentage of the PcTFL1.1 gene transgenic lines. In pear, PcTFL1.1 and PcTFL1.2 gene editing is essential for flowering. Mutations targeting *AcCEN4* and *AcCEN* genes through CRISPR/Cas9 system in kiwifruit transformed woody perennial into quick flower and fruit-developing plant (Varkonyi-Gasic et al., 2019).

Like field crops, growth, development and yield of horticulture crops are also affected by various biotic stresses caused by viruses, bacteria, fungi and nematodes. The importance of each crop varies, but generally increases production costs and causes significant losses. Resistant/tolerant cultivars are the most efficient and cost-effective substitutes to resolve these problems. CRISPR/Cas9-mediated resistance to these biotic stresses can have a vast impact on their production. *Citrus sinensis* lateral organ boundaries gene (*CsLOB1*) is a susceptibility gene for citrus canker, caused by bacterial pathogen *Xanthomonas citri* (Xcc) (Gottwald et al., 2002; Hu et al., 2014). *CsLOB1* gene has specific sequences in promoter region to which the Xcc pathogenicity factor *PthA4* binds, leading to *CsLOB1* gene induction. CRISPR/Cas9-mediated modified *PthA4* effector cis-elements in the promoter of *CsLOB1* reduced Xcc infection on the mutant plants (Jia et al., 2016). Biallelic promoter mutations of *CsLOB1* gene give a resistant plant, suggesting that the activation of a one allele *via* the *PthA4* element is adequate for disease development. In grapes, knocking out of *WRKY52* improved resistance to *Botrytis cinerea* (Wang et al., 2018a). Mutated transgenic biallelic lines were more resistant than monoallelic one.

Banana streak virus (BSV) is the key limitation in banana production. The genetic material of virus (dsDNA) combines into the host plant genome, producing viral particles under stress (Lockhart, 1995). Mutants were generated through CRISPR/Cas9 system in the BSV sequences of cultivar *Gonja manjaya*; 75% mutants have no symptom compared to non-mutant control plants under water stress (Tripathi et al., 2019). In banana, $MaGA_{20}ox_2$ gene regulates dwarfism. Mutations induced in the $MaGA_{20}ox_2$ gene using CRISPR/Cas9 gave seven mutant lines having semi-dwarf phenotype (Chen et al., 2016). Endogenous GA levels in the different tissues were in accordance

TABLE 10.1
List of Targeted Gene(s) and Trait Modified via the CRISPR/Cas9 System in Different Fruit Crops

S. No	Plant Species	Target Gene	Target Trait	Delivery Methods	References
1	Apple	DIPM-1, DIPM-2, DIPM-4	Fire blight disease resistance	PEG-mediated protoplast transfection	Malnoy et al. (2016)
2		PDS	Albino phenotypes	Agrobacterium-mediated transformation	Nishitani et al. (2016)
3	Apple and pear	PDS TLF1	Albino phenotypes Early flowering	Vacuum infiltration in a suspension of *Agrobacterium tumefaciens*	Charrier et al. (2019)
4	Banana	PDS	Albino phenotype	Agrobacterium-mediated suspension cell transformation	Naim et al. (2018)
5		PDS	Albino phenotypes	Agrobacterium-mediated embryogenic cell suspension culture transformation	Kaur et al. (2018)
6		MaGA20ox2	Semi-dwarfing size	Agrobacterium-mediated suspension cell transformation	Shao et al. (2020)
7		eBSV control of virus	Pathogenesis	Agrobacterium-mediated	Tripathi et al. (2019)
8	Cacao	TcNPR3	*Phytophthora tropicalis* resistance	Agrobacterium-mediated transient leaf transformation	Fister et al. (2018)
9	Citrus (*Carrizo citrange*)	PDS	Albino phenotypes	Agrobacterium-mediated epicotyl transformation	Zhang et al. (2017)
10	Citrus (grape fruit)	CsLOB1	Canker disease resistance	Agrobacterium-mediated epicotyl transformation	Jia et al. (2017)
11		PDS	Albino phenotype	Agrobacterium-mediated epicotyl transformation	Nakajima et al. (2017)
12	Citrus (sweet orange)	DMR6	Huanglongbing resistance	Agrobacterium-mediated epicotyl transformation	Osakabe et al. (2018)
13	Grape	L-Idonate dehydrogenase gene (IdnDH)	Tartaric acid content	Agrobacterium-mediated suspension cell transformation	Ren et al. (2016)
14		VvWRKY52	*Botrytis cinerea* resistance	Agrobacterium-mediated somatic embryo transformation	Wang et al. (2018a)
15	Kiwifruit	PDS	Albino phenotype	Agrobacterium-mediated transformation	Wang et al. (2018b)

with the observed phenotype. GA levels significantly change in roots and leaves of mutants compared to the control (Shao et al., 2020). This approach can be used to develop dwarf or semi-dwarf cultivars of banana. Crops modified using the CRISPR/Cas9 system are summarized in Table 10.1.

10.3 ROLE OF GENOME EDITING IN FRUIT RIPENING

The main challenge for fruit producers/traders is to supply a fruit having a high nutritive value, good flavour, taste and sufficient fruit shelf life to keep good quality prior to consumption. This is very important for climacteric fruits like mango, tomato etc., which are highly sensitive to ethylene, and non-climacteric fruits like strawberry, which become tasteless and uneatable. Therefore, improvement in fruit shelf life and delayed ripening have been a centre of interest for many researchers using conventional breeding and genetic modifications. However, genetic modifications depend upon the transgenic generation, having a restricted commercial use due to the current government policies and consumers sought for safety. Moreover, transgenic strategies sometime lead to incomplete and/or temporary knockdown effects, give non-target effects and background noise (Martin and Caplen, 2007). However, the CRISPR/Cas9 system offers a new prospect to beat these challenges.

Several studies have reported the role of various TFs involved in fruit ripening. Among those, the most studied ripening TF is RIPENING-INHIBITOR gene (RIN), a MADS-box gene family member. RIN TF is induced early during ripening and regulates ethylene-independent/dependent ripening pathways (Karlova et al., 2014). Due to the apparent phenotype and *rin* mutant importance, this locus has been chosen to validate the inheritance and functionality of mutations using CRISPR/Cas9 system. In tomato, mutations in the *RIN* locus mediated by CRISPR/Cas9 resulted in fruit ripening inhibition in T_0 generation (Ito et al., 2017). Fruits of homozygous *rin* mutants were pale red in colour. Physiological and molecular characterizations of these mutant lines demonstrate that ripening parameters in *rin* mutant were less affected. The use of CRISPR/Cas9 has allowed generating alternative knockout alleles, which have changed our current model of fruit ripening with RIN being necessary to initiate this process and *RIN* being a loss-of-function mutation.

Fruit softening is caused by cell-wall modifying proteins, *i.e.* expansin, pectin methylesterase, endoglucanase, β-galactosidase and polygalacturonase. In tomato, CRISPR/Cas9-induced pectate lyase mutants affect fruit firmness without changing colour and total soluble solids (TSS) content (Uluisik et al., 2016). Self-pruning 5G (SP5G) works as a flowering repressor under long-day conditions (Cao et al., 2015). CRISPR/Cas9 mutant plants in SP5G gene bear earlier flowering under long-day conditions. Long non-coding RNAs (lncRNAs) in plants are also the key gene regulators via interacting with RNA and DNA (Zhu and Wang, 2012). In tomato, lncRNA1840 and lncRNA1459 play an important role in ripening of fruit (Zhu et al., 2015). Knocking out of tomato lncRNA1459 using CRISPR/Cas9 system gives mutant lines of CR-lncRNA1459 with a delayed fruit ripening (Li et al., 2018).

10.4 CHALLENGES OF CRISPR/Cas9 GENOME EDITING

CRISPR/Cas9 system is a potential, innovative tool in crops biological research, which involves a directed modification in the crop genome. For the modification of target genes, NHEJ gene repair is mostly utilized in horticulture plants. In most cases, phenotype shows the usefulness of the CRISPR/Cas9 tool or has shown a gene function. Major obstacles of CRISPR/Cas9 system are genome sequence information of target organism and checking non-target editing of genes. It is not easy to perform gene editing in a crop lacking a complete genome sequence information. New DNA sequencing and genome assembling tools have been developed for many horticulture plant species like apple, peach, banana, papaya and grape. New sequencing platforms make possible economical and deep sequencing of complete genome of many horticulture and fruits tree, and in-depth understanding of their gene function, pathways as well as genome structure. Polyploidy is a well-known attribute in many plant genomes. Polyploidy genome nature of banana and strawberry makes their breeding as well genome editing very difficult. In polyploidy crops, to get a desired phenotype, it is very essential to mutation of two or more target gene copies. Hence, a more efficient genome editing method for generating biallelic or multiallelic mutations in one generation is necessary for the potential improvement of polyploidy crops (Naim et al., 2018; Wilson et al., 2019). To increase the induced mutation frequency, a suitable promoter, single/multiple gRNA for single gene and target gene selection are the key factors that must be addressed properly (Hashimoto et al., 2018; Labuhn et al., 2017).

Another major limitation of CRISPR/Cas9 system is an efficient protocol for delivering CRISPR/Cas9 components into plant cells and *in vitro* plants regeneration from transformed cells. With the advancement of plant transformation technologies in the past 30 years for many crops, standardized transformation and regeneration protocols have been developed in many crops. Therefore, the advancement of efficient and reproducible delivery systems, *in vitro* selection and regeneration protocols would be crucial for using CRISPR/Cas9 regularly in horticultural crops. The novel gene editing techniques are more accurate than earlier genetic engineering tools developed earlier. There is some apprehension regarding the effect on non-target gene mutations, leading to undesired/unknown effects, which occur in species with complex and large genome size (Zhao and Wolt, 2017). However, gene editing sometime gives undesired effects even when mutations occur in desired genes. Small deletion or insertion in DNA sequence alters the gene codon, preventing protein production/altered proteins having an unidentified effect. These unwanted mutations must be identified before the use of crop cultivars, although the methods for finding undesirable mutations still face shortfalls.

REFERENCES

Cao, K., L. Cui, X. Zhou, et al. "Four tomato Flowering Locus T-like proteins act antagonistically to regulate floral initiation." *Front Plant Sci* 6 (2015): 1213.

Charrier, A., E. Vergne, N. Dousset, et al. "Efficient targeted mutagenesis in apple and first time edition of pear using the CRISPR-Cas9 system." *Front Plant Sci* 10 (2019): 40.

Chen, J., J. Xie, Y. Duan, et al. "Genome-wide identification and expression profiling reveal tissue-specific expression and differentially-regulated genes involved in gibberellin metabolism between Williams banana and its dwarf mutant." *BMC Plant Biol* 16 (2016): 123.

Christian, M., T. Cermak, E. L. Doyle, et al. "Targeting DNA double-strand breaks with TAL effector nucleases." *Genetics* 186 (2010): 757–761.

Dhaliwal, A. K., A. Mohan, G. Sidhu, et al. "An ethylmethane sulfonate mutant resource in pre-green revolution hexaploid wheat." *PLoS One* 10 (2015):e0145227.

Fister, A. S., L. Landherr, S. N. Maximova, et al. "Transient expression of CRISPR/Cas9 machinery targeting TcNPR3 enhances defense response in *Theobroma cacao*." *Front Plant Sci* 9 (2018): 268.

Gaj, T., C. A. Gersbach, C. F. Barbas, et al. "ZFN, TALEN, and CRISPR/Cas-based methods for genome engineering." *Trends Biotechnol* 31 (2013): 397–405.

Gottwald, T. R., J. H. Graham, T. S. Schubert, et al. "Citrus canker: The pathogen and its impact." *Plant Health Prog* 3 (2002): 15.

Hashimoto, R., R. Ueta, C. Abe, et al. "Efficient multiplex genome editing induces precise and self-ligated type mutations in tomato plants." *Front Plant Sci* 9 (2018): 916.

Hu, Y., J. Zhang, H. Jia, et al. "Lateral organ boundaries 1 is a disease susceptibility gene for citrus bacterial canker disease." *Proc Natl Acad Sci U S A* 111 (2014): E521–E529.

Ito, Y., A. Nishizawa-Yokoi, M. Endo, et al. "Re-evaluation of the *rin* mutation and the role of RIN in the induction of tomato ripening." *Nat Plants* 3 (2017): 866–874.

Jia, H., V. Orbovic, J. B. Jones, et al. "Modification of the *PthA4* effector binding elements in Type I CsLOB1promoter using Cas9/sgRNA to produce transgenic Duncan grapefruit alleviating XccDpthA4:dCsLOB1.3 infection." *Plant Biotech J* 14 (2016): 1291–1301.

Jia, H., Y. Zhang, V. Orbovi´c, et al. "Genome editing of the disease susceptibility gene CsLOB 1 in citrus confers resistance to citrus canker." *Plant Biotech J* 15 (2017): 817–823.:

Jiang, W., H. Zhou, H. Bi, et al. "Demonstration of CRISPR/Cas9/sgRNA-mediated targeted gene modification in Arabidopsis, tobacco, sorghum and rice." *Nucleic Acids Res* 41 (2013): e188.

Karlova, R., N. Chapman, K. David, et al. "Transcriptional control of fleshy fruit development and ripening." *J Exp Bot* 65 (2014): 4527–4541.

Kaur, N., A. Alok, N. K. Shivani, et al. "CRISPR/Cas9-mediated efficient editing in phytoene desaturase demonstrates precise manipulation in banana cv. Rasthali genome." *Funct Integr Genomics* 18 (2018): 89–99.

Khatodia, S., K. Bhatotia, N. Passricha, et al. "The CRISPR/Cas genome-editing tool: Application in improvement of crops." *Front Plant Sci* 7 (2016): 506.

Labuhn, M., F. F. Adams, S. Knoess, et al. "Refined sgRNA efficacy prediction improves large-and small-scale CRISPR–Cas9 applications." *Nucleic Acids Res* 46 (2017): 1375–1385.

Li, T., S. Huang, W. Z. Jiang, et al. "TAL nucleases (TALNs): Hybrid proteins composed of TAL effectors and FokI DNA-cleavage domain." *Nucleic Acids Res* 39 (2010): 359–372.

Li, R., D. Fu, B. Zhu, et al. "CRISPR/Cas9-mediated mutagenesis of lncRNA1459 alters tomato fruit ripening." *Plant J* 160 (2018): 595.

Liljegren, S. J., C. Gustafson-Brown, A. Pinyopich, et al. "Interactions among Apetala1, leafy and terminal Flower1 specify meristem fate." *Plant Cell* 11 (1999): 1007–1018.

Liu, X., S. Wu, J. Xu, et al. "Application of CRISPR/Cas9 in plant biology." *Acta Pharm Sin B* 7 (2017): 292–302.

Lockhart, B. *"Banana Streak Badna Virus Infection in Musa: Epidemiology, Diagnosis and Control."* ASPAC Food & Fertilizer Technology Center: Taipei, Taiwan (1995): 143.

Malnoy, M., R. Viola, M. H. Jung, et al. "cDNA-free genetically edited grapevine and apple protoplast using CRISPR/Cas9 ribo-nucleoproteins." *Front Plant Sci* 7 (2016): 1904.

Martin, S. E., and N. J. Caplen, "Applications of RNA interference in mammalian systems." *Annu Rev Genomics Hum Genet* 8 (2007): 81–108.

Naim, F., B. Dugdale, J. Kleidon, et al. "Gene editing the phytoene desaturase alleles of Cavendish banana using CRISPR/Cas9." *Transgenic Res* 27 (2018): 451–460.

Nakajima, I., Y. Ban, A. Azuma, et al. "CRISPR/Cas9 mediated targeted mutagenesis in grape". *PLoS One* 12 (2017): e0177966.

Nishitani, C., N. Hirai, S. Komori, et al. "Efficient genome editing in apple using a CRISPR/Cas9 system." *Sci Rep* 6 (2016): 31481.

Oladosu, Y., M. Y. Rafii, N. Abdullah, et al. "Principle and application of plant mutagenesis in crop improvement: A review." *Biotechnol Equip* 30 (2016): 1–16.

Osakabe, Y., Z. Liang, C. Ren, et al. "CRISPR–Cas9-mediated genome editing in apple and grapevine." *Nat Protoc* 13 (2018): 2844.

Pacher, M., and H. Puchta, "From classical mutagenesis to nuclease-based breeding directing natural DNA repair for a natural end-product." *Plant J* 90 (2017): 819–833.

Pillitteri, L. J., C. J. Lovatt, L. L. Walling et al. "Isolation and characterization of a terminal flower homolog and its correlation with juvenility in citrus." *Plant Physiol* 135 (2004): 1540–1551.

Puchta, H. "The repair of double-strand breaks in plants: Mechanisms and consequences for genome evolution." *J Exp Bot* 56 (2004): 1–14.

Qin, G., H. Gu, L. Ma, et al. "Disruption of phytoene desaturase gene results in albino and dwarf phenotypes in *Arabidopsis* by impairing chlorophyll, carotenoid, and gibberellin biosynthesis." *Cell Res* 17 (2007): 471–482.

Ren, C., X. Liu, Z. Zhang, et al. "CRISPR/Cas9-mediated efficient targeted mutagenesis in Chardonnay (*Vitis vinifera L.*)." *Sci Rep* 6 (2016): 32289.

Shao, X., S. Wu, T. Dou, et al. "Using CRISPR/Cas9 genome editing system to create MaGA20ox2 gene-modified semi-dwarf banana." *Plant Biotech J* 18 (2020): 17–19.

Sikora, P., A. Chawade, M. Larsson, et al. "Mutagenesis as a tool in plant genetics, functional genomics, and breeding. "*Int J Plant Genomics* 2011 (2011): 314829.

Song, G., M. Jia, K. Chen, et al. "CRISPR/Cas9: A powerful tool for crop genome editing." *Crop J* 4 (2016): 75–82.

Tripathi, J. N., V. O. Ntui, M. Ron, et al. "CRISPR/Cas9 editing of endogenous banana streak virus in the B genome of *Musa* spp. overcomes a major challenge in banana breeding." *Commun Biol* 2 (2019): 46.

Uluisik, S., N. H. Chapman, R. Smith, et al. "Modulation of tomato fruit texture by silencing cell wall structure related genes." *Nat Biotech* 34 (2016): 950–952.

Varkonyi-Gasic, E., T. Wang, C. Voogd, et al. "Mutagenesis of kiwifruit CENTRORADIALIS-like genes transforms a climbing woody perennial with long juvenility and axillary flowering into a compact plant with rapid terminal flowering." *Plant Biotech J* 17 (2019): 869–880.

Voytas, D. F. "Plant genome engineering with sequence-specific nucleases." *Annu Rev Plant Biol* 64 (2013): 327–350.

Wang, X., M. Tu, D. Wang, et al. "CRISPR/Cas9-mediated efficient targeted mutagenesis in grape in the first generation." *Plant Biotech J* 16 (2018a): 844–855.

Wang, Z., S. Wang, D. Li, et al. "Optimized paired-sgRNA/Cas9 cloning and expression cassette triggers high efficiency multiplex genome editing in kiwifruit." *Plant Biotech J* 16 (2018b): 1424–1433.

Weinthal, D., A. Tovkach, V. Zeevi, et al. "Genome editing in plant cells by zinc finger nucleases." *Trends Plant Sci* 15 (2010): 308–321.

Wilson, F. M., K. Harrison, A. D. Armitage, et al. "CRISPR/Cas9-mediated mutagenesis of phytoene desaturase in diploid and octoploid strawberry." *Plant Methods* 15 (2019): 45.

Wright, D. A., T. Li, B. Yang, et al. "TALEN-mediated genome editing: Prospects and perspectives." *Biochem J* 462 (2014): 15–24.

Zhang, F., C. LeBlanc, V. F. Irish, et al. "Rapid and efficient CRISPR/Cas9 gene editing in Citrus using the YAO promoter." *Plant Cell Rep* 36 (2017): 1883–1887.

Zhao, H., and J. D. Wolt, "Risk associated with off-target plant genome editing and methods for its limitation." *Emerg Top Life Sci* 1 (2017): 231–240.

Zhu, Q.-H., and M.-B. Wang, "Molecular functions of long non-coding RNAs in plants." *Genes* 3 (2012): 176–190.

Zhu, B., Y. Yang, R. Li, et al. "RNA sequencing and functional analysis implicate the regulatory role of long non-coding RNAs in tomato fruit ripening." *J Exp Bot* 66 (2015): 4483–4495.

Zhu, C., X. Zheng, Y. Huang, et al. "Genome sequencing and CRISPR/Cas9 gene editing of an early flowering Mini-Citrus (*Fortunella hindsii*)." *Plant Biotech J* 17 (2019): 2199–2210.

11 Genome Editing for Metabolic Engineering in Plants

Sandeep Jaiswal
ICAR-Research Complex for North-Eastern Hill Region

Ashok Kumar
ICAR-Directorate of Onion and Garlic Research

Jyoti Priya
ICAR-Indian Agricultural Research Institute

Sivalingam Anandhan
ICAR-Directorate of Onion and Garlic Research

Shweta Singh
University of Minho (UMINHO)

Rupesh Kumar Singh
Universidade *de* Trás-os-Montes e Alto Douro (UTAD)

CONTENTS

11.1 INTRODUCTION

Metabolic pathway manipulation has been at the heart to produce a plethora of metabolites having commercial importance, such as amino acids, vitamins, antibiotics, solvents and various therapeutic secondary metabolites. Certainly, methods have relied upon the strain improvement of microbes through chemical mutagens and a rigorous selection for desirable traits. This random mutagenesis led to desirable changes, which were never characterized for the metabolic profile of microbes mainly due to the limitation of techniques available for metabolic flux analysis (MFA) and other molecular biology tools. Development in the tools of flux analysis and molecular biology has provided a new dimension to the efforts of metabolic pathway engineering as well as expanded the horizon from microbes to different other organism, including plants.

Various terms have been used to encompass this technology such as *in vitro* evolution by Timmis et al. (1988), pathway engineering by MacQuitty (1988) and Tong et al. (1991), cellular engineering by Nerem (1991), and the most promising and accepted term 'metabolic engineering' by Stephanopoulos and Vallino (1991).

Metabolic engineering has been defined in several ways conveying the similar meaning and goals; here, we define it as a powerful methodology which involves the molecular characterization of biochemical pathways and processes, its careful modification with the aim of developing novel/modified product, new biological pathway or changing cellular physiology and ultimately phenotypes. It comprises two comprehensive steps: *analysis* and *synthesis*. Analysis provides the information about the control architecture of a metabolic network, which enables to design the strategies for the modification of rational targets, which may involve the expression of novel gene(s), amplification of endogenous enzymes at transcripts as well as the protein level, elimination of certain gene(s) and modulation of certain enzymatic activity. Selection of ad hoc targets without the analysis of integrated metabolic network will be a futile exercise owing to several concerns such as thermodynamic feasibility, flux-dependent regulation of biological pathways. Hence, a combination of analytical methods for the characterization of metabolic flux and its control, and molecular biological tools for the requisite modification of rational targets is the spirit of metabolic engineering.

Metabolic pathway is any sequence of possible and observable biochemical reactions connecting a set of inputs, intermediates and output metabolites. Flux is the rate at which input metabolites are processed into product at the steady state; it is one of the key parameters of metabolic pathways that determines the cell physiology. MFA is a measure for the state of engagement of various pathways in overall metabolic process and cellular functions. As the knowledge of metabolic flux control is at the heart of metabolic engineering, a framework of metabolic control analysis (MCA) and quantitative representation of the degree of flux control exercised by enzymes, metabolites, effectors etc. involved in the pathway under consideration was developed (Kacser et al., 1995; Heinrich & Rapoport, 1974). MCA is a powerful methodology for the calculation of metabolic fluxes of the major intracellular reactions using the stoichiometric model, and it provides a metabolic flux map. It also provides many additional insights into the metabolic pathways such as nodal rigidity (branch point control), non-measured extracellular fluxes and maximum theoretical yields, and identifies the alternate pathways.

Plant metabolism in a highly complex, small model plant Arabidopsis can produce more than 2000 types of compounds, and there are more than 200,000 chemical compounds existing in the whole plant kingdom, which includes many highly complex and interconnected biochemical reactions (Pichersky & Gang, 2000). Among these numerous metabolites, each has a crucial function in the plant growth and development. Apart from being involved in primary life functions, some of these metabolites may have some special functions like defence from various biotic or abiotic stresses. Because of their indispensable nature, they are also known as secondary metabolites. Plant metabolites not only ensure the plant's survival but also sustain the organisms, which are dependent upon plants for their food source.

Plants provide almost every essential element and compound required for the human development. Apart from the primary food components like carbohydrates, fat, proteins and vitamins, plant-based foods also supply macro- and micronutrients and compounds like secondary metabolites to our body. It is essential to mention that these nutrients and compounds are unevenly distributed in the plant parts; for example, rice endosperm is rich in carbohydrate, but the quantity of essential compounds like folate, iron, and vitamin E and vitamin A is insufficient. Hence, it becomes desirable to manipulate plants in such a way that modified plants can produce desirable plant products at the specific tissue or organ. This is possible through metabolic engineering, and it has numerous commercial implications. Metabolic engineering is the application of modern genetic engineering along with computational tools to manipulate the biosynthetic pathways in such a manner that it enhances the metabolic capabilities of the target heterologous host organism. It can be developing a new strain/line for molecular designing and synthesis of pharmaceuticals, for novel compounds or for the enhanced ability of bioconversion by the host.

Recent development of genetic engineering, reverse genetics approach such as genome editing and genome sequencing tools along with the development of '–omics' techniques has revolutionized the study of the biosynthetic pathway and found the ways to manipulate it (Grusak & DellaPenna, 1999; DellaPenna, 2001). It becomes convenient to add, remove or replace any gene or a set of genes into the host organism to engineer the metabolism to the host. One of the famous examples of plant metabolic engineering is golden rice, with the ability to produce beta-carotene in the endosperm. Meanwhile, the emergence of modern genetic engineering tools like genome editing revolutionizes the genetic manipulation of organisms to get the desirable phenotype. Due to their precise nature of targeted mutagenesis and more extensive applicability, genome editing tools like TALEN and CRISPR/Cas9 become tools of choice by the researchers. As malnutrition is a global problem, affecting billions of people worldwide, the application of metabolic engineering can help in alleviating this problem and can develop biofortified plants rich in micronutrients, vitamins, protein, fatty acids and other essential compounds at the desired concentration and composition. Moreover, the increased understanding of stress physiology and metabolic fluxes in plants has allowed to develop metabolically altered plants with higher tolerance and increased performance under various biotic and abiotic stress conditions.

11.2 DEVELOPMENT OF METABOLIC ENGINEERING AS A SCIENCE

Initially, metabolic engineering was confined to the selection of improved strains. Random mutagenesis, screening of large mutant population, identification of desirable mutant and selection of improved strains were the methods of choice for metabolic engineering, especially in unicellular organisms like a bacterial system. The discovery of restriction enzyme enables the development of first genetically engineered *E. coli* in 1973 by Cohen, Boyer and co-workers. This enabled the genetic manipulation of an organism for strain improvement and was used for the development of a recombinant protein 'human insulin', from *E. coli*. In 1991, Bailey termed this newly emerging field as 'metabolic engineering' and defined it as 'the improvement of cellular activities by manipulations of enzymatic, transport, and regulatory functions of the cell with the use of recombinant DNA technology' (Nielsen, 2001). This definition implies that metabolic engineering uses recombinant DNA technology to restructure the metabolic network in order to improve the metabolite and protein production. The application of analytical techniques and computing power enhances the power of metabolic engineering. Modulation and modification of rational target to achieve the aim of metabolic engineering becomes much easier with the recent advancement in the techniques of molecular biology and recombinant DNA technology. Stephanopoulos (1999) comprehensively redefined the metabolic engineering as the directed improvement of cellular properties or product formation through the modification or introduction of specific biochemical reactions with the use of recombinant DNA technology.

According to Stephanopoulos, the early use of metabolic engineering was confined to microorganism, and emphasis was given to the five major areas: yield improvement of the native products in a microorganism, expanding the substrate range for metabolite synthesis, synthesis of heterologous metabolites, production of novel biological compounds and improving the adaptation of the microorganism to the new environment. In the early days, most of the metabolic engineering work was centred on the microbial cell manipulation for pharmaceutical-related applications. The primary objective of metabolic engineering is the introduction or modification in the biochemical pathway of the host organism to get the desired phenotype at the biochemical, physiological or morphological level. Incorporation of tools of molecular biology, computer-based mathematical models and various analytical tools has shifted the thought of 'single enzyme engineering' towards the integration of the whole biosynthesis system for getting desired changes.

11.3 APPROACHES OF METABOLIC ENGINEERING

Metabolic engineering can be achieved by modulating the endogenous metabolic pathway or using heterologous factors into the host. The purpose of this modulation can be:

1. Enhancing the production of the desired metabolite
2. Elimination of unwanted compound from the system
3. Modification of the metabolic flux to increase the availability, production and stability of the target compound.

These objectives of modulation in the metabolic pathway can be achieved by introducing desirable changes in the structure of gene and its regulatory networks (Figure 11.1). It has become much easier with the tools available with molecular biology and recombinant DNA technology.

11.3.1 Regulatory Genes

Regulatory genes are involved in coordinating the expression and regulation of the major structural genes involved in the biosynthetic pathway. Structural genes encode enzymes for the biosynthetic pathway, while the regulatory genes encode another group of regulatory proteins known as transcription factors. These transcription factors interact with the promoters of structural genes and modulate the gene expression (Ranish & Hahn, 1996). These transcription factors are activated by various biotic and abiotic signals, e.g. hormones, microbial elicitors, UV light, heat, etc. (Memelink et al., 2001; Vom Endt, 2002). It has been demonstrated in various reports that the ectopic expression of these transcription factors can change the metabolic profile of the host. The expression level of the structural gene is dependent upon various factors such as the binding affinity of a transcription factor to the promoter, interaction between internal and introduced transcription factor, and level of stimulation in transcription factor to the stimulus (Quattrocchio et al. 1993; Grotewold, 2006). These parameters can be altered by the use of genome editing tools as targeted sequence changes in the regulatory genes can change the way it interacts with the structural genes, and this can ultimately lead to a change in the metabolic profile of the host. Hence, genome editing can modulate the metabolism by optimizing the transcript level of the structural genes.

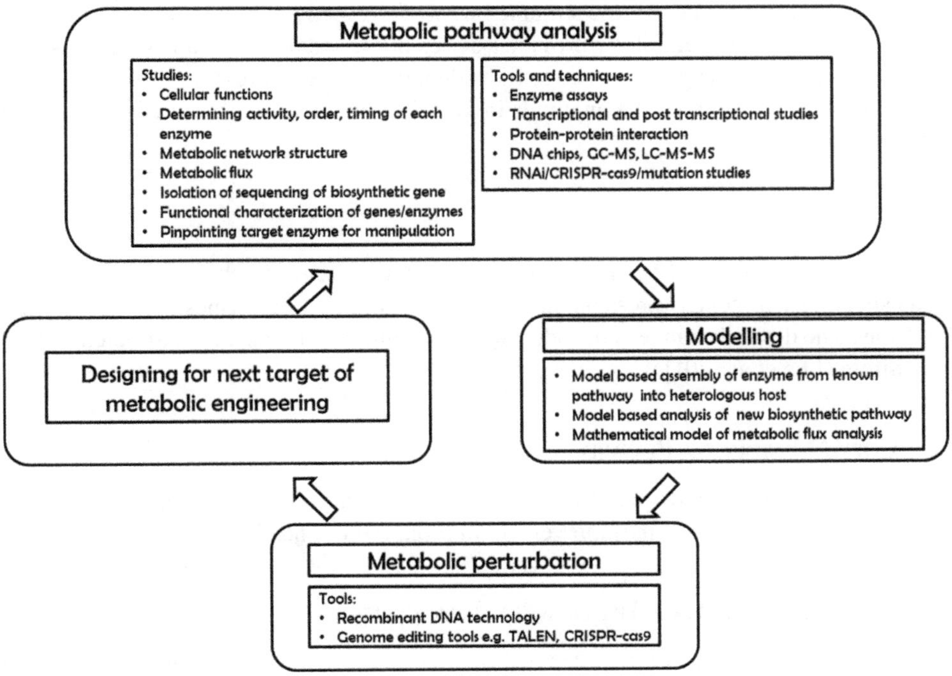

FIGURE 11.1 Scheme of metabolic engineering in plants.

11.3.2 Modification in Structural Genes

Structural genes encode proteins that may involve in the biosynthetic pathway. In order to modulate the metabolism, it is possible to introduce, remove, modify or inactivate the structural genes. A completely novel pathway, or branch of the pathway or additional steps can be added into the biosynthetic pathway if the corresponding structural gene has been identified and functionally characterized. Moreover, metabolic engineering tools can alter the DNA sequence of the structural gene so that desired changes in the structure, affinity or activity of the enzymes involved in the biosynthetic pathway can be done. Recent genome editing tools can perform a precise DNA sequence change at the promoter or coding region of the structural gene, and can fine-tune the gene expression level so that desirable change in metabolism can be made.

11.3.3 Blocking Specific Steps in the Pathway

MFA may necessitate to block the specific biochemical reaction in the endogenous biosynthesis pathway for desirable characteristics. These biochemical pathways might be involved in siphoning off the essential intermediates or producing undesirable products. RNA interference was the tool of choice to down-regulate the expression of specific structural genes but limitations like incomplete silencing are not always desirable. Modified genome effectors like dCas/aCas in association with CRISPR can deactivate/activate the target gene in a highly precise and sequence-specific manner. Similarly, some CRISPR effectors like Cas13a have been identified to target RNA transcript of the target *genes* (Abudayyeh et al., 2017).

11.4 TOOLS OF METABOLIC ENGINEERING

Metabolic engineering relies heavily upon methods and tools that can perform a desirable change in the genome, biochemical approaches that can measure metabolic fluxes, and cellular biology as well as molecular tools that can analyse the state of the cell. All these tools decipher the cell's network architecture; hence, the suitable targets for genetic manipulation can be identified. Molecular biology and biotechnology provide a series of techniques that can be harnessed to perform a precise and desirable genetic change like silencing or overexpression of the target genes in the host so that the activities of the enzymes involved in the target pathway can be manipulated in the desired manner.

11.4.1 Tuneable Promoters

These promoters are developed by changing the promoter sequence to change the timing, strength and tissue specificity of gene expression, and hence, the activity of enzymes involved in the metabolic pathway. The desired change is introduced by conventional mutagenesis, directed evolution or genome editing (Lee et al., 2011).

11.4.2 Copy Number of a Plasmid

Increasing the copy number of plasmid can increase the expression level of the inserted gene. This tool is applicable to the strain improvement in prokaryotes and single-celled eukaryotes.

11.4.3 Engineering the Half-Life of RNA Transcripts

The half-life of the RNA transcript is decided by its sequence, post-transcriptional modifications and secondary structure, etc. Genetic engineering tools can increase/decrease the stability of the RNA transcript; hence, the level of expression of a particular gene involved in a metabolic pathway can be regulated.

11.4.4 High-Efficiency Transformations

Plant transformation techniques like Agrobacterium-mediated transformation and biolistic methods are most commonly used. Higher efficiency of plant transformation techniques also contributes to the success of metabolic engineering.

11.4.5 Transposon Mutagenesis

It involves random mapping and mutagenesis of the genome by transposable elements introduced into the host, followed by the selection of the mutant with the desired metabolic change.

11.4.6 Gene Shuffling

Advancement in PCR enabled scientists to shuffle the gene sequence for its improved expression. Directed evolution of genes leads to the development of genes having desired sequence change and improved performance.

11.4.7 Positional Isotopomers

Use labelled isotopes for MFA and metabolic network analysis, particularly in plant systems (Fernie & Morgan, 2013).

11.4.8 Transcription Profiling

Using DNA microarrays determines the global change in gene expression of an organism at a particular stage, time and tissue.

11.4.9 Metabolite Profiling

This tool determines the global change in the level of the metabolite of an organism. It includes a spatial abundance of metabolites at the intermediate step or the final metabolic product.

11.4.10 Statistical Modelling

Complex computational model and algorithm are being used for the estimation of flux coupling, the analysis of metabolic flux and the possible lethality of overproducing a particular metabolite. Modelling in relation to metabolic engineering focuses on the three primary goals: identifying mistakes in a metabolic model, correcting metabolic models and optimizing experiments to gain the maximum yield by identifying targets for genetic manipulation (Zomorrodi et al., 2013).

11.4.11 RNA-Antisense Technique

This technique has been used in various metabolic engineering experiments to down-regulate the target gene expression in a sequence-specific manner at the transcriptional or post-transcriptional level.

11.4.12 Genome Editing

It involves a group of specially designed nuclease for the sequence-specific modification of the target gene or specific locus in the genome. The most commonly used genome editing tool in the present time is CRISPR/Cas9, which is an RNA-guided nuclease.

11.4.13 Fluxome

It assess the response of the system to environmental stimulus as well as metabolic engineering. It identifies the regulatory mechanism controlling metabolic flux qualitatively and quantitatively as well.

11.5 CATEGORIES OF METABOLIC ENGINEERING

There are following seven categories of metabolic engineering based on the objective.

11.5.1 Production of Desirable Protein in Heterologous Host

It involves the identification, isolation and transfer of gene of interest into a suitable host, followed by its expression so that the desirable metabolic change can be obtained. Conventionally, proteins of pharmaceutical importance or novel enzyme activity have been produced by metabolic engineering. However, the use of genome editing for such a category of modification is not applicable since genome editing tools manipulate the genome without inserting genes for the trait development from a foreign source.

11.5.2 Extension of Substrate Range

This category of metabolic engineering involves genetic modification in order to develop the ability in the host for utilizing new substrate or more efficient utilization of the raw material.

11.5.3 Pathways Leading to New Products

The existing biosynthetic pathway of an organism can be extended in order to get a new product. This can be accomplished by recruiting a heterologous enzyme or activating/modifying the

TABLE 11.1
Metabolic Engineering in Plants for Novel Characteristics

Traits	Crop	Genome Editing Tool Used	Modification	References
Fragrance	Rice (*Oryza sativa*)	CRISPR/Cas9	SF3B1	Butt et al. (2018)
Modification of flower colour	A model plant torenia (*Torenia fournieri L.*)	CRISPR/Cas9	Mutations in the torenia flavanone 3-hydroxylase (F3H) gene encoding a key enzyme involved in flavonoid biosynthesis	Nishihara et al. (2018)
Alkaloid content	Opium poppy	CRISPR-Cas9	• Biosynthesis of BIAs (e.g. morphine, the baine) was significantly reduced • A novel uncharacterized alkaloid was observed	Alagoz et al. (2016)

existing pathway. Gene shuffling-directed evolution or genome editing tools can be used to modify the pathway for the new product formation. A few examples are available for such genome editing, which are listed in Table 11.1.

11.5.4 Metabolic Engineering for Bioremediation

The plant metabolism can be engineered for the degradation of xenobiotics either by inserting a heterologous pathway or by modifying the existing pathway. The transgenic approach has successfully been used to develop a plant with bioremediation abilities for various heavy metals and xenobiotic compounds. Till now, genome editing tools have not been applied for bioremediation.

11.5.5 Modification of Plant's Physiology and Molecular Biology

Genetic manipulation with genetic engineering and genome editing tools can be utilized for changing the physiology of the plants. Metabolic engineering using these recent tools has been used to develop plants with novel and desirable agronomic traits like improved tolerance to biotic and abiotic stress conditions, and plant architecture (Table 11.2).

11.5.6 Elimination or Reduction of Undesirable/By-Product Formation

By-product formation may be an undesirable metabolic event as it can cause the expense of extra carbon and energy. Moreover, these by-products can cause toxicity or interfere with other pathways. These by-products can be eliminated by the inactivation of the responsible gene. However, in some cases, the by-product is crucial for some cellular functions, so gene inactivation can be undesirable. In those cases, molecular tools can be used to reduce the by-product formation up to the optimum state (Table 11.3).

11.5.7 Improvement of Yield, Quality or Productivity

Yield is one of the most desirable traits for crop improvement. Yield enhancement can be achieved in terms of grain or particular metabolite. As the grain yield is a polygenic trait, some major QTLs have been modified using genome editing tools to improve the grain yield in crops like rice. The yield of some important metabolites like carotenoids in tomato has been improved by simply manipulating the one or few enzymes of the biosynthetic pathway. Though, the manipulation of complex pathways is difficult; hence, analysis of the flux regulation as well as engineering central carbon metabolism is critical (Table 11.4).

TABLE 11.2
Metabolic Engineering of Plant Physiology for Improved Fitness of Plants in Desirable Environmental Conditions

Traits	Crop	Genome Editing Tool Used	Modification	References
Dwarf phenotype	Tomato	CRISPR/Cas9	Generation of dominant dwarf PROCERA (DELLA protein-encoding) allele	Tomlinson et al. (2019)
Broad-spectrum disease resistance	Tomato	CRISPR/Cas9	Small deletions in the *Sl*DMR6-1 gene, which resulted in increased salicylic acid levels	de Toledo Thomazella et al. (2016)
Powdery mildew resistant	Tomato	CRISPR/Cas9	Targeted the SlMlo1 locus using the double sgRNA strategy	Nekrasov et al. (2017)
Tomato yellow leaf curl virus	Tomato	CRISPR/Cas9	Targeting the TYLCV genome with Cas9-single guide RNA at the sequences encoding the coat protein (CP) or replicate (Rep)	Tashkandi et al. (2018)
Parasitic weed resistance	Tomato	CRISPR/Cas9-mediated	CRISPR/Cas9-mediated mutagenesis of the CCD8 gene in tomato to develop host resistance to the parasitic weed *Phelipanche aegyptiaca*	Bari et al. (2019)
Parthenocarpic plants	Tomato	CRISPR/Cas9	Somatic mutation into SlIAA9 – a key gene controlling parthenocarpy	Ueta et al. (2017)
Facultative parthenocarpy	Tomato	CRISPR/Cas9	CRISPR/Cas9-mediated mutation in SlAGAMOUS-LIKE 6 (SlAGL6)	Klap et al. (2017)
Plant architecture	Rice (*Oryza sativa*)	CRISPR/Cas9	Targeted disruption of CAROTENOID CLEAVAGE DIOXYGENASE 7 (CCD7), which controls a key step in strigolactone biosynthesis	Butt et al. (2018)
BLB resistance	Rice (*Oryza sativa*)	CRISPR/Cas9	Knockout of rice Os8N3 (also known as OsSWEET11), which belongs to the Sugar Will Eventually be Exported Transporters (SWEET) family of sugar transporters	Kim, Moon, and& Park (2019)
Rice blast resistance	Rice (*Oryza sativa*)	CRISPR/Cas9	Targeting the OsERF922 gene in rice	Wang et al. (2016)
Herbicide resistance	Rice (*Oryza sativa*)	CRISPR/CAS9	ALS	Sun et al. (2016)
Powdery mildew resistance	Wheat (*Triticum aestivum L.*)	CRISPR/Cas9	Generated Taedr1 wheat plants by a simultaneous modification of the three homoeologs of wheat EDR1	Zhang et al. (2017)
Herbicide resistance	Wheat (*Triticum aestivum L.*)	CRISPR/CAS9	ALS	Zhang et al. (2017)

(Continued)

TABLE 11.2 (*Continued*)
Metabolic Engineering of Plant Physiology for Improved Fitness of Plants in Desirable Environmental Conditions

Traits	Crop	Genome Editing Tool Used	Modification	References
Male sterile	Wheat (*Triticum aestivum L.*)	CRISPR-Cas9	Mutagenesis of a strictosidine synthase-like enzyme encoded by male sterile 45 (Ms45) gene	Singh et al. (2018)
Salinity tolerance	Rice (*Oryza sativa*)	CRISPR/Cas9	Mutagenesis of the OsRR22 gene (involved in both cytokinin signal transduction and metabolism)	Zhang et al. (2019b)
Delayed flowering time	Soya bean (*Glycine max*)	CRISPR/Cas9	Targeted mutagenesis of GmFT2a	Cai et al. (2018, 2020)
Potyvirus resistance (turnip mosaic virus (tumv))	*Arabidopsis sp.*	CRISPR/Cas9	Introduce the sequence-specific deleterious point mutations at the eIF(iso)4E locus in *Arabidopsis thaliana.*	Pyott, Sheehan, and Molnar (2016)
Drought stress tolerance	*Arabidopsis sp.*	CRISPR/dCas9 fusion with a histone acetyl transferase	CRISPRa dCas9^{HAT} improves drought stress tolerance through the positive regulation of AREB1	Roca Paixão et al. (2019)
Drought stress tolerance	Maize	CRISPR/Cas9	ARGOS8 variants generated by CRISPR-Cas9 increase ARGOS8 transcript level and hence improve the yield under drought stress	Shi et al. (2017)
Transgene-free male sterile lines	Maize	CRISPR/Cas9	Targeting the MS8 gene	Chen et al. (2018)
Herbicide resistance	*Zea mays* (maize)	CRISPR/CAS9	ALS2	Svitashev et al. (2015)
Resistance to citrus canker	Citrus	CRISPR/Cas9	Modify the canker susceptibility gene CsLOB1 in Duncan grapefruit	Jia et al. (2017)
Cassava brown streak disease		CRISPR/Cas9	Editing of cassava eIF4E isoforms nCBP-1 and nCBP-2-ncbp-1/ncbp-2 mutants displayed delayed and attenuated CBSD aerial symptoms, as well as reduced severity and incidence of storage root necrosis	Gomez et al. (2019)
Banana streak virus	Banana	CRISPR/Cas9	Inactivation of endogenous banana streak virus (eBSV) from B-genome of the banana	Tripathi et al. (2019)

(*Continued*)

TABLE 11.2 (*Continued*)
Metabolic Engineering of Plant Physiology for Improved Fitness of Plants in Desirable Environmental Conditions

Traits	Crop	Genome Editing Tool Used	Modification	References
Citrus canker	Wanjincheng orange	CRISPR/Cas9	Editing of CsWRKY22 reduces susceptibility to *Xanthomonas citri* subsp. citri	Wang et al. (2019a)
Cotton leaf curl Multan virus	*Nicotiana benthamiana*	CRISPR/Cas9	Targeting the IR region and the C1 coding region	Yin et al. (2019)
Early flowering	Chines cabbage	CRISPR/Cas9	*FLOWERING LOCUS C* (BraFLC2 and BraFLC3) double knockout	Jeong et al. (2019)
Enhanced flower longevity	Petunia cv. mirage rose	CRISPR/Cas9	CRISPR/Cas9-mediated editing of 1-aminocyclopropane-1-carboxylate oxidase1	Xu et al. (2020)
Herbicide resistance	*Solanum tuberosum* (potato)	CRISPR/CAS9	ALS1	Butler et al. (2016)
Herbicide resistance	Watermelon	CRISPR/CAS9	ALS	Tian et al. (2018)
Overcoming self-incompatibility	Diploid potato	CRISPR-Cas9	Target conserved exonic regions of the S-RNase gene and generate targeted knockouts (KOs)	Enciso-Rodriguez et al. (2019)

TABLE 11.3
Metabolic Engineering of Plant for Eliminating Certain Toxic/Undesirable Metabolites

Traits	Crop	Genome Editing Tool Used	Modification	References
Gamma-aminobutyric acid levels (GABA)	Tomato	CRISPR/Cas9	Single-site guide RNAs to target the tomato *phytoene desaturase* gene	Li et al. (2018c)
Removal of toxic factor (steroidal glycoalkaloids)	Potato	CRISPR/Cas9	Editing of the St16DOX (2-oxoglutarate-dependent dioxygenase) gene	Nakayasu et al. (2018)

TABLE 11.4
Metabolic Engineering of Plants for Improving Yield and Its Quality

Traits	Crop	Genome Editing Tool Used	Modification	References
Lycopene content	Tomato	CRISPR/Cas9-mediated multiplex genome editing	Five genes involved in the carotenoid metabolic pathway in tomato were selected and edited simultaneously	Li et al. (2018d)
Fruit colour	Tomato	CRISPR/Cas9	Generation of pink-fruited tomato plants via disruption of the SlMYB12 gene	Deng et al. (2018)
Fruit ripening	Tomato	CRISPR-Cas9	• Loss-of-function mutants of lncRNA1459 • Ethylene production and lycopene accumulation were largely repressed in lncRNA1459 mutants	Li et al. (2018a)
Increased lycopene content	Tomato	CRISPR/Cas9	Multiplex genome editing of genes involved in lycopene synthesis	Li et al. (2018c)
Low-gluten content	Wheat	CRISPR/Cas9	Mutation of 33-mer in the alpha-gliadin genes (the main immunodominant peptide in patients with coeliac disease)	Sánchez-León et al. (2018)
Fatty acid composition	Soya bean	TALEN	Mutations in FAD2-1A and FAD2-1B	Haun et al. (2014)
Increase in oleic acid content	*Arabidopsis thaliana*	CRISPR/Cas9	CRISPR/Cas9-mediated mutation of the FAD2 gene	Jiang et al. (2017)
Isoflavone content and resistance to soya bean mosaic virus	Soya bean	CRISPR/Cas9-mediated multiplex gene editing	Simultaneously target GmF3H1, GmF3H2 and GmFNSII-1 in soya bean hairy roots and plants	Zhang et al. (2019a)
Starch quality	Sweet potato (*Ipomoea batatas*)	CRISPR/Cas9	Mutagenesis of starch biosynthetic genes, IbGBSSI (encoding granule-bound starch synthase I) and IbSBEII (encoding starch-branching enzyme II) for producing low-amylose starch	Wang et al. (2019b)
Oleic acid content increase	*Brassica napus*	CRISPR/Cas9	FAD2.a (FAD2_Aa)	Okuzaki et al. (2018)
Increase in oleic acid content	*Camelina sativa*	CRISPR/Cas9	Mutation of FAD2 gene	Jiang et al. (2017)
Increase in the oleic acid content	*Arachis hypogaea*	TALEN	Mutagenesis of fatty acid desaturase 2 (AhFAD2)	Wen et al. (2018)
Protein quality	Sorghum	CRISPR/Cas9	Reduction in α-kafirins by mutating conserved region of the α-kafirin-encoding gene	Li et al. (2018b)

11.6 CONCLUSION

Product metabolite of any specific biological pathways may be desirable but it has an influence on other metabolic pathway. Therefore, there is a need to divert the attention from individual pathway to the analysis of integrated metabolic network in order to identify the rational target for modification. Recent development in molecular biology must be integrated with the coherent development of high-throughput and cost-effective analytical tools for the efficient MFA and thermodynamically feasible cellular changes.

Rapidly changing climate has immensely impacted the agriculture in various forms but spirit of human kind can't surrender to this advent. We need to pace up the crop improvement programmes for better adaptability along with high quality and productivity. Although efforts have been made but at large, these are centric to single gene lacking the systemic characterization of the metabolic network. There is a need to engineer a robust metabolic network having an inherent capability of maintaining homeostasis under changing climatic condition without compromising on yield and quality. It is worthy to mention that plants also provide a cost-effective alternative to microbes for the production of commercial metabolites. It has a better scalability potential with a minimal cost of maintenance.

REFERENCES

Abudayyeh, O. O., Gootenberg, J. S., Essletzbichler, P., Han, S., Joung, J., Belanto, J. J., ... Zhang, F. (2017). RNA targeting with CRISPR–Cas13. *Nature*, *550*(7675), 280–284. doi: 10.1038/nature24049.

Alagoz, Y., Gurkok, T., Zhang, B., & Unver, T. (2016). Manipulating the biosynthesis of bioactive compound alkaloids for next-generation metabolic engineering in opium poppy using CRISPR-Cas 9 genome editing technology. *Scientific Reports*, *6*(1), 30910. doi: 10.1038/srep30910.

Bailey, J. (1991). Toward a science of metabolic engineering. *Science*, *252*(5013), 1668–1675. doi: 10.1126/science.2047876.

Bari, V. K., Nassar, J. A., Kheredin, S. M., Gal-On, A., Ron, M., Britt, A., ... Aly, R. (2019). CRISPR/Cas9-mediated mutagenesis of CAROTENOID CLEAVAGE DIOXYGENASE 8 in tomato provides resistance against the parasitic weed Phelipanche aegyptiaca. *Scientific Reports*, *9*(1), 1–12. doi: 10.1038/s41598-019-47893-z.

Butler, N. M., Baltes, N. J., Voytas, D. F., & Douches, D. S. (2016). Geminivirus-mediated genome editing in potato (*Solanum tuberosum L.*) using sequence-specific nucleases. *Frontiers in Plant Science*, *7*, 1–13. doi: 10.3389/fpls.2016.01045.

Butt, H., Jamil, M., Wang, J. Y., Al-Babili, S., & Mahfouz, M. (2018). Engineering plant architecture via CRISPR/Cas9-mediated alteration of strigolactone biosynthesis. *BMC Plant Biology*, *18*(1), 174. doi: 10.1186/s12870-018-1387-1.

Cai, Y., Chen, L., Liu, X., Guo, C., Sun, S., Wu, C., ... Hou, W. (2018). CRISPR/Cas9-mediated targeted mutagenesis of GmFT2a delays flowering time in soya bean. *Plant Biotechnology Journal*, *16*(1), 176–185. doi: 10.1111/pbi.12758.

Cai, Y., Wang, L., Chen, L., Wu, T., Liu, L., Sun, S., ... Hou, W. (2020). Mutagenesis of GmFT2a and GmFT5a mediated by CRISPR/Cas9 contributes for expanding the regional adaptability of soybean. *Plant Biotechnology Journal*, *18*(1), 298–309. doi: 10.1111/pbi.13199.

Chen, R., Xu, Q., Liu, Y., Zhang, J., Ren, D., Wang, G., & Liu, Y. (2018). Generation of transgene-free maize male sterile lines using the CRISPR/Cas9 system. *Frontiers in Plant Science*, *9*(September), 1–8. doi: 10.3389/fpls.2018.01180.

de Toledo Thomazella, D. P., Brail, Q., Dahlbeck, D., & Staskawicz, B. J. (2016). CRISPR-Cas9 mediated mutagenesis of a DMR6 ortholog in tomato confers broad-spectrum disease resistance. In bioRxiv. doi: 10.1101/064824.

DellaPenna, D. (2001). Plant metabolic engineering. *Plant Physiology*, *125*(1), 160–163. doi: 10.1104/pp.125.1.160.

Deng, L., Wang, H., Sun, C., Li, Q., Jiang, H., Du, M., ... Li, C. (2018). Efficient generation of pink-fruited tomatoes using CRISPR/Cas9 system. *Journal of Genetics and Genomics*, *45*, 51–54. doi: 10.1016/j.jgg.2017.10.002.

Enciso-Rodriguez, F., Manrique-Carpintero, N. C., Nadakuduti, S. S., Buell, C. R., Zarka, D., & Douches, D. (2019). Overcoming self-incompatibility in diploid potato using CRISPR-Cas9. *Frontiers in Plant Science*, *10*(April), 1–12. doi: 10.3389/fpls.2019.00376.

Fernie, A. R., & Morgan, J. A. (2013). Analysis of metabolic flux using dynamic labelling and metabolic modelling. *Plant, Cell & Environment*, *36*(9), 1738–1750. doi: 10.1111/pce.12083.

Gomez, M. A., Lin, Z. D., Moll, T., Chauhan, R. D., Hayden, L., Renninger, K., … Bart, R. S. (2019). Simultaneous CRISPR/Cas9-mediated editing of cassava eIF4E isoforms nCBP-1 and nCBP-2 reduces cassava brown streak disease symptom severity and incidence. *Plant Biotechnology Journal*, *17*(2), 421–434. doi: 10.1111/pbi.12987.

Grotewold, E. (2006). The genetics and biochemistry of floral pigments. *Annual Review of Plant Biology*, *57*(1), 761–780. doi: 10.1146/annurev.arplant.57.032905.105248.

Grusak, M. A., & DellaPenna, D. (1999). Improving the nutrient composition of plants to enhance human nutrition and health. *Annual Review of Plant Physiology and Plant Molecular Biology*, *50*(1), 133–161. doi: 10.1146/annurev.arplant.50.1.133.

Haun, W., Coffman, A., Clasen, B. M., Demorest, Z. L., Lowy, A., Ray, E., … Zhang, F. (2014). Improved soybean oil quality by targeted mutagenesis of the fatty acid desaturase 2 gene family. *Plant Biotechnology Journal*, *12*(7), 934–940. doi: 10.1111/pbi.12201.

Heinrich, R., & Rapoport, T. A. (1974). A linear steady-state treatment of enzymatic chains: General properties, control and effector strength. *European Journal of Biochemistry*, *42*(1), 89–95. doi: 10.1111/j.1432-1033.1974.tb03318.x.

Jeong, S. Y., Ahn, H., Ryu, J., Oh, Y., Sivanandhan, G., Won, K. H., … Kim, S. G. (2019). Generation of early-flowering Chinese cabbage (Brassica rapa spp. pekinensis) through CRISPR/Cas9-mediated genome editing. *Plant Biotechnology Reports*, *13*(5), 491–499. doi: 10.1007/s11816-019-00566-9.

Jia, H., Zhang, Y., Orbović, V., Xu, J., White, F. F., Jones, J. B., & Wang, N. (2017). Genome editing of the disease susceptibility gene CsLOB1 in citrus confers resistance to citrus canker. *Plant Biotechnology Journal*, *15*(7), 817–823. doi: 10.1111/pbi.12677.

Jiang, W. Z., Henry, I. M., Lynagh, P. G., Comai, L., Cahoon, E. B., & Weeks, D. P. (2017). Significant enhancement of fatty acid composition in seeds of the allohexaploid, Camelina sativa, using CRISPR/Cas9 gene editing. *Plant Biotechnology Journal*, *15*(5), 648–657. doi: 10.1111/pbi.12663.

Kacser, H., Burns, J. A., Kacser, H., & Fell, D. A. (1995). The control of flux. *Biochemical Society Transactions*, *23*(2), 341–366. doi: 10.1042/bst0230341.

Kim, Y.-A., Moon, H., & Park, C.-J. (2019). CRISPR/Cas9-targeted mutagenesis of Os8N3 in rice to confer resistance to Xanthomonas oryzae pv. oryzae. *Rice*, *12*(1), 67. doi: 10.1186/s12284-019-0325-7.

Klap, C., Yeshayahou, E., Bolger, A. M., Arazi, T., Gupta, S. K., Shabtai, S., … Barg, R. (2017). Tomato facultative parthenocarpy results from SlAGAMOUS-LIKE 6 loss of function. *Plant Biotechnology Journal*, *15*(5), 634–647. doi: 10.1111/pbi.12662.

Lee, J. W., Kim, T. Y., Jang, Y.-S., Choi, S., & Lee, S. Y. (2011). Systems metabolic engineering for chemicals and materials. *Trends in Biotechnology*, *29*(8), 370–378. doi: 10.1016/j.tibtech.2011.04.001.

Li, R., Fu, D., Zhu, B., Luo, Y., & Zhu, H. (2018a). CRISPR/Cas9-mediated mutagenesis of lncRNA1459 alters tomato fruit ripening. *Plant Journal*, *94*(3), 513–524. doi: 10.1111/tpj.13872.

Li, A., Jia, S., Yobi, A., Ge, Z., Sato, S. J., Zhang, C., … Holding, D. R. (2018b). Editing of an alpha-kafirin gene family increases digestibility and protein quality in sorghum. *Plant Physiology*, *177*(4), 1425–1438. doi: 10.1104/pp.18.00200.

Li, R., Li, R., Li, X., Fu, D., Zhu, B., Tian, H., … Zhu, H. (2018c). Multiplexed CRISPR/Cas9-mediated metabolic engineering of γ-aminobutyric acid levels in Solanum lycopersicum. *Plant Biotechnology Journal*, *16*(2), 415–427. doi: 10.1111/pbi.12781.

Li, X., Wang, Y., Chen, S., Tian, H., Fu, D., Zhu, B., … Zhu, H. (2018d). Lycopene is enriched in tomato fruit by CRISPR/Cas9-mediated multiplex genome editing. *Frontiers in Plant Science*, *9*(April), 1–12. doi: 10.3389/fpls.2018.00559.

MacQuitty, J. J. (1988). *The Impact of Chemistry on Biotechnology* (M. Phillips, S. P. Shoemaker, R. D. Middlekauff, & R. M. Ottenbrite, Eds.). doi: 10.1021/bk-1988-0362.

Memelink, J., Kijne, J. W., van der Heijden, R., & Verpoorte, R. (2001). Genetic modification of plant secondary metabolite pathways using transcriptional regulators. *Advances in Biochemical Engineering/Biotechnology*, 103–125. doi: 10.1007/3-540-45302-4_4.

Nakayasu, M., Akiyama, R., Lee, H. J., Osakabe, K., Osakabe, Y., Watanabe, B., … Mizutani, M. (2018). Generation of α-solanine-free hairy roots of potato by CRISPR/Cas9 mediated genome editing of the St16DOX gene. *Plant Physiology and Biochemistry*, *131*(February), 70–77. doi: 10.1016/j.plaphy.2018.04.026.

Nekrasov, V., Wang, C., Win, J., Lanz, C., Weigel, D., & Kamoun, S. (2017). Rapid generation of a transgene-free powdery mildew resistant tomato by genome deletion. *Scientific Reports*, *7*(1), 1–6. doi: 10.1038/s41598-017-00578-x.

Nerem, R. M. (1991). Cellular engineering. *Annals of Biomedical Engineering, 19*(5), 529–545. doi: 10.1007/BF02367396.

Nielsen, J. (2001). Metabolic engineering. *Applied Microbiology and Biotechnology, 55*(3), 263–283. doi: 10.1007/s002530000511.

Nishihara, M., Higuchi, A., Watanabe, A., & Tasaki, K. (2018). Application of the CRISPR/Cas9 system for modification of flower color in Torenia fournieri. *BMC Plant Biology, 18*(1), 1–9. doi: 10.1186/s12870-018-1539-3.

Okuzaki, A., Ogawa, T., Koizuka, C., Kaneko, K., Inaba, M., Imamura, J., & Koizuka, N. (2018). CRISPR/Cas9-mediated genome editing of the fatty acid desaturase 2 gene in Brassica napus. *Plant Physiology and Biochemistry, 131*(April), 63–69. doi: 10.1016/j.plaphy.2018.04.025.

Pichersky, E., & Gang, D. R. (2000). Genetics and biochemistry of secondary metabolites in plants: An evolutionary perspective. *Trends in Plant Science, 5*(10), 439–445. doi: 10.1016/S1360-1385(00)01741-6.

Pyott, D. E., Sheehan, E., & Molnar, A. (2016). Engineering of CRISPR/Cas9-mediated potyvirus resistance in transgene-free Arabidopsis plants. *Molecular Plant Pathology, 17*(8), 1276–1288. doi: 10.1111/mpp.12417.

Quattrocchio, F., Wing, J. F., Leppen, H. T. C., Mol, J. N. M., & Koes, R. E. (1993). Regulatory genes controlling anthocyanin pigmentation are functionally conserved among plant species and have distinct sets of target genes. *Plant Cell, 5*(11), 1497–1512. doi: 10.1105/tpc.5.11.1497.

Ranish, J. A., & Hahn, S. (1996). Transcription: Basal factors and activation. *Current Opinion in Genetics & Development, 6*(2), 151–158.

Roca Paixão, J. F., Gillet, F. X., Ribeiro, T. P., Bournaud, C., Lourenço-Tessutti, I. T., Noriega, D. D., … Grossi-de-Sa, M. F. (2019). Improved drought stress tolerance in Arabidopsis by CRISPR/dCas9 fusion with a Histone AcetylTransferase. *Scientific Reports, 9*(1), 1–9. doi: 10.1038/s41598-019-44571-y.

Sánchez-León, S., Gil-Humanes, J., Ozuna, C. V., Giménez, M. J., Sousa, C., Voytas, D. F., & Barro, F. (2018). Low-gluten, nontransgenic wheat engineered with CRISPR/Cas9. *Plant Biotechnology Journal, 16*(4), 902–910. doi: 10.1111/pbi.12837.

Shi, J., Gao, H., Wang, H., Lafitte, H. R., Archibald, R. L., Yang, M., … Habben, J. E. (2017). ARGOS8 variants generated by CRISPR-Cas9 improve maize grain yield under field drought stress conditions. *Plant Biotechnology Journal, 15*(2), 207–216. doi: 10.1111/pbi.12603.

Singh, M., Kumar, M., Albertsen, M. C., Young, J. K., & Cigan, A. M. (2018). Concurrent modifications in the three homeologs of Ms45 gene with CRISPR-Cas9 lead to rapid generation of male sterile bread wheat (*Triticum aestivum L.*). *Plant Molecular Biology, 97*(4–5), 371–383. doi: 10.1007/s11103-018-0749-2.

Stephanopoulos, G. (1999). Metabolic fluxes and metabolic engineering. *Metabolic Engineering, 1*(1), 1–11. doi: 10.1006/mben.1998.0101.

Stephanopoulos, G., & Vallino, J. (1991). Network rigidity and metabolic engineering in metabolite overproduction. *Science, 252*(5013), 1675–1681. doi: 10.1126/science.1904627.

Sun, Y., Zhang, X., Wu, C., He, Y., Ma, Y., Hou, H., … Xia, L. (2016). Engineering herbicide-resistant rice plants through CRISPR/Cas9-mediated homologous recombination of acetolactate synthase. *Molecular Plant, 9*(4), 628–631. doi: 10.1016/j.molp.2016.01.001.

Svitashev, S., Young, J. K., Schwartz, C., Gao, H., Falco, S. C., & Cigan, A. M. (2015). Targeted mutagenesis, precise gene editing, and site-specific gene insertion in maize using Cas9 and guide RNA. *Plant Physiology, 169*(2), 931–945. doi: 10.1104/pp.15.00793.

Tashkandi, M., Ali, Z., Aljedaani, F., Shami, A., & Mahfouz, M. M. (2018). Engineering resistance against tomato yellow leaf curl virus via the CRISPR/Cas9 system in tomato. *Plant Signaling and Behavior, 13*(10). doi: 10.1080/15592324.2018.1525996.

Tian, S., Jiang, L., Cui, X., Zhang, J., Guo, S., Li, M., … Xu, Y. (2018). Engineering herbicide-resistant watermelon variety through CRISPR/Cas9-mediated base-editing. *Plant Cell Reports, 37*(9), 1353–1356. doi: 10.1007/s00299-018-2299-0.

Timmis, K. N., Rojo, F., & Ramos, J. L. (1988). Prospects for laboratory engineering of bacteria to degrade pollutants. *Basic Life Sciences, 45*, 61–79. doi: 10.1007/978-1-4899-0824-7_4.

Tomlinson, L., Yang, Y., Emenecker, R., Smoker, M., Taylor, J., Perkins, S., … Jones, J. D. G. (2019). Using CRISPR/Cas9 genome editing in tomato to create a gibberellin-responsive dominant dwarf DELLA allele. *Plant Biotechnology Journal, 17*(1), 132–140. doi: 10.1111/pbi.12952.

Tong, I. T., Liao, H. H., & Cameron, D. C. (1991). 1,3-propanediol production by *Escherichia coli* expressing genes from the Klebsiella pneumoniae dha regulon. *Applied and Environmental Microbiology, 57*(12), 3541–3546. doi: 10.1128/AEM.57.12.3541-3546.1991.

Tripathi, J. N., Ntui, V. O., Ron, M., Muiruri, S. K., Britt, A., & Tripathi, L. (2019). CRISPR/Cas9 editing of endogenous banana streak virus in the B genome of Musa spp. overcomes a major challenge in banana breeding. *Communications Biology*, *2*(1), 1–11. doi: 10.1038/s42003-019-0288-7.

Ueta, R., Abe, C., Watanabe, T., Sugano, S. S., Ishihara, R., Ezura, H., … Osakabe, K. (2017). Rapid breeding of parthenocarpic tomato plants using CRISPR/Cas9. *Scientific Reports*, *7*(1), 1–8. doi: 10.1038/s41598-017-00501-4.

Vom Endt, D. (2002). Transcription factors controlling plant secondary metabolism: What regulates the regulators? *Phytochemistry*, *61*(2), 107–114. doi: 10.1016/S0031-9422(02)00185-1.

Wang, F., Wang, C., Liu, P., Lei, C., Hao, W., Gao, Y., … Zhao, K. (2016). Enhanced rice blast resistance by CRISPR/Cas9-targeted mutagenesis of the ERF transcription factor gene OsERF922. *PLoS One*, *11*(4), 1–18. doi: 10.1371/journal.pone.0154027.

Wang, L., Chen, S., Peng, A., Xie, Z., He, Y., & Zou, X. (2019a). CRISPR/Cas9-mediated editing of CsWRKY22 reduces susceptibility to Xanthomonas citri subsp. citri in Wanjincheng orange (Citrus sinensis (L.) Osbeck). *Plant Biotechnology Reports*, *13*(5), 501–510. doi: 10.1007/s11816-019-00556-x.

Wang, H., Wu, Y., Zhang, Y., Yang, J., Fan, W., Zhang, H., … Zhang, P. (2019b). CRISPR/Cas9-based mutagenesis of starch biosynthetic genes in sweet potato (ipomoea Batatas) for the improvement of starch quality. *International Journal of Molecular Sciences*, *20*(19). doi: 10.3390/ijms20194702.

Wen, S., Liu, H., Li, X., Chen, X., Hong, Y., Li, H., … Liang, X. (2018). TALEN-mediated targeted mutagenesis of fatty acid desaturase 2 (FAD2) in peanut (*Arachis hypogaea L.*) promotes the accumulation of oleic acid. *Plant Molecular Biology*, *97*(1–2), 177–185. doi: 10.1007/s11103-018-0731-z.

Xu, J., Kang, B. C., Naing, A. H., Bae, S. J., Kim, J. S., Kim, H., & Kim, C. K. (2020). CRISPR/Cas9-mediated editing of 1-aminocyclopropane-1-carboxylate oxidase1 enhances Petunia flower longevity. *Plant Biotechnology Journal*, *18*(1), 287–297. doi: 10.1111/pbi.13197.

Yin, K., Han, T., Xie, K., Zhao, J., Song, J., & Liu, Y. (2019). Engineer complete resistance to cotton leaf curl Multan virus by the CRISPR/Cas9 system in Nicotiana benthamiana. *Phytopathology Research*, *1*(1), 1–9. doi: 10.1186/s42483-019-0017-7.

Zhang, Y., Bai, Y., Wu, G., Zou, S., Chen, Y., Gao, C., & Tang, D. (2017). Simultaneous modification of three homoeologs of TaEDR1 by genome editing enhances powdery mildew resistance in wheat. *Plant Journal*, *91*(4), 714–724. doi: 10.1111/tpj.13599.

Zhang, P., Du, H., Wang, J., Pu, Y., Yang, C., Yan, R., … Yu, D. (2019a). Multiplex CRISPR/Cas9-mediated metabolic engineering increases soya bean isoflavone content and resistance to soya bean mosaic virus. *Plant Biotechnology Journal*, 1–12. doi: 10.1111/pbi.13302.

Zhang, A., Liu, Y., Wang, F., Li, T., Chen, Z., Kong, D., … Luo, L. (2019b). Enhanced rice salinity tolerance via CRISPR/Cas9-targeted mutagenesis of the OsRR22 gene. *Molecular Breeding*, *39*(3). doi: 10.1007/s11032-019-0954-y.

Zomorrodia, A. R., Suthers, P. F., Ranganathan, S., Maranasa, C. D. (2012). Mathematical optimization applications in metabolic networks. *Metabolic Engineering*, *14*(6), 672–686.

12 Genome Editing in Algae: Perspectives and Challenges

Kavita Rani, Jitendra Kumar, and Taruneet Kaur
ICAR-National Dairy Research Institute

Vikas Kumar
Maharishi Markandeshwar (Deemed to be University)

Om Prakash Gupta
ICAR-Indian Institute of Wheat and Barley Research

CONTENTS

12.1 INTRODUCTION

The food or fodder that has been consumed by humans, large and small animals comes from plants, animals, fungi and microorganisms that have been grown, bred and improved by human beings over thousands of years. Apart from the breeding method, nanofertilizers like zinc oxide are being used for a sustainable approach to enhance plant productivity. Through the centuries, by using different breeding methods, we have managed to select the best-performing individuals and strains of crop plants, animals and fungi with enhanced growth rates and productivities, along with improved nutritional composition and health. As a thumb rule, the improved organisms arose from naturally occurring variations in the genetic make-up of those plants, animals, fungi or natural partial- or full-genome duplications or breeding (crosses) between closely related organisms. For the past 90 years, gamma ray-induced changes in genomes (Eriksson, 2018) have been used as a method to increase the development of improved organisms for application to human beings. Biotechnology-based techniques have become the possible tools to modify the genetic make-up of living cells and organisms since the 1970s.

Engineered crops or animals are produced through the manipulation of the genetic material by the introduction of a specific piece of DNA called a transgene (generally considered to be a gene acquired from a various, unrelated species, like a bacterial gene introduced into a higher plant or even a synthetic gene) to give the organism a new and different property like resistance to a disease or a specific herbicide, better food quality or more crop productivity. In order to achieve such goals, targeted genome modifications are pivotal for genetic analyses and genetic engineering, and compared to various forms of a random integration of cloned genes for over expression, specific alterations of the eukaryotic genome have been considerable challenges for life scientist. Gene targeting (GT) was initially developed in recombinogenic lower eukaryotes by introducing a homologous transgene into the cell, and by utilizing homologous recombination (HR), scientists were able to knock out or replace genes of interest (Hinnen et al., 1978). GT has been successfully demonstrated in animals (Doetschman et al., 1987; Thomas and Capecchi, 1987) and plants (Lee et al., 1990). However, GT in these higher organisms has been very difficult, because they are not recombinogenic (Puchta and Fauser, 2013). The engineered nucleases include zinc finger nucleases (ZFNs), transcription activator-like effector nucleases (TALENs) and clustered regularly interspaced palindromic sequences CRISPR-Cas9 (CRISPR-associated protein 9) (Kim, 2014). These sequence-specific nucleases have enabled researchers to cleave genomic DNA and to obtain mutations of a gene resulting from the faulty repair of the cleaved DNA.

Microalgae have emerged as important platforms for the production of biofuel and other biomolecules, and genetic engineering of microalgae is, thus, one of the fastest-growing biotechnology fields (Jeon et al., 2017). In addition to the over expression of genes of interest, genome editing is essential for the suppression of genes interfering with the production of target molecules. However, progress in this field has been hampered by multiple layers of difficulties inherent to microalgae. This chapter will describe what has been achieved in microalgal genome editing, examine in detail the problems associated with microalgal genome editing and suggest the possible solutions. Genome editing has many applications that have been shown in other organisms, and the possibility of these applications in microalgae will also be evaluated. Because the over expression of genes shares the same technical problems, we will also include a brief description of microalgal transformation technology.

12.2 MICROALGAE TRANSFORMATION

Genetic modifications are designed into the native genome of a given organism through the manipulation technology to achieve the specific phenotype. Since reports of successful transfection of animal cells with isolated viral DNAs within the early 1970s (Nicolson and McAllister, 1972; Graham and van der Eb, 1973), it took over 10 years to attain the transformation of plants through Agrobacteria that are capable of remodelling plants in nature and are considered as a natural genetic engineer (Tepfer, 1984). The transformation of the microalgae *Chlamydomonas* was identified in the 1980s (Debuchy, Purton, and Rochaix, 1989). The long delay in achieving microalgae transformation wasn't simply because of a smaller microalgal community but also because of technical problems inherent to microalgae. The identical delay was seen for genome editing techniques in microalgae. As known lately, successful genome editing should be supported by the solid transformation techniques for an organism (Altpeter et al., 2016).

Chlamydomonas reinhardtii used as transformation of the model microalga was achieved by the transformation techniques used for plants, which were done for the nuclear and chloroplast genomes. Transformation by glass bead agitation was uniquely developed for *C. reinhardtii*, where protoplast cells were agitated vigorously with glass beads (Nelson et al., 1994). These pioneering works led to the development of the transformation toolbox in *Chlamydomonas* and other microalgae. In general, the delivery of genetic material in microalgae is taken into account much less efficient as compared to that of plants. On the other hand, once transformants are obtained, microalgae are mostly single cells and may be maintained somatically, not requiring the tedious (and recalcitrant in some species) regeneration steps in plants (Altpeter et al., 2016). The event of the transformation toolbox in microalgae, excluding *Chlamydomonas*, presented another layer of difficulties mainly because of the paucity of selection markers that are developed for plants and other microbes. This is probably caused by their evolutionary divergence, where their cellular machineries have been differentiated so much that existing herbicides and antibiotics cannot be used for selection purposes. This is often particularly true for the model 'industrial microalgae'. Nannochloropsis is tolerant to most antibiotics and herbicides, which leaves only phleomycin and its derivative *zeocin* for selection by the *ble* gene from *Streptoalloteichus hindustanus* (Kang et al., 2015). Exceptionally, *N. oceanica* CCMP1779 is sensitive to hygromycin that's selectable with *aphVII* (Vieler et al., 2012).

However, algae sharing lineages with plants offer more options for selection, including hygromycin and other herbicides and antibiotics, which are the same as those of plants (Leon and Fernandez, 2007). Microalgal transformation still suffers from a very low transformation efficiency, even compared to plants, which requires drastically improved transformation techniques. Lately, semipermeable membrane removal and PEG-mediated protoplast transformation are reported in *Chlorella* and *Cyanidioschyzon* (Yang et al., 2015). PEG-mediated protoplast transformation in plants achieves higher transformation efficiency without the priority of the host range of Agrobacteria or expensive equipment (Masani et al., 2014). *Chlorella* is gaining popularity in many global consortia of algal biofuel production, and it is, thus, interesting to determine how this technology develops not only in *Chlorella* but also in other microalgae. Bacterial conjugation is routinely employed for the delivery of episomal vectors into microalgal cells, which is claimed to be more efficient than conventional transformation techniques in diatoms (Karas et al., 2015). Optimized DNA transfer techniques may provide solutions for these problems of delivery techniques in microalgae, which can be used for genome editing in microalgae. Development of the transformation toolboxes also provided opportunities for reverse genetic techniques in microalgae, facilitated by the genomic sequencing of *C. reinhardtii* and other microalgae. These boxes including RNA interference (RNAi) and artificial microRNAs (miRNAs) supported the findings that microalgae are capable of RNA silencing by small interfering RNAs, siRNA, miRNAs in *C. reinhardtii* and other microalgae (Vieler et al., 2012; Yamasaki et al., 2016). In addition to the RNA-based knockdown techniques, GT via HR was also introduced in *Chlamydomonas and Chlorella* (Molnár et al., 2007; Dawson et al., 1997) however, theses outcomes were not reproducible possibly due to the non-recombinogenic characteristics of microalgae. Similar difficulties were also observed in nearly all the upper eukaryotes, including animals and plants, because GT was originally developed in recombinogenic yeasts. Difficulties or inefficiencies of the above-mentioned reverse genetic techniques entailed a more efficient and precise modification of DNA, which led to genome editing, also called genome engineering, using engineered nucleases. In particular if we talk about plants, the Agrobacterium-mediated transformation established a strong base to utilize the application of genetic engineering. Obviously some structural differences in the transformation of plants lagged behind animals were due to the presence of the protective cell wall. Initially, some attempts were done with protoplasts, which did not provide enough success. Instead, plant geneticists come up with an alternative technique using the natural transformation machinery of Agrobacteria, which circumvented the removal of the cell wall (Kempken and Jung, 2009; Tepfer, 1984). Moving on from plants to microalgal transformation, the model microalga *Chlamydomonas reinhardtii* was achieved by the transformation techniques used for plants, which were done for nuclear (Fernández et al., 1989; Debuchy et al., 1989) and chloroplast genomes (Boynton et al., 1988). Later, transformation by other delivery measures such as glass bead agitation was specifically developed for *C. reinhardtii*, where protoplast cells were vigorously agitated with glass beads (Nelson et al., 1994). The continuous research progress from the latest work in this area led to the development of the transformation toolbox in *Chlamydomonas* and other microalgae. However, the manipulation of genome in microalgae is still considered much less efficient compared to that of plants. On the contrary, once transformants are obtained, microalgae are mostly single cells and can be maintained somatically, not requiring the tedious (and recalcitrant in some species) regeneration steps in plants (Altpeter et al., 2016). Despite this ease of experimentation, attempts of genome editing in microalgae had suffered from a limited success with only a handful of results. Hence, it would be interesting to explore and evolve different ways of working of these nucleases in microalgae.

In this chapter, we attempted to explore the gene editing techniques used in microalgae along with associated challenges and future prospects. We will also cover the significance of gene-edited microalgae in a wide scope of industrial applications such as pharmaceuticals, nutraceuticals, biofuels etc.

12.3 CRISPR-Cas9, TALENs AND ZFNs TECHNOLOGY FOR GENOME EDITING

In recent years, the emergence of highly versatile genome editing technologies has enabled the scientists with an ability to rapidly and economically introduce sequence-specific alterations into the genetic material of a broad spectrum of cell types and organisms. The most commonly used genome editing technologies are (1) CRISPR- Cas9, (2) TALENs, (3) ZFNs and (4) homing endonuclease or meganucleases. These molecular scissors make a targeted DNA double-stranded break that is repaired by either of the two basic mechanisms: non-homologous end-joining, which causes a change in gene sequences by inducing mutations at the disrupted gene site, or HR, which induces a gene insertion or gene replacement using exogenous DNA templates with homology to the targeted site. In particular, the ease and continuous successful outcomes associated with CRISPR-Cas9 and TALENs make them more suitable to identify new genomic sequences in different organisms, which ultimately have driven a revolution in genome editing. This revolution further speeds up the scientific breakthroughs and discoveries in disciplines as diverse as synthetic biology, neuroscience, human gene therapy, disease modelling, environment management, drug discovery, agricultural and food sciences. ZFNs are chimeric proteins consisting of a synthetic zinc finger-based DNA-binding domain and a DNA cleavage domain. ZFNs, which are fusions between a custom-designed Cys2-His2 zinc finger protein and the cleavage domain of the FokI restriction endonuclease (Kim et al., 1996), were the first genomic editing tool to achieve widespread use (Porteus and Carroll, 2005; Urnov et al., 2010). ZFNs function as dimeric proteins, with each monomer identifying a specific 'half site' sequence typically 9–18 base pairs (bps) of DNA via the zinc finger DNA-binding domain. Dimerization of the ZFN proteins is initiated by the FokI cleavage domain, which cleaves DNA within a 5- to 7-bp spacer sequence that makes two flanking zinc finger binding sites (Smith et al., 2000). Each ZFN is particularly composed of three or four zinc finger domains, with each individual domain having 30 amino acid residues that are constituted in a bba motif (Pavletich and Pabo, 1991). The residues that facilitate DNA recognition are located within the a-helical domain and typically interact with 3 bps of DNA, with an occasional overlap from an adjacent domain (Wolfe et al., 2000). TALE proteins are bacterial effectors and used efficiently for genomic editing in unicellular as well as multicellular organisms widely. In 2009, the nucleotide sequence used by TALE proteins to recognize DNA was discovered (Boch et al., 2009; Moscou and Bogdanove 2009). With a short interval of time, this discovery facilitated the synthesis of custom TALENs capable of altering nearly any gene. Similar like ZFNs, TALENs are specific in form and function, consisting of an amino-terminal TALE DNA-binding domain attached to a carboxy-terminal FokI cleavage domain (Christian et al., 2010; Miller et al., 2011). Also like ZFNs, dimerization of

TALEN proteins is facilitated by the FokI cleavage domain, which cleaves within a 12- to 19-bp spacer sequence that separates each TALE binding site (Miller et al., 2011). TALEs are typically organized to recognize between 12 and 20 bps of DNA, with more nucleotide bases particularly resulting in higher genome editing specificity (Guilinger et al., 2014). The TALE binding domain includes a series of repeat domains, each 34 bases in length. Each repeat communicates DNA via the amino acid residues at positions 12 and 13, called as the repeat variable diresidues (Boch et al., 2009; Moscou and Bogdanove 2009). Unlike zinc fingers, which recognize DNA triplets, each TALE repeat recognizes only a single base pair, associated with no target site overlap from nearby domains (Deng et al., 2012; Mak et al., 2012).

12.3.1 Homing Endonuclease

Homing endonuclease, also known as meganuclease, is defined as the final member in the series of targeted nuclease family. These enzymes have been reviewed at length elsewhere (Stoddard, 2014) but, briefly, members of the LAGLIDADG family of endonucleases, so named for the conserved amino acid motif located within these enzymes that interacts with DNA, are a collection of naturally existing enzymes that identify and cut long DNA sequences (14–40 bps). These enzymes build extensive sequence-specific associations with their DNA substrate (Stoddard, 2011) and thus typically show a unique specificity. The binding and cleavage domains in homing endonuclease are not modular, unlike ZFNs and TALENs. This overlap in structure and functionality makes their repurposing challenging, thus leading to their less suitability for more routine applications of genome editing. Recent studies report about megaTALs fusions of a rare-cleaving homing endonuclease to a TALE-binding domain, which ultimately induce highly specific gene alterations (Boissel et al., 2014; Lin et al., 2015). These endonucleases increase the possibility of integration of antitumor and anti-HIV factors into the human CCR5 gene in both primary T cells and hematopoietic stem/progenitor cells (Sather et al., 2015), and cause the disruption of endogenous T-cell receptor elements in T cells (Osborn et al., 2016), indicating their tremendous potential for improving and enhancing immunotherapy.

12.3.2 CRISPR-Cas9

The CRISPR-Cas9 system, which evolves as an adaptive immunity system in bacteria (Marraffini and Sontheimer 2010), is the most recent addition to the genome editing toolbox. In bacteria, the type-II CRISPR system provides defence against DNA from invaders like viruses and plasmids via RNA-guided DNA cleavage by Cas9 proteins (Wiedenheft et al., 2012). Foreign DNA in the form of short segments are intercalated within the CRISPR locus and transcribed into CRISPR RNA (crRNA), which then join to trans-activating crRNA (tracrRNA) to direct a sequence-specific digestion of pathogenic DNA by the Cas9 protein (Jinek et al., 2012). Jinek et al. (2012) find that target recognition by the Cas9 protein only needs a seed sequence integrated within the crRNA and a conserved protospacer-adjacent motif (PAM) upstream of the crRNA binding site. Progressive scientific efforts with time are simplified for genome engineering (Cho et al., 2013; Cong et al., 2013; Jinek et al., 2013; Mali et al., 2013) and now contain only the Cas9 nuclease and a single guide RNA (gRNA) consisting of the essential crRNA and tracrRNA parts. Because the target site recognition is done completely by the gRNA, CRISPR-Cas9 has evolved as the most flexible, efficient and user-friendly tool for genome editing, thus deleting the need for engineering new proteins to find out each new target site. The only major problem for Cas9 target site recognition is the presence of PAM motif, which is recognized by the Cas9 proteins and is mandatory for DNA cleavage that is located just downstream of the gRNA target site.

Several scientific reports available now have described in detail about the structural basis of DNA recognition by Cas9, explaining that the heteroduplex generated by the gRNA and its complementary strand of DNA is located in a positively charged groove between the two nuclease domains (RuvC and HNH) within the Cas9 protein (Nishimasu et al., 2014) and that PAM recognition is carried out by an arginine-rich motif located in Cas9 (Anders et al., 2014). Doudna and colleagues have since proposed that a change in DNA strand position leads to a structural reorganization within the Cas9 protein that carries the non-target DNA strand into the RuvC active site, which then keeps the HNH domain near the target DNA (Jiang et al., 2016), enabling Cas9-directed cleavage of both DNA strands.

12.3.3 DNA Delivery Is a Key Step of Genome Editing

DNA introduction into the cell is the crucial step for targeted and efficient gene editing. The transformation efficiency depends upon DNA delivery method. Various methods are available for performing algal transformation, like Agrobacterium-mediated co-culture electroporation, biolistic microparticle bombardment and shaking with glass beads (Ahmad et al., 2015; Gimpel et al., 2015).

12.3.3.1 Electroporation

The pores were created in the cell membrane through electric pulses that facilitate the delivery of exogenous DNA into the cell. There are many factors involved in a better and efficient DNA delivery like number of cell, DNA quantity, intensity of electric pulses, duration and frequency of electric pulses provided. The process is assisted through electroporation buffer (HEPES or Tris, calcium chloride) along with electrolytes sucrose, mannitol and sorbitol (Coll, 2006). Electroporation conditions have been used for efficient DNA delivery into *Chlamydomonas sp.* (Shin et al., 2016), *Nannochloropsis sp.* (Wang et al., 2016) and *P. tricornutum* (Niu et al., 2013).

12.3.3.2 Biolistic Method

It is a gene gun-based DNA delivery method, which has been optimized for chloroplast transformation in algae and nuclear genome transformation. This method of transformation was achieved in model microalga *C. reinhardtii* (Boynton et al., 1988), *Volvox carteri* and *P. tricornutum* (Coll, 2006). The plasmid DNA is deposited on the tungsten or gold microparticles and accelerated toward cells incubated in the centre of the agar petri dish using standardized parameters like cell density, nature and dimensions of microparticles, the speed with which microparticles are accelerating, distance to the microalgae and the DNA content.

12.3.3.3 Agitation with Glass Beads

In this method, microalgal cell was agitated with glass beads and the foreign DNA. PEG acts as a mediator for the delivery of DNA inside the cells. The cells are vortexed to break up the cell wall or the use of cell wall-degrading enzymes like cellulolysin, hemicelluloses, macerase and cellulose to facilitate the cell wall disruption. The efficiency and accuracy of this method depend upon various factors such as microalgae cell size, presence or absence of cell wall, cell wall composition and strength along with duration and speed of agitation and concentrations of enzymes and PEG (Coll, 2006).

12.3.3.4 Agrobacterium-Mediated Transformation

Gene transformation into the plants is mediated by this method. Integration of DNA into the nucleus at a large scale is achieved by Agrobacterium. In this process, natural Agrobacterium infection having T-DNA oncogenes and opine catabolism-related genes is replaced with genes of interest in the targeted cell (Tzfira and Citovsky, 2006). The limitation of this method is less effective in microalgae C. *reinhardtii* (Kumar et al., 2004), *Haematococcus pluvialis* (Kathiresan et al., 2009), *Chlorella vulgaris* (San Cha et al., 2012) and *Parachlorella kessleri* (Rathod et al., 2013). Glass bead-based agitation is the most commonly used method for DNA delivery due to simple and handiness, and does not need any sophisticated instrumentation. Electroporation and biolistic particle bombardment both require sophisticated instrumentation, and it uses brute force for the delivery of DNA in the cell, which in turn causes cell damage. Despite possessing some limitations of cell damage and sophisticated instrumentation, it is still opted as the preferred choice over the agitation method. Microalgae having thick and tough walls are to be transformed DNA by this method (Coll, 2006).

12.3.4 Selectable Marker

For the selection of transformed cells from the wild type, a selectable marker (whether endogenous or exogenous antibiotic-expressing gene) must be required. The resistance confirmation was done by inhibition, knockout of an endogenous gene and heterologous expression of exogenous genes. Changes that were observed in the expression profile of endogenous genes certainly lead to the production of auxotrophic mutants with increased sensitivity/ resistance towards a specific metabolite (Jiang et al., 2014; Shin et al., 2016). The commonly used antibiotic selection markers are hygromycin, kanamycin, zeocin and nourseothricin, among others (Bashir et al., 2016).

12.4 APPLICATION OF CRISPR-Cas9 IN MICROALGAE

CRISPR-Cas9 tools have been widely used for genome editing purposes from animals to plants. CRISPR-Cas9 generates more efficient and more applied outcomes. In recent years, researcher uses CRISPR-Cas9 editing tool in algal research area. Previous studies showed the functionality and working of CRISPR by targeting the genes; mutations in genes gave visible phenotypic changes or auxotrophs. The optimized protocols were used in subsequent studies to create mutants with better molecular characteristics. Continuous research efforts make a number of different Cas9 expression modes, and different combinations of selection systems have been upgraded. This section will focus specifically on the use of CRISPR-Cas9 system genome editing studies carried out in microalgae. The different transformation and editing tools for targeted genes in microalgae and macroalgae are enlisted in Table 12.1.

12.4.1 Genome Editing in *C. reinhardtii* by Cas9 Nuclease

C. reinhardtii have been used as model algae for genome editing. Cas9 and sgRNA were expressed transiently in the cells to mutate exogenous GFP, hygromycin and Gluc (Gaussia luciferase). The induced mutations were studied after 24 hours' post-electroporation in cells amplifying the target genes and checking mutations by sequencing and loss of restriction cutting sites. Endogenous gene peptidyl-prolyl cis-trans isomerase (FKB12) has 168 genome engineering via CRISPR-Cas9 system for targeting to induce mutations leading to the production of mutants algae, which have the ability to tolerance of rapamycin. Recovery of only one mutant out of 16 transformations indicates that the constitutive expression of Cas9 causes toxicity to cells (Jiang et al., 2014). Cas9 and target-specific sgRNA were joined together and introduced into the cells along with hygromycin selection vector, and the capability of sgRNAs to cut the target DNA was checked by *in vitro* cleavage assay. The concentration and ratio of Cas9 and sgRNA is playing a significant role for efficient mutagenesis, and 21 tryptophan auxotrophs with mutations at the target site were obtained. Knockout of phospholipase A2 gene increased by 64.25% lipid, which resulted in the wild-type organism (Shin et al., 2019). Cas9 RNPs were used to create single and double knockouts of CpFTSY (chloroplast signal recognition particle receptor protein) and ZEP (zeaxanthin epoxidase) genes. The mutants were generated by editing Chlamydomonas cells with activated RNP complex possessing 200 mg of Cas9 nuclease and 140 mg of *in vitro* transcribed of sgRNA. Unlike previous reports, it was a DNA free editing as there was not involvement of any selection marker vector. Indel mutation frequencies of 0.56%, 0.46% and 1.1% were found for CpFTSY, ZEP loci and double mutants, respectively (Baek et al., 2016).

TABLE 12.1
Genome Editing in Microalgae and Macroalgae

Sr. No.	Microalgae/ Macroalgae	Nucleases	Method of Delivery	Target Gene	Outcome	References
			Microalgae			
1	*C. reinhardtii* strain CC4350	ZFN	Glass beads	*COP3*	Targeted gene knockout	Sizova et al. (2013)
2	*P. tricornutum* CCMP2561	TALEN	Bombardment	*UGPase*	Increase in triacylglycerol accumulation	Daboussi et al. (2014)
3	*C. reinhardtii* CC503	CRISPR/Cas9	Electroporation	*FKB12*	Targeted gene modification was achieved	Jiang et al. (2014)
4	*P. tricornutum* CCMP2561	CRISPR/Cas9	Bombardment	*CpSRP54*	Stable targeted gene mutation	Nymark et al. (2016)
5	*C. reinhardtii* CC124	CRISPR/Cas9	Electroporation	*MAA7, CpSRP43* and *ChlM*	Targeted gene knockout	Shin et al. (2016)
6	*C. reinhardtii* CC4349	CRISPR/Cas9	Electroporation	*ZEP and CpFTSY*	Production of two gene-knockout mutants	Baek et al. (2016)
7	*C. reinhardtii* CC400	CRISPR/dCas9	Glass beads	*PEPC1* and *RFP*	Lipid production was enhanced	Kao and Ng (2017)
8	*N. oceanica* IMET1	CRISPR/Cas9	Electroporation	*Nitrate reductase gene*	Targeted gene knockout in nannochloropsis	Tong et al. (2015)
9	*N. gaditana* CCMP1894	Cas9 editor lines	Electroporation	*ZnCys TF BSD*	Knockout and attenuation of ZnCys in nannochloropsis	Ajjawi et al. (2017)
			Macroalgae			
1	*Synechococcus elongatus* UTEX 2973	CRISPR/Cas9	Conjugation	*nblA*	Rapid and efficient genetic engineering of Synechococcus strain was done	Wendt et al. (2016)
2	*Synechococcus elongates* PCC 7942	CRISPR/Cas9	Conjugation	*glgc*	Increase in succinate production	Li et al. (2016)
3	*Synechococcus* UTEX 2973	CRISPR/Cpf1	Conjugation	*psbA1, nblA*	Genome editing using CRISPR/Cpf1-based nucleases	Ungerer and Pakrasi (2016)
4	*Synechocystis* 6803 Anabaena 7120	CRISPR/Cpf1	Conjugation	*nblA, nifH*	Genome editing using CRISPR/Cpf1-based nucleases	Ungerer and Pakrasi (2016)

12.4.2 CRISPR-Cas9 Mechanism in Microalgae

A typical CRISPR-Cas9 mechanism for understanding is presented in a simple form (Figure 12.1). They can be divided into the following steps along with the showing how genomic editing with different endonuclease leads to a better utilization of these small unicellular plants.

12.4.2.1 Selection of Target Gene

Target gene can be picked up from the published literature, while the real-time gene expression experiments and data from the next-generation sources.

12.4.2.2 Selection of Cas9 Expression System

It depends upon the experiment which of the variants of Cas9 should be chosen. For example, the production of knockout Cas9 nuclease can be a modular selection. Cas9 nickase creates a nick in the DNA. dCas9 is a favoured choice for gene repression studies. Cpf1 generates staggered cuts, locates a T-rich PAM (targets having T-rich sequences) and prompts HDR for a repair template.

12.4.2.3 Single Guide RNA (sgRNA) Suitability to Reduce Off-Target Effects

Guide RNAs (gRNA) should be target-specific without or very few associated off-target site. Sequence prediction was done by software that is available both online and offline for efficiency and even off-targets for the input sequence. Commonly used and accessible software packages are RGEN software suites having tools such as Cas-OFFinder (Bae et al., 2014a,b), Cas-Designer (Park et al., 2015), Cas-Analyzer (Park et al., 2017), CHOPCHOP (Montague et al., 2014; Labun et al., 2016), sgRNA design tool dBroad Institute (Doench et al., 2016), CRISPR-P (Lei et al., 2014) and CCTop-CRISPR/Cas9 target online predictor (Stemmer et al., 2015).

12.4.2.4 Transformation of Target Cells with Cas9

Transformation of target cells with Cas9 was achieved by the introduction of Cas9 and components into the cells. In the case of RNPs, the concentration of Cas9 endonuclease and sgRNA must be essentially optimized to give the sufficient mutagenesis frequency. Selection marker must be chosen very carefully to check the occurrence of transformation in cells. As a test, the preliminary trials should be carried out by testing the mutagenesis efficiency of Cas9 against endogenous selectable phenotypes.

12.4.2.5 Screening of Mutants Post-Transformation

Large number of methods are available to screen the mutants. The mutants can be screened effectively using band shift assay, surveyor assay, T_7 endonuclease I assay, Sanger sequencing, PCR/RE assay, HRM analysis, targeted genome sequencing and NGS.

12.4.2.6 Characterization of Mutants

To check out the effect of mutation in the mutants, it is essential to study its physiology. It basically involves the analysis of mutants to evaluate the effect of mutations in their physiological attributes like growth rate, chlorophyll content, accumulation of lipids, photosynthesis efficiency, metabolite types and their generation rate, starch and other metabolites, omics, and metabolic flux studies.

Figure 12.1 depicts the well-defined role of endonuclease in genome editing in microalgae.

12.5 APPLICATION OF GENE-EDITED MICROALGAE

The application of genetic metabolic engineering tools in microalgae has the potential to create important sources of metabolites, pharmaceuticals and nutraceuticals produced, renewable fuel with food production in a sustainable industry, sustainable agriculture and ecological economics, which are summarized in Table 12.2.

12.6 FUTURE PERSPECTIVES AND CHALLENGES

Efficient and accurate genome editing tools are necessary for specific mutagenesis or gene editing without producing off-targeting site, which has not been well known in the field of microalgae. Selection of Cas9 expression system is a vital factor for estimating the genome editing efficiency. Vector-based expression systems are preferred to choose as a better tool to carry out genome editing studies, providing higher rates of expression of Cas9 and sgRNA. A major disadvantage is the occurrence of Cas9-induced toxicity (Jiang et al., 2014), which increases the possibility of offsite

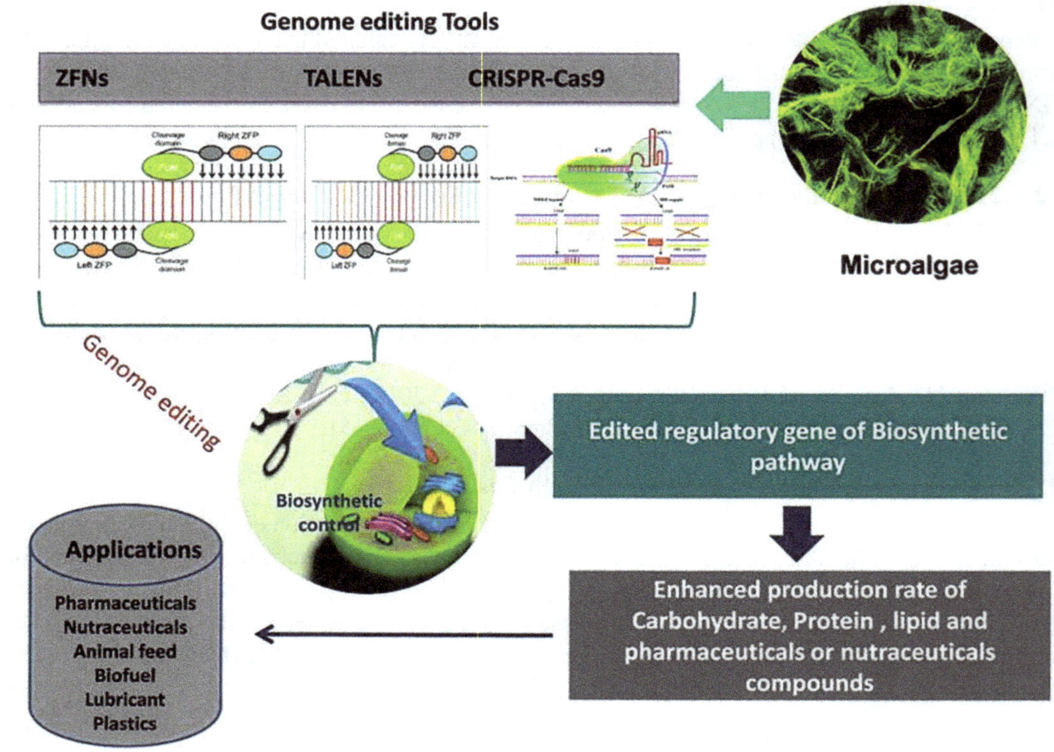

FIGURE 12.1 Diagram depicting the well-defined role of endonuclease in genome editing in microalgae.

TABLE 12.2
Application of Genome Editing in Microalgae

Sr. No.	Microalgae	Genetic Tools	Gene-Edited Pathway/ Bioactive Compounds	Enhanced Production	References
1.	*C. reinhardtii*	CRISPR/Cas9 system for nuclear genetic manipulation	Enzyme β-carotene ketolase of *H. pluvialis* has been expressed in *C. reinhardtii*	Carotenoids keto-lutein and adonixanthin	Leon et al. (2007)
2.	*P. tricornutum*	CRISPR/Cas9, TALE-based system	Stable gene knockouts	Industrial applications/' iLB1027_lipid' model	Khraiwesh et al. (2015)
3.	*Chlorella spp*	Genes for active enzymes cloned	Lipids, human growth hormone	Biomass, biofuel and value-added chemical production	Gao et al. (2014)
4.	*Chlorella minutissima*	Cas9 system for knockdown	Acyltransferases	Lipid contents have twofold increase	Hsieh et al. (2012)
5.	*Nannochloropsis gaditana*	Cas9 system (knockdown of a single transcription regulator ZnCys)	Lipid biosynthesis pathway	Enhanced lipid yield	Ajjawi et al. (2017)

mutations and strict GM regulations coming due to the integration of Cas9 genes into the cellular genomes. Instances of Cas9-induced toxicity may be reduced by using inducible promoters and terminators, managing Cas9 expression to a particular lifecycle stage and more effectively decreasing off-targets (Stukenberg et al., 2018; Verruto et al., 2018). CRISPR-Cas9 and allied genome editing systems have revolutionized the entire world. Its easy adaptability and versatile nature, efficiency and associated applications have significantly contributed to its meteoric rise as a favoured genomic editing tool. Though, its success rate as a strain improvement technique in microalgae is not much fast, still the developments point toward a promising trend. To fullfil this challenging task, more research is required.

12.7 CONCLUSIONS

Genome editing is required for creating mutations of target genes enabled by recombinant nucleases with sequence specificity. The latest endonucleases associated with the CRISPR systems are far better than the previous versions in terms of outcome generated, accuracy and efficiency. This improved version of CRISPR technology can be used in creating mutations, editing of genes, and targeted integration of overexpression cassettes. Application can be further extended for many other purposes, including attenuation of gene expression, elimination of transgenic markers, etc., and many more to be in the list. The microalgal community is catching up with this advanced and exciting technology but not is able to come up to a required level of improvement as that of the mainstream technical developments in animals and plants. Hence, it's essentially urgent to first solve the fundamental issues in microalgae, which is the inefficient transfer of genetic materials into the cell. In spite of this big issue, many more new researches have been reported with CRISPR compared to the previously available techniques, which surely will set a new age of genome editing in microalgae.

REFERENCES

Ahmad, Irshad, Anil K. Sharma, Henry Daniell, and Shashi Kumar. "Altered lipid composition and enhanced lipid production in green microalga by introduction of brassica diacylglycerol acyltransferase 2." *Plant Biotechnology Journal* 13, no. 4 (2015): 540–550.

Ajjawi, Imad, John Verruto, Moena Aqui, Leah B. Soriaga, Jennifer Coppersmith, Kathleen Kwok, Luke Peach, et al. "Lipid production in Nannochloropsis gaditana is doubled by decreasing expression of a single transcriptional regulator." *Nature Biotechnology* 35, no. 7 (2017): 647–652.

Altpeter, Fredy, Nathan M. Springer, Laura E. Bartley, Ann E. Blechl, Thomas P. Brutnell, Vitaly Citovsky, Liza J. Conrad, et al. "Advancing crop transformation in the era of genome editing." *The Plant Cell* 28, no. 7 (2016): 1510–1520.

Altpeter, Fredy, Nathan M. Springer, Laura E. Bartley, Ann E. Blechl, Thomas P. Brutnell, Vitaly Citovsky, Liza J. Conrad, et al. "Advancing crop transformation in the era of genome editing." *The Plant Cell* 28, no. 7 (2016): 1510–1520.

Anders, Carolin, Ole Niewoehner, Alessia Duerst, and Martin Jinek. "Structural basis of PAM-dependent target DNA recognition by the Cas9 endonuclease." *Nature* 513, no. 7519 (2014): 569–573.

Bae, Sangsu, Jiyeon Kweon, Heon Seok Kim, and Jin-Soo Kim. "Microhomology-based choice of Cas9 nuclease target sites." *Nature Methods* 11, no. 7 (2014a): 705–706.

Bae, Sangsu, Jeongbin Park, and Jin-Soo Kim. "Cas-OFFinder: A fast and versatile algorithm that searches for potential off-target sites of Cas9 RNA-guided endonucleases." *Bioinformatics* 30, no. 10 (2014b): 1473–1475.

Baek, Kwangryul, Duk Hyoung Kim, Jooyeon Jeong, Sang Jun Sim, Anastasios Melis, Jin-Soo Kim, EonSeon Jin, and Sangsu Bae. "DNA-free two-gene knockout in Chlamydomonas reinhardtii via CRISPR-Cas9 ribonucleoproteins." *Scientific Reports* 6 (2016): 30620.

Bashir, Khawaja Muhammad Imran, Moo-Sang Kim, Ulf Stahl, and Man-Gi Cho. "Microalgae engineering toolbox: Selectable and screenable markers." *Biotechnology and Bioprocess Engineering* 21, no. 2 (2016): 224–235.

Boch, Jens, Heidi Scholze, Sebastian Schornack, Angelika Landgraf, Simone Hahn, Sabine Kay, Thomas Lahaye, Anja Nickstadt, and Ulla Bonas. "Breaking the code of DNA binding specificity of TAL-type III effectors." *Science* 326, no. 5959 (2009): 1509–1512.

Boissel, Sandrine, Jordan Jarjour, Alexander Astrakhan, Andrew Adey, Agnes Gouble, Philippe Duchateau, Jay Shendure, et al. "megaTALs: A rare-cleaving nuclease architecture for therapeutic genome engineering." *Nucleic Acids Research* 42, no. 4 (2014): 2591–2601.

Boynton, John E., Nicholas W. Gillham, Elizabeth H. Harris, Jonathan P. Hosler, Anita M. Johnson, Allan R. Jones, Barbara L. Randolph-Anderson, Dominique Robertson, Ted M. Klein, and Katherine B. Shark. "Chloroplast transformation in Chlamydomonas with high velocity microprojectiles." *Science* 240, no. 4858 (1988): 1534–1538.

Cho, Seung Woo, Sojung Kim, Jong Min Kim, and Jin-Soo Kim. "Targeted genome engineering in human cells with the Cas9 RNA-guided endonuclease." *Nature Biotechnology* 31, no. 3 (2013): 230–232.

Christian, Michelle, Tomas Cermak, Erin L. Doyle, Clarice Schmidt, Feng Zhang, Aaron Hummel, Adam J. Bogdanove, and Daniel F. Voytas. "Targeting DNA double-strand breaks with TAL effector nucleases." *Genetics* 186, no. 2 (2010): 757–761.

Coll, J. M. "Methodologies for transferring DNA into eukaryotic microalgae." *Spanish Journal of Agricultural Research* 4 (2006): 316–330.

Cong, Le, F. Ann Ran, David Cox, Shuailiang Lin, Robert Barretto, Naomi Habib, Patrick D. Hsu, et al. "Multiplex genome engineering using CRISPR/Cas systems." *Science* 339, no. 6121 (2013): 819–823.

Daboussi, Fayza, Sophie Leduc, Alan Maréchal, Gwendoline Dubois, Valérie Guyot, Christophe Perez-Michaut, Alberto Amato, et al. "Genome engineering empowers the diatom Phaeodactylum tricornutum for biotechnology." *Nature Communications* 5, no. 1 (2014): 1–7.

Dawson, Hana N., Richard Burlingame, and Andrew C. Cannons. "Stable transformation of Chlorella: Rescue of nitrate reductase-deficient mutants with the nitrate reductase gene." *Current Microbiology* 35, no. 6 (1997): 356–362.

Debuchy, Robert, Saul Purton, and Jean-David Rochaix. "The argininosuccinate lyase gene of Chlamydomonas reinhardtii: An important tool for nuclear transformation and for correlating the genetic and molecular maps of the ARG7 locus." *The EMBO Journal* 8, no. 10 (1989): 2803–2809.

Deng, D., Yan, C., Pan, X., Mahfouz, M., Wang, J., Zhu, J.K., Shi, Y. and Yan, N., 2012. Structural basis for sequence-specific recognition of DNA by TAL effectors. *Science*, 335, no. 6069: 720–723.

Doench, John G., Nicolo Fusi, Meagan Sullender, Mudra Hegde, Emma W. Vaimberg, Katherine F. Donovan, Ian Smith, et al. "Optimized sgRNA design to maximize activity and minimize off-target effects of CRISPR-Cas9." *Nature Biotechnology* 34, no. 2 (2016): 184–191.

Doetschman, Thomas, Ronald G. Gregg, Nobuyo Maeda, Martin L. Hooper, David W. Melton, Simon Thompson, and Oliver Smithies. "Targetted correction of a mutant HPRT gene in mouse embryonic stem cells." *Nature* 330, no. 6148 (1987): 576–578.

Eriksson, Dennis. "Recovering the original intentions of risk assessment and management of genetically modified organisms in the European Union." *Frontiers in Bioengineering and Biotechnology* 6 (2018): 52.

Fernández, Emilio, Rogene Schnell, Laura P.W. Ranum, Susanne C. Hussey, Carolyn D. Silflow, and Paul A. Lefebvre. "Isolation and characterization of the nitrate reductase structural gene of Chlamydomonas reinhardtii." *Proceedings of the National Academy of Sciences of the United States of America* 86, no. 17 (1989): 6449–6453.

Gao, Chunfang, Yun Wang, Yue Shen, Dong Yan, Xi He, Junbiao Dai, and Qingyu Wu. "Oil accumulation mechanisms of the oleaginous microalga Chlorella protothecoides revealed through its genome, transcriptomes, and proteomes." *BMC Genomics* 15, no. 1 (2014): 582.

Gimpel, Javier A., Vitalia Henríquez, and Stephen P. Mayfield. "In metabolic engineering of eukaryotic microalgae: Potential and challenges come with great diversity." *Frontiers in Microbiology* 6 (2015): 1376.

Graham, Frank L., and Alex J. van der Eb. "A new technique for the assay of infectivity of human adenovirus 5 DNA." *Virology* 52, no. 2 (1973): 456–467.

Guilinger, John P., Vikram Pattanayak, Deepak Reyon, Shengdar Q. Tsai, Jeffry D. Sander, J. Keith Joung, and David R. Liu. "Broad specificity profiling of TALENs results in engineered nucleases with improved DNA-cleavage specificity." *Nature Methods* 11, no. 4 (2014): 429–435.

Hinnen, Albert, James B. Hicks, and Gerald R. Fink. "Transformation of yeast." *Proceedings of the National Academy of Sciences of the United States of America* 75, no. 4 (1978): 1929–1933.

Hsieh, Hsin-Ju, Chia-Hung Su, and Liang-Jung Chien. "Accumulation of lipid production in Chlorella minutissima by triacylglycerol biosynthesis-related genes cloned from Saccharomyces cerevisiae and Yarrowia lipolytica." *Journal of Microbiology* 50, no. 3 (2012): 526–534.

Jeon, Seungjib, Byeong-ryool Jeong, and Yong Keun Chang. Chemicals and fuels from microalgae: Consequences of microbial interactions with hydrocarbons, oils, and lipids: Production of fuels and chemicals. In Sang Yup Lee, ed. *Handbook of Hydrocarbon and Lipid Microbiology*. Cham: Springer International Publishing (2017) pp. 1–21.

Jiang, Wenzhi, Andrew J. Brueggeman, Kempton M. Horken, Thomas M. Plucinak, and Donald P. Weeks. "Successful transient expression of Cas9 and single guide RNA genes in Chlamydomonas reinhardtii." *Eukaryotic Cell* 13, no. 11 (2014): 1465–1469.

Jiang, Fuguo, David W. Taylor, Janice S. Chen, Jack E. Kornfeld, Kaihong Zhou, Aubri J. Thompson, Eva Nogales, and Jennifer A. Doudna. "Structures of a CRISPR-Cas9 R-loop complex primed for DNA cleavage." *Science* 351, no. 6275 (2016): 867–871.

Jinek, Martin, Krzysztof Chylinski, Ines Fonfara, Michael Hauer, Jennifer A. Doudna, and Emmanuelle Charpentier. "A programmable dual-RNA–guided DNA endonuclease in adaptive bacterial immunity." *Science* 337, no. 6096 (2012): 816–821.

Jinek, Martin, Alexandra East, Aaron Cheng, Steven Lin, Enbo Ma, and Jennifer Doudna. "RNA-programmed genome editing in human cells." *eLife* 2 (2013): e00471.

Kang, Nam Kyu, Seungjib Jeon, Sohee Kwon, Hyun Gi Koh, Sung-Eun Shin, Bongsoo Lee, Gang-Guk Choi, Ji-Won Yang, Byeong-ryool Jeong, and Yong Keun Chang. "Effects of overexpression of a bHLH transcription factor on biomass and lipid production in Nannochloropsis salina." *Biotechnology for Biofuels* 8, no. 1 (2015): 1–13.

Kao, Pei-Hsun, and I.-Son Ng. "CRISPRi mediated phosphoenolpyruvate carboxylase regulation to enhance the production of lipid in Chlamydomonas reinhardtii." *Bioresource Technology* 245 (2017): 1527–1537.

Karas, Bogumil J., Rachel E. Diner, Stephane C. Lefebvre, Jeff McQuaid, Alex PR Phillips, Chari M. Noddings, John K. Brunson, et al. "Designer diatom episomes delivered by bacterial conjugation." *Nature Communications* 6, no. 1 (2015): 1–10.

Kathiresan, S., A. Chandrashekar, G. A. Ravishankar, and R. Sarada. "Agrobacterium-mediated transformation in the green alga Haematococcus pluvialis (Chlorophyceae, Volvocales) 1." *Journal of Phycology* 45, no. 3 (2009): 642–649.

Kempken, Frank, and Christian Jung, eds. *Genetic Modification of Plants: Agriculture, Horticulture and Forestry*, Vol. 64. Berlin, Germany: Springer Science & Business Media (2009).

Khraiwesh, Basel, Kenan Jijakli, Joseph Swift, Amphun Chaiboonchoe, Rasha Abdrabu, Pei-Wen Chao, Laising Yen, and Kourosh Salehi-Ashtiani. "Prospective applications of synthetic biology for algal bioproduct optimization." In Navid R. Moheimani, Mark P. McHenry, Karne de Boer, Parisa A. Bahri, eds. *Biomass and Biofuels from Microalgae*. Cham: Springer (2015) 137–154.

Kim, Hyongbum, and Jin-Soo Kim. "A guide to genome engineering with programmable nucleases." *Nature Reviews Genetics* 15, no. 5 (2014): 321–334.

Kim, Yang-Gyun, Jooyeun Cha, and Srinivasan Chandrasegaran. "Hybrid restriction enzymes: Zinc finger fusions to FokI cleavage domain." *Proceedings of the National Academy of Sciences of the United States of America* 93, no. 3 (1996): 1156–1160.

Kumar, Sasidharanpillai Vinod, Rachel William Misquitta, Vanga Siva Reddy, Basuthkar Jagadeeswar Rao, and Manchikatla Venkat Rajam. "Genetic transformation of the green alga: Chlamydomonas reinhardtii by Agrobacterium tumefaciens." *Plant Science* 166, no. 3 (2004): 731–738.

Labun, Kornel, Tessa G. Montague, James A. Gagnon, Summer B. Thyme, and Eivind Valen. "CHOPCHOP v2: A web tool for the next generation of CRISPR genome engineering." *Nucleic Acids Research* 44, no. W1 (2016): W272–W276.

Lee, Kathleen Y., Peter Lund, Keith Lowe, and Pamela Dunsmuir. "Homologous recombination in plant cells after Agrobacterium-mediated transformation." *The Plant Cell* 2, no. 5 (1990): 415–425.

Lei, Yang, Li Lu, Hai-Yang Liu, Sen Li, Feng Xing, and Ling-Ling Chen. "CRISPR-P: A web tool for synthetic single-guide RNA design of CRISPR-system in plants." *Molecular Plant* 7, no. 9 (2014): 1494–1496.

Leon, Rosa, and Emilio Fernandez. "Nuclear transformation of eukaryotic microalgae." In León Rosa, Aurora Galván, Fernández Emilio, eds. *Transgenic Microalgae as Green Cell Factories*. New York: Springer (2007) pp. 1–11.

Leon, Rosa, Inmaculada Couso, and Emilio Fernández. "Metabolic engineering of ketocarotenoids biosynthesis in the unicelullar microalga Chlamydomonas reinhardtii." *Journal of Biotechnology* 130, no. 2 (2007): 143–152.

Li, Hung, Claire R. Shen, Chun-Hung Huang, Li-Yu Sung, Meng-Ying Wu, and Yu-Chen Hu. "CRISPR-Cas9 for the genome engineering of cyanobacteria and succinate production." *Metabolic Engineering* 38 (2016): 293–302.

Lin, Jianfei, He Chen, Ling Luo, Yongrong Lai, Wei Xie, and Kehkooi Kee. "Creating a monomeric endonuclease TALE-I-SceI with high specificity and low genotoxicity in human cells." *Nucleic Acids Research* 43, no. 2 (2015): 1112–1122.

Mak, Amanda Nga-Sze, Philip Bradley, Raul A. Cernadas, Adam J. Bogdanove, and Barry L. Stoddard. "The crystal structure of TAL effector PthXo1 bound to its DNA target." *Science* 335, no. 6069 (2012): 716–719.

Mali, Prashant, Luhan Yang, Kevin M. Esvelt, John Aach, Marc Guell, James E. DiCarlo, Julie E. Norville, and George M. Church. "RNA-guided human genome engineering via Cas9." *Science* 339, no. 6121 (2013): 823–826.

Marraffini, Luciano A., and Erik J. Sontheimer. "CRISPR interference: RNA-directed adaptive immunity in bacteria and archaea." *Nature Reviews Genetics* 11, no. 3 (2010): 181–190.

Masani, Mat Yunus Abdul, Gundula A. Noll, Ghulam Kadir Ahmad Parveez, Ravigadevi Sambanthamurthi, and Dirk Prüfer. "Efficient transformation of oil palm protoplasts by PEG-mediated transfection and DNA microinjection." *PLoS One* 9, no. 5 (2014): e96831.

Miller, Jeffrey C., Siyuan Tan, Guijuan Qiao, Kyle A. Barlow, Jianbin Wang, Danny F. Xia, Xiangdong Meng, et al. "A TALE nuclease architecture for efficient genome editing." *Nature Biotechnology* 29, no. 2 (2011): 143–148.

Molnár, Attila, Frank Schwach, David J. Studholme, Eva C. Thuenemann, and David C. Baulcombe. "miRNAs control gene expression in the single-cell alga Chlamydomonas reinhardtii." *Nature* 447, no. 7148 (2007): 1126–1129.

Montague, Tessa G., José M. Cruz, James A. Gagnon, George M. Church, and Eivind Valen. "CHOPCHOP: A CRISPR/Cas9 and TALEN web tool for genome editing." *Nucleic Acids Research* 42, no. W1 (2014): W401–W407.

Moscou, Matthew J., and Adam J. Bogdanove. "A simple cipher governs DNA recognition by TAL effectors." *Science* 326, no. 5959 (2009): 1501.

Nelson, J. A., Paul B. Savereide, and Paul A. Lefebvre. "The CRY1 gene in Chlamydomonas reinhardtii: Structure and use as a dominant selectable marker for nuclear transformation." *Molecular and Cellular Biology* 14, no. 6 (1994): 4011–4019.

Nicolson, Margery O., and Robert M. McAllister. "Infectivity of human adenovirus-1 DNA." *Virology* 48, no. 1 (1972): 14–21.

Nishimasu, Hiroshi, F. Ann Ran, Patrick D. Hsu, Silvana Konermann, Soraya I. Shehata, Naoshi Dohmae, Ryuichiro Ishitani, Feng Zhang, and Osamu Nureki. "Crystal structure of Cas9 in complex with guide RNA and target DNA." *Cell* 156, no. 5 (2014): 935–949.

Niu, Ying-Fang, Meng-Han Zhang, Da-Wei Li, Wei-Dong Yang, Jie-Sheng Liu, Wei-Bin Bai, and Hong-Ye Li. "Improvement of neutral lipid and polyunsaturated fatty acid biosynthesis by overexpressing a type 2 diacylglycerol acyltransferase in marine diatom Phaeodactylum tricornutum." *Marine Drugs* 11, no. 11 (2013): 4558–4569.

Nymark, Marianne, Amit Kumar Sharma, Torfinn Sparstad, Atle M. Bones, and Per Winge. "A CRISPR/Cas9 system adapted for gene editing in marine algae." *Scientific Reports* 6 (2016): 24951.

Osborn, Mark J., Beau R. Webber, Friederike Knipping, Cara-lin Lonetree, Nicole Tennis, Anthony P. DeFeo, Amber N. McElroy, et al. "Evaluation of TCR gene editing achieved by TALENs, CRISPR/Cas9, and megaTAL nucleases." *Molecular Therapy* 24, no. 3 (2016): 570–581.

Park, Jeongbin, Sangsu Bae, and Jin-Soo Kim. "Cas-Designer: A web-based tool for choice of CRISPR-Cas9 target sites." *Bioinformatics* 31, no. 24 (2015): 4014–4016.

Park, Jeongbin, Kayeong Lim, Jin-Soo Kim, and Sangsu Bae. "Cas-analyzer: An online tool for assessing genome editing results using NGS data." *Bioinformatics* 33, no. 2 (2017): 286–288.

Pavletich, Nikola P., and Carl O. Pabo. "Zinc finger-DNA recognition: Crystal structure of a Zif268-DNA complex at 2.1 A." *Science* 252, no. 5007 (1991): 809–817.

Porteus, Matthew H., and Dana Carroll. "Gene targeting using zinc finger nucleases." *Nature Biotechnology* 23, no. 8 (2005): 967–973.

Puchta, Holger, and Friedrich Fauser. Gene targeting in plants: 25 years later. *International Journal of Developmental Biology* 57, no. (6–7–8) (2013): 629–637.

Rathod, Jayant Pralhad, Gunjan Prakash, Reena Pandit, and Arvind M. Lali. "Agrobacterium-mediated transformation of promising oil-bearing marine algae Parachlorella kessleri." *Photosynthesis Research* 118, no. 1–2 (2013): 141–146.

San Cha, Thye, Willy Yee, and Ahmad Aziz. "Assessment of factors affecting Agrobacterium-mediated genetic transformation of the unicellular green alga, Chlorella vulgaris." *World Journal of Microbiology and Biotechnology* 28, no. 4 (2012): 1771–1779.

Sather, Blythe D., Guillermo S. Romano Ibarra, Karen Sommer, Gabrielle Curinga, Malika Hale, Iram F. Khan, Swati Singh, et al. "Efficient modification of CCR5 in primary human hematopoietic cells using a megaTAL nuclease and AAV donor template." *Science Translational Medicine* 7, no. 307 (2015): 307.

Shin, Sung-Eun, Jong-Min Lim, Hyun Gi Koh, Eun Kyung Kim, Nam Kyu Kang, Seungjib Jeon, Sohee Kwon, et al. "CRISPR/Cas9-induced knockout and knock-in mutations in Chlamydomonas reinhardtii." *Scientific Reports* 6, no. 1 (2016): 1–15.

Shin, Ye Sol, Jooyeon Jeong, Thu Ha Thi Nguyen, Jaoon Young Hwan Kim, EonSeon Jin, and Sang Jun Sim. "Targeted knockout of phospholipase A2 to increase lipid productivity in Chlamydomonas reinhardtii for biodiesel production." *Bioresource Technology* 271 (2019): 368–374.

Sizova, Irina, Andre Greiner, Mayanka Awasthi, Suneel Kateriya, and Peter Hegemann. "Nuclear gene targeting in C hlamydomonas using engineered zinc-finger nucleases." *The Plant Journal* 73, no. 5 (2013): 873–882.

Smith, Jeff, Marina Bibikova, Frank G. Whitby, A. R. Reddy, Srinivasan Chandrasegaran, and Dana Carroll. "Requirements for double-strand cleavage by chimeric restriction enzymes with zinc finger DNA-recognition domains." *Nucleic Acids Research* 28, no. 17 (2000): 3361–3369.

Stemmer, Manuel, Thomas Thumberger, Maria del Sol Keyer, Joachim Wittbrodt, and Juan L. Mateo. "CCTop: An intuitive, flexible and reliable CRISPR/Cas9 target prediction tool." *PLoS One* 10, no. 4 (2015): e0124633.

Stoddard, Barry L. "Homing endonucleases: From microbial genetic invaders to reagents for targeted DNA modification." *Structure* 19, no. 1 (2011): 7–15.

Stoddard, Barry L. "Homing endonucleases from mobile group I introns: Discovery to genome engineering." *Mobile DNA* 5, no. 1 (2014): 7.

Stukenberg, Daniel, Stefan Zauner, Gianluca Dell'Aquila, and Uwe G. Maier. "Optimizing CRISPR/Cas9 for the diatom Phaeodactylum tricornutum." *Frontiers in Plant Science* 9 (2018): 740.

Tepfer, David. "Transformation of several species of higher plants by Agrobacterium rhizogenes: sexual transmission of the transformed genotype and phenotype." *Cell* 37, no. 3 (1984): 959–967.

Thomas, Kirk R. and Mario R. Capecchi. "Site-directed mutagenesis by gene targeting in mouse embryo-derived stem cells." *Cell* 51, no. 3 (1987): 503–512.

Tong, Yaojun, Pep Charusanti, Lixin Zhang, Tilmann Weber, and Sang Yup Lee. "CRISPR-Cas9 based engineering of actinomycetal genomes." *ACS Synthetic Biology* 4, no. 9 (2015): 1020–1029.

Tzfira, Tzvi, and Vitaly Citovsky. "Agrobacterium-mediated genetic transformation of plants: Biology and biotechnology." *Current Opinion in Biotechnology* 17, no. 2 (2006): 147–154.

Ungerer, Justin, and Himadri B. Pakrasi. "Cpf1 is a versatile tool for CRISPR genome editing across diverse species of cyanobacteria." *Scientific Reports* 6, no. 1 (2016): 1–9.

Urnov, Fyodor D., Edward J. Rebar, Michael C. Holmes, H. Steve Zhang, and Philip D. Gregory. "Genome editing with engineered zinc finger nucleases." *Nature Reviews Genetics* 11, no. 9 (2010): 636–646.

Verruto, John, Kristie Francis, Yingjun Wang, Melisa C. Low, Jessica Greiner, Sarah Tacke, Fedor Kuzminov, et al. "Unrestrained markerless trait stacking in Nannochloropsis gaditana through combined genome editing and marker recycling technologies." *Proceedings of the National Academy of Sciences of the United States of America* 115, no. 30 (2018): E7015–E7022.

Vieler, Astrid, Guangxi Wu, Chia-Hong Tsai, Blair Bullard, Adam J. Cornish, Christopher Harvey, Ida-Barbara Reca, et al. "Genome, functional gene annotation, and nuclear transformation of the heterokont oleaginous alga Nannochloropsis oceanica CCMP1779." *PLoS Genetics* 8, no. 11 (2012): e1003064.

Wang, Qintao, Yandu Lu, Yi Xin, Li Wei, Shi Huang, and Jian Xu. "Genome editing of model oleaginous microalgae Nannochloropsis spp. by CRISPR/Cas9." *The Plant Journal* 88, no. 6 (2016): 1071–1081.

Wendt, Kristen E., Justin Ungerer, Ryan E. Cobb, Huimin Zhao, and Himadri B. Pakrasi. "CRISPR/Cas9 mediated targeted mutagenesis of the fast growing cyanobacterium Synechococcus elongatus UTEX 2973." *Microbial Cell Factories* 15, no. 1 (2016): 115.

Wiedenheft, Blake, Samuel H. Sternberg, and Jennifer A. Doudna. "RNA-guided genetic silencing systems in bacteria and archaea." *Nature* 482, no. 7385 (2012): 331–338.

Wolfe, Scot A., Lena Nekludova, and Carl O. Pabo. "DNA recognition by Cys2His2 zinc finger proteins." *Annual Review of Biophysics and Biomolecular Structure* 29, no. 1 (2000): 183–212.

Yamasaki, Tomohito, Eun-Jeong Kim, Heriberto Cerutti, and Takeshi Ohama. "Argonaute3 is a key player in mi RNA-mediated target cleavage and translational repression in Chlamydomonas." *The Plant Journal* 85, no. 2 (2016): 258–268.

Yang, Bo, Jin Liu, Bin Liu, Peipei Sun, Xiaonian Ma, Yue Jiang, Dong Wei, and Feng Chen. "Development of a stable genetic system for Chlorella vulgaris: A promising green alga for CO_2 biomitigation." *Algal Research* 12 (2015): 134–141.

13 Plant Genome Editing in Basic Research to Understand Molecular Functions

P. T. K. Jagannadham
ICAR-Central Citrus Research Institute

Siddanna Savadi
ICAR-Directorate of Cashew Research

Senthilkumar K. Muthusamy
ICAR-Central Tuber Crops Research Institute

CONTENTS

13.1 INTRODUCTION

Being sessile, plants have evolved more sophisticated molecular machinery for growth and survival in response to the changing environment (Bressan et al., 2014; Lenka et al., 2019). Several biotic and abiotic stress factors influence the growth and development as well as the survival of the plants. To cope up the external pressure, plants have to modulate the expression using energy-efficient molecular machinery. Understanding the functions of the genes in response to biotic and abiotic stress conditions would help to develop climate smart crops (Muthusamy et al., 2017b; Lenka et al., 2018; Sharma et al., 2018). Unlike other organisms, plants offer many advantages as model organisms mainly because of their inherent plastic and modular nature, which makes them possible for tolerating many experimental conditions. In addition, the propagating materials, *viz.* seed, vegetative propagules, *etc.*, can be stored and shared between laboratories across different countries easily. The functions of many genes have been decoded using the model organisms *Arabidopsis* and rice (O'Malley and Ecker, 2010). The main advantage of these organisms is of smaller genome size; *i.e.*, *Arabidopsis* has a small genome size of 103 Mb comprising five chromosomes, whereas rice has a genome size of 389 Mb comprising 12 chromosomes. In addition, both have a huge collection of genetic and genomic resources and a well-established transformation system amenable for tissue culture (Woodward and Bartel, 2018). Moreover, the availability of the genetic and genomic resources of the plants eases and fastens the process of characterization of genes using reverse genetics approach (Provart et al., 2016; Lenka et al., 2018; Muthusamy et al., 2019). The large collection of mutant stocks are maintained and made available through Arabidopsis Biological Resource Center (ABRC) and the Nottingham Arabidopsis Stock Centre (NASC) (Provart et al., 2016). Many research groups have utilized the mutant resources to discover various functions of the plant genes. Thus, understanding the functions of the plant molecular machinery is the major requirement to harness their potential using modern molecular biology tools for the benefit of humans. Interestingly, many human disease-related putative genes had orthologs in *Arabidopsis*, which makes it as a potential model organism to understand the basic molecular mechanism of these genes (Jones et al., 2008; Provart et al., 2016; Zluhan-Martínez et al., 2020). Thus, various molecular tools available can be harnessed to study and understand the various molecular mechanisms of the plants (Bolser et al., 2016; Muthusamy et al., 2016). Recent advances in genomics, genetic engineering and functional genomics techniques ease the process of gene discovery and functional characterization of several processes (Kumar et al., 2018; Jagannadham et al., 2019).

13.2 STRATEGIES FOR INTRODUCING MUTATIONS IN CROPS

Chemical, physical and biological mutagens were successfully employed for creating variations in plants (Page and Grossniklaus, 2002; Muthusamy et al., 2019b). Each mutagen has its own advantages and limitations on its application. Chemical or physical mutagens result in wider variations in plant genomes, which can be identified by high-throughput screening methods such as TILLING (Targeting-Induced Local Lesions IN Genomes). Biological mutagens, including *Agrobacterium* and transposon-mediated chromosomal insertion vectors, are extensively used to create the mutant lines in plants.

13.2.1 Mutagen-Based Mutagenesis

Chemical agents, including ethyl methane sulphonate (EMS), N-methyl-N-nitrosourea (MNU) and sodium azide, were successfully employed in basic research for creating variations in plants. Several genes were identified and characterized in plants using the mutant populations created with chemical mutagens (reviewed in Provart et al., 2016; Cuperus et al., 2010). For instance, many genes – including a *chloroplast-targeted UMP kinase* (Zhu et al., 2016), *SPL28*, a clathrin-associated adaptor protein complex 1, medium subunit µ1 (AP1M1) regulating spotted leaf and early senescence trait (Qiao et al., 2010) and *cysteine-rich receptor-like kinase* involved in disease resistance (Zhou et al., 2017) – were characterized through map-based cloning using EMS mutagenized populations in rice (Viana et al., 2019).

Physical mutagens, including gamma-rays, fast neutron, UV and X-ray radiations, were widely employed for creating variations in plants. Although physical mutagens result in heavy damage of DNA, many genes have been successfully identified and characterized in plants. For instance, mutation in *OASA1* results in new allele, which leads to higher tryptophan content in rice (Chun et al., 2012; Viana et al., 2019).

13.2.2 Insertional Mutagenesis

Targeted mutagenesis is a key genetic tool for creating novel allelic variation in plants. Genetic elements such as T-DNA, retrotransposons, transposons and activator/dissociation (Ac/Ds) insertions have been successfully used for the development of mutants through insertional mutagenesis. The function of several genes has been discovered by creating, screening and characterizing the mutant libraries. The most important advantage of this strategy is that the insertion is almost random, and the flanking region of the insertion site can be easily tagged using the PCR techniques, like inverse PCR or thermal asymmetric interlaced PCR (TAIL-PCR). Moreover, these approaches can be easily used for studying the essential genes whose function is necessary for the survival of the plant, which is difficult in the knockout mutations created by chemical or physical mutagens. Additionally, insertional mutagenesis approach can be efficiently exploited to characterize a large number of genes using the activation tagging, promoter trap, enhancer trap and gene trap vectors (Pérez-Martín et al., 2017; Muthusamy et al., 2017a).

The main advantage of T-DNA-based mutagenesis is that it exhibits a low copy and stable inheritance (Wu et al., 2003), and T-DNA allows a gene detection or activation by allowing the insertion of gene activation elements such as Ds/dSpm element in the proximity of the genes. Recently, large populations of T-DNA insertion lines have been produced in model organisms, including rice, *Arabidopsis* and tomato (O'Malley and Ecker, 2010; O'Malley et al., 2015; Pérez-Martín et al., 2017; Muthusamy et al., 2019b). Several genes with diverse functions have been characterized in plants using T-DNA-based mutant resources (O'Malleyet al., 2015; Viana et al., 2019). For instance, *rid1* gene, a Cys2/His2-type zinc finger transcription factor, was identified and characterized as a master regulator switch related to flowering induction in rice using a never-flowering T-DNA-mutant rice phenotype, which never headed after growing for more than 500 days (Wu et al., 2008). Similarly, *OsGASR1*, a member of the GAST (gibberellin (GA)-stimulated transcript), regulates the seedling growth and α-amylase content in rice (Lee et al., 2017).

13.3 TARGETED MUTAGENESIS

Desirable genetic variations in the gene pools of crop species are often scarce due to a very low rate of spontaneous mutations. Artificial induction of mutations can efficiently create a large set of genetic variations in a species. However, the majority of the induced mutations in plants are randomly created using physical irradiation sources such as X-rays or γ-rays and chemical agents like EMS. Mutation breeding is found valuable for the genetic improvement of crop species. Recently, availability of the genome sequence information provides the resources and tools for the systematic discovery of mutations and screening for desirable mutants from a large population of mutants.

13.3.1 TILLING and Its Applications in Crop Plants

Identification of desirable mutants from a large population of mutants is a challenge to the researchers. The high-throughput screening technique called TILLING has become an important tool for identifying desirable mutants from a large mutant population. TILLING is a reverse genetic tool that facilitates the identification of a series of point mutations in genes. Being a high-throughput mutant screening method, TILLING facilitates a rapid and low-cost discovery of mutations in large populations. For the first time, TILLING approach was developed in *Arabidopsis* by Till et al. (2003). In TILLING, mutation detection involved the PCR amplification of particular loci using the pooled DNA of mutant population, followed by denaturation and reannealing of the pooled DNA PCR products and cleaving of the reannealed products using the CEL1 enzyme. The cleaved products are detected using a LICOR system, which gives information on the site of point mutation in the candidate gene (Kurowska et al., 2011). The importance of this methodology in high-throughput mutation detection was established by the success of the *Arabidopsis* TILLING Project (ATP) (Henikoff et al., 2004).

Generally, TILLING spots the single base pair changes or small deletions in the specific candidate gene (McCallum et al., 2000; Till et al., 2003). Further, the altered form of TILLING, known as EcoTILLING, allows the detection of SNPs or genetic variations in naturally occurring populations such as landraces and wild accessions (Haughn and Gilchrist, 2006). This tool can be employed in genetic mapping, and studying the gene structure, linkage disequilibrium and population structure in natural populations (Haughn and Gilchrist, 2006).

TILLING can play an important role in the improvement of crop plants. The useful mutations detected using the TILLING technique can be directly utilized in plant breeding programmes as TILLING is considered as a nongenetically modified (GM) technology (Uauy et al., 2009). Several groups have developed TILLING resources in a number of crop species such as rice (Wu et al., 2005), wheat (Rawat et al., 2012), barley (Lababidi et al., 2009), tomato (Minoia et al., 2010) and also in perennial tree crops (Table 13.1).

Recently, mutation screening efficiency of the TILLING technology has been improved using advanced molecular techniques (Parry et al., 2009). For instance, the improvement of TILLING platforms with the next-generation sequencing (NGS) technologies referred to as the TILLING by Sequencing (TbyS) accelerates the mutation detection rate by several times, and thus, TbyS is expected to enhance the outcomes of the crop functional genomics studies as well as to facilitate the efficient harnessing of induced genetic variations in crop breeding (Kumar et al., 2017). Further, a combination of recent advances in NGS and genome editing capabilities would add more power to the TILLING approach's applications (Kumar et al., 2017). Thus, the recent advances in genomics and TILLING would allow the discovery of mutants in a precise and high-throughput way, which is expected to enhance the utility of induced mutations in the breeding and functional genomics.

13.3.2 RNA Interference (RNAi)

RNA interference is a phenomenon occurring in eukaryotic organisms to suppress the functions of endogenous genes (de Framond et al., 2007). It is a post-transcriptional gene silencing (PTGS) mechanism, which is found to be conserved in plants, and it regulates different biological processes such as growth and development, genome stability and defence from the molecular parasites such as viruses and transposable elements (TEs) (Joseph and Osman, 2012).

RNAi mechanism was discovered accidentally in petunia plant while attempting to enhance the chalcone synthase (*chsA*) gene expression for the production of an enzyme involved in purple pigment biosynthesis. The chalcone synthase (*chsA*) gene overexpression in petunia had resulted in white colour flowers, which should have been purpler (Napoli et al., 1990). Thus, it formed the foundation of the discovery of RNAi. The mechanism involved in these effects was unknown until it was discovered by Fire et al. (1998). They found that the introduction of double-stranded RNA (sense and antisense) in *C. elegans* resulted in ten times greater gene silencing than the introduction of sense and antisense strands separately. The introduction of a double-stranded break (DSB) triggered the production of dsRNAs in the cell, which are recognized by DICER (enzyme) and cleaved into siRNAs (Tijsterman and Plasterk, 2004). These siRNAs are used by the RISC complex (Hammond et al., 2001) to target and degrade complementary mRNAs, resulting in the downregulation of gene expression. There are two types of small RNAs, viz. small interfering RNAs (siRNAs) and microRNAs (miRNAs) involved in the RNAi process. These two types of small RNAs are ~22 nucleotides long but are different in their function, specificity and origin. Generally, miRNAs are considered as endogenous, whereas siRNAs are considered to be exogenous in origin, derived from the virus, transposon or transgene trigger. Further, miRNAs are generated from stem loop precursors that are not wholly double-stranded, whereas siRNAs were found to be processed from long and completely double-stranded RNAs (dsRNAs) (Tomari and Zamore, 2005).

TABLE 13.1
TILLING Applications in Crops

Host Plant	TILLING Strategy	Target Genes/Region	Population Size	Outcome of the Study	References
Western black cottonwood (*Populus trichocarpa*)	Eco-TILLING with crude celery juice extract (CJE) digestion and SNPs confirmed by sequencing	Nine candidate genes: *PoptrTB1, PoptrLFY, PoptrMP, PoptrKNAT1, PoptrF5H, PoptrCHI, Poptr4CL3, PoptrNPR1* and *PoptrMPK4*	Individuals from 41 populations	Allowed a rapid discovery of natural genetic variations existing in a large pool of individuals of *P. trichocarpa* tree species. Variation in genes varied from 1 SNP in *PoptrTB1* to >23 SNPs in *PoptrLFY* in the 1-kbp region examined	Gilchrist et al. (2006)
Brassica rapa	TILLING	*CAX1a* encoding cadmium transporter	-	Enhanced tolerance to cadmium toxicity and thus possess phytoremediation potential	Navarro-León et al. (2020)
Triticum turgidum L.ssp. durum	TILLING	*Lycopene epsilon cyclase (LCYE)*	-	Identified mutants with enhanced β-carotene content in leaves	Richaud et al. (2018)
Rice	TILLING	*SBEI* encoding starch-branching enzymes (SBEs)	Cultivar Nipponbare (n = 4096)	A substantial change in amylose and protein content was found. In amylopectin chain-length distribution, mutant has a small but significant reduction in B3 chains, whereas a significant increase was found in B1 chains, which can be used in breeding of novel quality traits	Kim et al. (2018)
Populus nigra	TILLING by resequencing using NGS and Sanger sequencing	Five candidate genes for lignin biosynthesis: *CAD4, HCT1, C3H3, CCR7, 4CL3*	768 accessions	Identified 36 non-synonymous mutations in the lignin biosynthesis genes	Marroni et al. (2011b)
Populus nigra	Eco-TILLING	One candidate gene of lignin biosynthesis *CAD4*	360 accessions	Identified 45 SNPs, 6 of which caused an amino acid substitution, which could be utilized in breeding programmes	Marroni et al. (2011a)
Jatropha curcas	TILLING by genotyping-by-sequencing (GBS)	Genome sequence (400 million paired-end reads)	96 M_1V_6 plants	Identified 1452 SNPs in coding regions, of 995 genes and functional annotation showed that the total 16 SNPs were located in six KEGG pathways	Maghuly et al. (2015)

Since the discovery of RNAi, it has become an important tool for analysing the functions of the genes (Hannon and Rossi, 2004; Pradhan et al., 2015).

13.3.2.1 Applications of RNAi in Crop Improvement

RNAi technology has gained great importance in the improvement of crop plants with the advances in crop genomics and molecular biology. RNAi technology has been employed to improve different useful traits such as disease resistance, shelf life and qualitative traits in many crop plants (Segal et al., 2003; Konan et al., 2007; Dodo et al., 2008; Valarmathi et al., 2016; Yang et al., 2017).

Initially, RNAi was utilized to develop resistance to viruses in crop varieties. Engineered virus resistance strategies in plants are similar to the natural RNA silencing mechanisms. At first, RNAi technology was employed to develop potato virus Y-resistant plants by the suppression of a viral proteinase gene (Mansoor et al., 2006). Since then, resistance to several other viruses, viz. cucumber and tobacco mosaic virus (Hu et al., 2011), tomato spotted wilt virus (Mitter et al., 2013), bean golden mosaic virus (Bonfim et al., 2007), banana bract mosaic virus (Sankaranarayanan et al., 2020) and rice tungro bacilliform virus (Valarmathi et al., 2016), has been developed in plants. In addition, crop plants resistant to several fungal and other plant parasites have been developed (Table 13.2). RNAi has also been used to produce male sterility in crop plants. Male sterility is a valuable trait in hybrid seed production. Targeting of genes expressed exclusively in pollen tissues involved in pollen production using RNAi can result in male sterility. For instance, male-sterile tobacco lines were developed by the suppression of *TA29*, a gene essential for pollen development (Nawaz-Ul-Rehman et al., 2007). RNAi was also used to create cytoplasmic male sterility by the suppression of *Msh1* expression in tobacco and tomato, which resulted in rearrangements in the male sterility-associated genes of mitochondria (Sandhu et al., 2007). RNAi has also been used to modify quality traits such as nutrient content and reduced toxin production (Segal et al., 2003; Konan et al., 2007).

Although much progress has been made in the application of RNAi technology, the realization of full potential of RNAi in the improvement of crop plants remains incomplete. Further, understanding of complexities of the RNAi pathway, the molecular machinery and their role in different biological processes is yet to be completely elucidated.

13.3.3 Zinc Finger Nucleases (ZFNs)

ZFNs comprise artificial fusion proteins that combine a zinc finger DNA-binding domain to a nonspecific nuclease domain. ZFNs were initially constructed by Chandrasekaran and co-workers (Mani et al., 2005). ZFNs have the ability to cut DNA templates *in vitro as* well as in a cell on both the strands, thereby triggering homologous recombination (HR) (Bibikova et al., 2001). ZFNs contain a DNA-binding domain comprising three Cys_2His_2 zinc fingers and a nonspecific DNA cleavage domain derived from the endonuclease FokI (Kim et al., 1996).

13.3.3.1 ZFN Application in Crops

The successful demonstration of ZFN-based genome editing in maize using an *IPK1* gene has resulted in reduced phytate biosynthesis in seeds (Shukla et al., 2009). This results in an increased bioavailability of phosphorus in corn grains and reduces the adverse environmental impact. In another instance, ZFNs were used to induce a targeted mutagenesis of a transgene and nine endogenous soya bean (*Glycine max*) genes. ZFN was even successfully implemented in soya bean, which had duplicated paleopolyploid genome (Curtin et al., 2011; Bonawitz et al., 2019). Even in bread wheat, a allohexaploid ZFN-mediated DNA repair helped to introduce a specific single amino acid change into the coding sequence of acetohydroxamic synthase (AHAS), which conferred resistance to imidazolinone herbicides (Ran et al., 2018). In tomato, ZFN-mediated disruption of the *LEAFY COTYLEDON1-LIKE4 (L1L4)* gene was undertaken with an aim of understanding the molecular mechanisms that control pleiotropy in fruit-bearing species (Hilioti et al., 2016). Even though

TABLE 13.2
List of Studies Employing RNAi Technology for Developing Resistance to Different Fungal Pathogens in Crop Plants

Plant	Silencing Strategy	Target Gene	Gene Function	Result	References
Wheat	RNAi	*Fusarium graminearum* chitin synthase 3b gene *FgCh3b*	Synthesis of the majority of the chitin found in the fungal cell wall	Confers durable resistance to *Fusarium* head blight and seedling blight in wheat	Cheng et al. (2015)
Barley	*RNAi*	*B. graminis Avra10*	Encodes virulence effectors	Inhibited fungal development in the absence of the cognate R gene *MLO10*	Nowara et al. (2010)
Barley and wheat	*RNAi*	*PSTha12J12* (haustorial Pst transcript)	Demethylation of lanosterol to create a precursor of ergosterol, a fundamental component of fungal membranes	No obvious reductions in rust development or sporulation of *Puccinia striiformis*f. sp. *tritici*or *P. graminis*f. sp.*tritici*	Yin et al. (2011)
Wheat	*RNAi*	*PtMAPK1* (MAP kinase), *PtCYC1* (cyclophilin) and *PtCNB* (calcineurin B)	Encode effector proteins that facilitate the host infection by rust pathogens	Disease suppression, reduced fungal growth and sporulation *P. triticina, P. graminis* and *P. striiformis*	Panwar et al. (2013a,b)
Arabidopsis and barley	*RNAi*	*CYP51A, CYP51B* and *CYP51C*	Demethylation of lanosterol to create a precursor of ergosterol, a fundamental component of fungal membranes	Resistance to *Fusarium graminearum*	Koch et al. (2013)
Wheat	RNAi	*F. culmorum β-1, 3-glucan synthase* gene *Fc Gls1*	Glucosyltransferase enzyme involved in the generation of beta-glucan in fungi	Conferred resistance to *Fusarium* head blight to *Fusarium* head blight	Chen et al. (2016)
Wheat	Silencing	Wheat lipoxygenase gene*TaLpx-1*	Lipoxygenases that facilitate the infection of *F. graminearum*	Conferred resistance to *F. graminearum*	Nalam et al. (2015)

(Continued)

TABLE 13.2 (*Continued*)
List of Studies Employing RNAi Technology for Developing Resistance to Different Fungal Pathogens in Crop Plants

Plant	Silencing Strategy	Target Gene	Gene Function	Result	References
Arabidopsis and tomato	Silencing	*Botrytis cinerea dicer-like protein 1 (Bc-DCL1) and Bc-DCL2*	Dicer protein involved in production of *B. cineraria* RNA effectors that facilitate pathogenicity and growth	Attenuates fungal pathogenicity and growth	Wang et al. (2016)
Barley	RNAi	RACB-regulated signalling genes encoding a leucine-rich repeat (LRR) protein, a leucine-rich receptor-like kinase and an S-domain SD1-receptor-like kinase	Rho-like GTPases (called RACs or ROPs) function in transcriptional reprogramming of hormone response and innate immunity gene expression patterns	Enhanced resistance to powdery mildew	Schnepf et al. (2018)
Barley	RNAi	Spraying of long noncoding ds*CYP3*	Cytochrome P450 *lanosterol C-14α-demethylases*, required for the biosynthesis of fungal ergosterol	Suppression of *CYP51A*, *CYP51B* and *CYP51C* confers resistance to *F. graminearum*	Koch et al. (2016)
Barley	RNAi	Cellulose synthase-like D2 (*HvCslD2*)	Synthesis of cell wall components	Easy penetration by host-adapted, virulent and avirulent non-host races of the powdery mildew fungus	Douchkov et al. (2016)
Wheat	RNAi	Stripe rust (*P. striiformis*) effector candidate (PEC6)	Encodes a pathogenicity factor involved in the suppression of host defence	VIGS of *PEC6* expression decreased the number of rust pustules	Liu et al. (2016)

ZFN applications were widespread in animal cells, it led to genome editing in the plants for its improvement.

13.3.3.2 Potential Limitations of the ZFN Technology

Despite its initial success, the ZFN technology has its limitations. The assembly and testing process of ZFNs is exorbitant. The endogenous targets are limited due to a minimal number of modules for the context-dependent assembly platform. Another major constraint is restrictions imposed by the company Sangamo Biosciences to access the design tools (Sander et al., 2011). This technique has low targeting efficiency and specificity, and frequent off-target effects are partly due to nonspecific DNA binding (DeFrancesco, 2011).

13.3.4 Transcription Activator-Like Effector Nucleases (TALENs)

Transcription activator-like effector nucleases (TALENs) are sequence-specific nucleases consisting of a catalytic domain of the *FokI* endonuclease (Boch et al., 2009; Christian et al., 2010). The DNA-binding domain in TALE monomers is comprised of a central repeat domain (CRD) that directs DNA binding and host specificity. TALENs work in pairs: DSBs induced by a pair of TALENs activate DNA repair mechanisms either by HR or by non-homologous end-joining (NHEJ). NHEJ creates frameshift mutations with insertions or deletions (indels), resulting in gene disruption (Moscou and Bogdanove, 2009).

13.3.4.1 The Application of TALENs in Crops

The utilization of TALENs technology for crop improvement started with rice. The promoter region of the sucrose-efflux transporter gene *OsSWEET14* was designed and transformed into rice, which interferes with the binding of bacterial effector protein of *Xanthomonas oryzae*, a causal pathogen for blight disease. The stable transformants have a mutation in the effector binding site in the promoter region, which leads to silencing of the gene and consequent resistance to *X. oryzae* (Li et al., 2012). In wheat, three homoeoalleles of the MILDEW-RESISTANCE LOCUS (MLO) proteins were edited. Mutation of all three TaMLO homologs created a heritable broad-spectrum resistance to powdery mildew. In tomato, the promoter site of the barley phytase gene of the purple acid phosphatase (HvPAPhy) and a negative regulator of GA signalling PROCERA (PRO) gene were disrupted for the increased GA response (Lor et al., 2014). In soya bean, mutations were created in two fatty desaturase genes (FAD2-1A and FAD2-1B), which led to an increase in the linoleic acid content from 20% to 80% and a decrease in linoleic acid from 50% to under 4%. This also led to an increased oil shelf life and improved oxidative stability (Haun et al., 2014). There are several more studies that utilized the TALENs for the crop improvement.

13.3.4.2 Potential Limitations of the TALEN Technology

The target sites of TALENs are limited by their prerequisite presence of a thymidine nucleotide at the 5′ position (Doyle et al., 2012). The conventional TALENs are unable to cleave DNA containing 5-methylcytosine, to counter a repeat that recognizes cytosine which can be substituted for one that binds to thymine. But it compromises the target specificity (Valton et al., 2012). The presence of repetitive sequences in TALE array makes things difficult while handling the PCR (Holkers et al., 2013). The Golden Gate platform provides a robust, economical and user-friendly protocol for TALEN assembly. The major hindrance in using the TALEN system is in preparing the vector with suitable monomers for binding the target DNA (Weber et al., 2011).

13.3.5 The CRISPR/Cas9

CRISPR is an acronym for clustered regularly interspaced short palindromic repeat. CRISPR sequences are a crucial component of the immune systems of these simple life forms. The CRISPR/Cas systems can be used in multiple ways like DNA cleavage, gene silencing, gene activation,

SNP detection and homology-mediated gene integration (Ghogare et al., 2020). Compared to ZFNs and TALENs, CRISPR genome editing is less complex and needs designing of a simple guide RNA (gRNA) of about 20 nucleotides complementary to the DNA stretch within the target gene. The CRISPR cleavage requires (1) a short synthetic gRNA sequence of 20 nucleotides that bind to the target DNA and (2) Cas9 nuclease enzyme that cleaves three to four bases after the protospacer-adjacent motif (PAM) (Jinek et al., 2012). The technology being most adapted and studied is in constant improvement to increase its specificity and reduce the off-targets (Lee et al., 2016; Hu et al., 2018).

13.3.5.1 Application of CRISPR/Cas9 in Crops

CRISPR/CAS9 is predominantly used and found successful in the animal cells. Its efficiency to induce mutations was successfully demonstrated in various crops like rice, wheat and maize (Shan et al., 2013; Zong et al., 2017). With its ease of application, established protocols and availability of genome information, the utilization of the technology increased manifold within a short span. In rice, CRISPR/Cas9 system enhanced salinity tolerance by mutating OsRR22 gene that is involved in both cytokinin signal transduction and metabolism (Zhang et al., 2019). The CRISPR/Cas9 system was used to edit *Xa13* gene promoter, such that only pathogen-induced expression is impaired without affecting the other roles (Li et al., 2020). With such level of ease of CRISPR/Cas9 system, multiplex editing has been successfully implemented in Durum wheat to reduce the allergen proteins (Camerlengo et al., 2020). In cucumber, CRISPR/Cas9 system disrupted the function of the recessive *eIF4E* (eukaryotic translation initiation factor 4E) gene to develop broad virus resistance (Chandrasekaran et al., 2016). In tomato, full knockout of the tomato susceptibility (S) gene *powdery mildew resistance* 4 (*PMR4)* was carried out to provide resistance against powdery mildew. With several protocols being standardized, ironing out all the inconsistencies, the technology is on its way to revolutionize the crop improvement for the betterment of the world.

13.3.5.2 Potential Limitations of the CRISPR/Cas9 Technology

Every technology has its pros and cons, and the same applies to this technology. The major concern regarding the CRISPR/Cas system is its relatively high off-target (Hahn and Nekrasov, 2019). The target site selection is also dependent on the presence of the PAM sites (Jinek et al., 2012). Besides target sites, the size of catalytic window of cytosine deaminase base editors ranges from four to five nucleotides; efforts have been made to narrow this window to increase the accuracy and efficiency (Tang et al., 2019).

13.4 CONCLUSION

Genome editing or gene editing techniques change an organism's DNA in several ways like addition and deletion of alteration. Over the years, several technologies have spawned in search of flawless genome editing. ZFN and TALEN have improved the precision but have the limitation of target and operational cost. CRISPR /Cas9 technology has come to close to the perfection in terms of editing and ease of use. The CRISPR/Cas9 system derived from the prokaryotic system has been successfully executed across the genera. The utilization of CRISPR has left to the imagination of the scientist with its myriad ways and uses. The off-target is one drawback of the CRISPR, but with several researchers handling the technology, there is significant improvement in the technology over the years. Compared to developed world in developing several successful genome edited crop, speed of adoption of such technologies in developing world is of paramount significance in breeding new genotypes.

REFERENCES

Bibikova, M., D. Carroll, D. Segal, et al. "Stimulation of homologous recombination through targeted cleavage by chimeric nucleases." *Molecular and Cellular Biology* 21, no. 1 (2001): 289–297.

Bibikova, M., K. Beumer, J. K. Trautman, and D. Carroll. "Enhancing gene targeting with designed zinc finger nucleases." *Science* 300, no. 5620 (2003): 764–764.

Boch, J., H. Scholze, S. Schornack, et al. "Breaking the code of DNA binding specificity of TAL-type III effectors." *Science* 326, no. 5959 (2009): 1509–1512.

Bolser, D., D. M. Staines, E. Pritchard and P. Kersey. "Ensembl plants: integrating tools for visualizing, mining, and analyzing plant genomics data". In *Plant Bioinformatics,* edited by D. Edwards 115–40. Humana Press, New York, NY, 2016.

Bonawitz, N. D., W. M. Ainley, A. Itaya, et al. "Zinc finger nuclease-mediated targeting of multiple transgenes to an endogenous soybean genomic locus via non-homologous end joining." *Plant Biotechnology Journal* 17, no. 4 (2019): 750–761.

Bonfim, K., J. C. Faria, E. O. Nogueira, É. A. Mendes and F. J. Aragão. "RNAi-mediated resistance to Bean golden mosaic virus in genetically engineered common bean (*Phaseolus vulgaris*)." *Molecular Plant-Microbe Interactions* 20, no. 6 (2007): 717–726.

Bressan, R. A., J. K. Zhu, M. J. Van Oosten, A. Maggio, H. J. Bohnert and V. Chinnusamy. "Epigenetics connects the genome to its environment." *Plant Breeding Reviews* 38 (2014): 69–142.

Camerlengo, F., A. Frittelli, C. Sparks, et al. "CRISPR-Cas9 multiplex editing of the α-amylase/trypsin inhibitor genes to reduce allergen proteins in durum wheat." *Frontiers in Sustainable Food Systems* 4 (2020): 104.

Chandrasekaran, J., M. Brumin, D. Wolf, et al. "Development of broad virus resistance in non-transgenic cucumber using CRISPR/Cas9 technology." *Molecular Plant Pathology* 17, no. 7 (2016): 1140–1153.

Chen, W., C. Kastner, D. Nowara, et al. "Host-induced silencing of Fusarium culmorum genes protects wheat from infection." *Journal of Experimental Botany* 67, no. 17 (2016): 4979–4991

Cheng, W., X. S. Song, H. P. Li, et al. "Host-induced gene silencing of an essential chitin synthase gene confers durable resistance to F usarium head blight and seedling blight in wheat." *Plant Biotechnology Journal* 13, no. 9 (2015): 1335–1345.

Christian, M., T. Cermak, E. L. Doyle, et al. "Targeting DNA double-strand breaks with TAL effector nucleases." *Genetics* 186, no. 2 (2010): 757–761.

Chun, J. B., B. K. Ha, D. S. Jang, et al. "Identification of mutations in OASA1 gene from a gamma-irradiated rice mutant population." *Plant Breeding* 131, no. 2 (2012): 276–281.

Cuperus, J. T., T. A. Montgomery, N. Fahlgren, et al. "Identification of MIR390a precursor processing-defective mutants in Arabidopsis by direct genome sequencing." *Proceedings of the National Academy of Sciences of the United States of America* 107, no. 1 (2010): 466–471.

Curtin, S. J., F. Zhang, J. D. Sander, et al. "Targeted mutagenesis of duplicated genes in soybean with zinc-finger nucleases." *Plant Physiology* 156, no. 2 (2011): 466–473.

de Framond, A., P. J. Rich, J. McMillan and G. Ejeta. "Effects on Striga parasitism of transgenic maize armed with RNAi constructs targeting essential *S. asiatica* genes." In *Integrating New Technologies for Striga Control: Towards Ending the Witch-Hunt*, edited by: G. Ejeta and J. Gressel. 185–196. World Scientific Publishing Co Pte Ltd, 2007.

DeFrancesco, L. "Move over ZFNs: a new technology for genome editing may put the zinc finger nuclease franchise out of business, some believe. Not so fast, say the finger people." *Nature Biotechnology* 29, no. 8 (2011): 681–685.

Dodo, H. W., K. N. Konan, F. C. Chen, M. Egnin and O. M. Viquez. "Alleviating peanut allergy using genetic engineering: the silencing of the immunodominant allergen Ara h 2 leads to its significant reduction and a decrease in peanut allergenicity." *Plant Biotechnology Journal* 6, no. 2 (2008): 135–145.

Douchkov, D., S. Lueck, G. Hensel, et al. "The barley (*Hordeum vulgare*) cellulose synthase-like D2 gene (HvCslD2) mediates penetration resistance to host-adapted and nonhost isolates of the powdery mildew fungus." *New Phytologist* 212, no. 2 (2016): 421–433.

Doyle, E. L., N. J. Booher, D. S. Standage, et al. "TAL Effector-Nucleotide Targeter (TALE-NT) 2.0: tools for TAL effector design and target prediction." *Nucleic Acids Research* 40.W1 (2012): W117–W122.

Fire, A., S. Xu, M. K. Montgomery, et al. "Potent and specific genetic interference by double-stranded RNA in *Caenorhabditis elegans*." *Nature* 391, no. 6669 (1998): 806–811.

Ghogare, R., B. Williamson-Benavides, F. Ramírez-Torres and A. Dhingra, et al. "CRISPR-associated nucleases: the Dawn of a new age of efficient crop improvement." *Transgenic Research* 29, no. 1 (2020): 1–35.

Gilchrist, E. J., G. W. Haughn, C. C. Ying, et al. "Use of Ecotilling as an efficient SNP discovery tool to survey genetic variation in wild populations of *Populus trichocarpa*." *Molecular Ecology* 15, no. 5 (2006): 1367–1378.

Hahn, F. and V. Nekrasov. "CRISPR/Cas precision: do we need to worry about off-targeting in plants?" *Plant Cell Reports* 38, no. 4 (2019): 437–441.

Hammond, S. M., S. Boettcher, A. A. Caudy, R. Kobayashi, and G. J. Hannon, "Argonaute2, a link between genetic and biochemical analyses of RNAi." *Science* 293, no. 5532 (2001): 1146–1150.

Hannon, G. J., and J. J. Rossi. "Unlocking the potential of the human genome with RNA interference." *Nature* 431, no. 7006 (2004): 371–378.

Haughn, G. W., and E. J. Gilchrist "TILLING in the Botanical Garden: a reverse genetic technique feasible for all plant species." In *Floriculture, Ornamental and Plant Biotechnology: Advances and Topical Issues*, edited by J. A. da Silva, pp. 476–482. Global Science Books, London (2006).

Haun, W., A. Coffman, B. M. Clasen, et al. "Improved soybean oil quality by targeted mutagenesis of the fatty acid desaturase 2 gene family." *Plant Biotechnology Journal* 12, no. 7 (2014): 934–940.

Henikoff, S., B. J. Till, and L. Comai. "TILLING. Traditional mutagenesis meets functional genomics." *Plant Physiology* 135, no. 2 (2004): 630–636.

Hilioti, Z., I. Ganopoulos, S. Ajith, I. Bossis, and A. Tsaftaris "A novel arrangement of zinc finger nuclease system for in vivo targeted genome engineering: the tomato LEC1-LIKE4 gene case." *Plant Cell Reports* 35, no. 11 (2016): 2241–2255.

Holkers, M., I. Maggio, J. Liu, et al. "Differential integrity of TALE nuclease genes following adenoviral and lentiviral vector gene transfer into human cells." *Nucleic Acids Research* 41, no. 5 (2013): e63.

Hu, Q., Y. Niu, K. Zhang, Y. Liu and X. Zhou. "Virus-derived transgenes expressing hairpin RNA give immunity to tobacco mosaic virus and cucumber mosaic virus." *Virology Journal* 8, no. 1 (2011): 41.

Hu, X., X. Meng, Q. Liu, J. Li and K. Wang. "Increasing the efficiency of CRISPR-Cas9-VQR precise genome editing in rice." *Plant Biotechnology Journal* 16, no. 1 (2018): 292–297.

Jagannadham, P. T. K., S. K. Muthusamy and P. Chidambaranathan. "Micromics: A novel approach to understand the molecular mechanisms in plant stress tolerance." In *Recent Approaches in Omics for Plant Resilience to Climate Change*, edited by S. Wani, pp. 93–108. Springer, Cham (2019).

Jinek, M., K. Chylinski, I. Fonfara, M. Hauer, J. A. Doudna and E. Charpentier. "A programmable dual-RNA–guided DNA endonuclease in adaptive bacterial immunity." *Science* 337, no. 6096 (2012): 816–821.

Jones, A. M., J. Chory, J. L. Dangl, et al. "The impact of Arabidopsis on human health: diversifying our portfolio." *Cell* 133, no. 6 (2008): 939–943.

Joseph, T. T., and R. Osman "Convergent transmission of RNAi guide-target mismatch information across argonaute internal allosteric network." *PLoS Computational Biology* 8, no. 9 (2012): e1002693.

Katiyar, A., S. Smita, S. K. Muthusamy, V. Chinnusamy, D. M. Pandey and K. C. Bansal. "Identification of novel drought-responsive microRNAs and trans-acting siRNAs from *Sorghum bicolor* (L.) Moench by high-throughput sequencing analysis." *Frontiers in Plant Science* 6 (2015): 506.

Kim, Y. G., J. Cha and S. Chandrasegaran. "Hybrid restriction enzymes: Zinc finger fusions to FokI cleavage domain." *Proceedings of the National Academy of Sciences of the United States of America* 93, no. 3 (1996): 1156–1160.

Kim, H. J., H. J. Lee, H. Kim, S. W. Cho and J. S. Kim. "Targeted genome editing in human cells with zinc finger nucleases constructed via modular assembly." *Genome Research* 19, no. 7 (2009): 1279–1288.

Kim, H., M. R. Yoon, A. Chun and T. H. Tai "Identification of novel mutations in the rice starch branching enzyme I gene via TILLING by sequencing." *Euphytica* 214, no. 6 (2018): 94.

Koch, A., N. Kumar, L. Weber, H. Keller, J. Imani and K. H. Kogel "Host-induced gene silencing of cytochrome P450 lanosterol C14α-demethylase–encoding genes confers strong resistance to Fusarium species." *Proceedings of the National Academy of Sciences of the United States of America* 110, no. 48 (2013): 19324–19329.

Koch, A., D. Biedenkopf, A. Furch, et al. "An RNAi-based control of *Fusarium graminearum* infections through spraying of long dsRNAs involves a plant passage and is controlled by the fungal silencing machinery." *PLoS Pathogens* 12, no. 10 (2016): e1005901.

Konan, K. N., O. M. Viquez, F. C. Chen and H. W. Dodo. "Success towards alleviating peanut allergy: The major allergen Ara h 2 is silenced via RNA interfere (RNAi)." In *Biotechnology and Sustainable Agriculture 2006 and Beyond*, edited by Z. Xu et al., pp. 261–264. Springer, Dordrecht (2007).

Kumar, A. P., P. C. McKeown, A. Boualem, et al. "TILLING by Sequencing (TbyS) for targeted genome mutagenesis in crops." *Molecular Breeding* 37, no. 2 (2017): 14.

Kumar, S., S. K. Muthusamy, C. N. Mishra, V. Gupta and K. Venkatesh. "Importance of genomic selection in crop improvement and future prospects." In *Advanced Molecular Plant Breeding: Meeting the Challenge of Food Security*, edited by D. N. Bharadwaj, pp. 275–96. CRC Press, Boca Raton, FL (2018).

Kurowska, M., A. Daszkowska-Golec, D. Gruszka, et al. "TILLING-a shortcut in functional genomics." *Journal of Applied Genetics* 52, no. 4 (2011): 371.

Lababidi, S., N. Mejlhede, S. K. Rasmussen, et al. "Identification of barley mutants in the cultivar 'Lux'at the Dhn loci through TILLING." *Plant Breeding* 128, no. 4 (2009): 332–336.

Le, L. Q., V. Mahler, Y. Lorenz, et al. "Reduced allergenicity of tomato fruits harvested from Lyc e 1–silenced transgenic tomato plants." *Journal of Allergy and Clinical Immunology* 118, no. 5 (2006): 1176–1183.

Lee, C. M., T. J. Cradick and G. Bao. "The Neisseria meningitidis CRISPR-Cas9 system enables specific genome editing in mammalian cells." *Molecular Therapy* 24, no. 3 (2016): 645–654.

Lee, S. C., S. J. Kim, S. K. Han, G. An and S. R. Kim "A gibberellin-stimulated transcript, *OsGASR1*, controls seedling growth and α-amylase expression in rice." *Journal of Plant Physiology* 214 (2017): 116–122.

Lenka, S. K., S. K. Muthusamy, V. Chinnusamy and K. C. Bansal "Ectopic expression of rice *PYL3* enhances cold and drought tolerance in *Arabidopsis thaliana*." *Molecular Biotechnology* 60, no. 5 (2018): 350–361.

Lenka, S. K., A. K. Singh, S. K. Muthusamy, S. Smita, V. Chinnusamy and K. C. Bansal "Heterologous expression of rice RNA-binding glycine-rich (RBG) gene *OsRBGD3* in transgenic *Arabidopsis thaliana* confers cold stress tolerance." *Functional Plant Biology* 46, no. 5 (2019): 482–491.

Li, T., B. Liu, M. H. Spalding, D. P. Weeks and B. Yang "High-efficiency TALEN-based gene editing produces disease-resistant rice." *Nature Biotechnology* 30, no. 5 (2012): 390.

Li, C., W. Li, Z. Zhou, H. Chen, C. Xie and Y. Lin "A new rice breeding method: CRISPR/Cas9 system editing of the Xa13 promoter to cultivate transgene-free bacterial blight-resistant rice." *Plant Biotechnology Journal* 18, no. 2 (2020): 313–315.

Lieber, M. R. "The mechanism of double-strand DNA break repair by the nonhomologous DNA end-joining pathway." *Annual Review of Biochemistry* 79 (2010): 181–211.

Liu, C., C. Pedersen and T. Schultz-Larsen. "The stripe rust fungal effector PEC 6 suppresses pattern-triggered immunity in a host species-independent manner and interacts with adenosine kinases." *New Phytologist* (2016). doi: 10.1111/nph.14034.

Lor, V. S., C. G. Starker, D. F. Voytas, D. Weiss and N. E. Olszewski. "Targeted mutagenesis of the tomato PROCERA gene using transcription activator-like effector nucleases." *Plant Physiology* 166, no. 3 (2014): 1288–1291.

Maghuly, F., S. Pabinger, J. Krainer and M. Laimer "Facilitation of *Fusarium graminearum* infection by 9-lipoxygenases in Arabidopsis and wheat." *Molecular Plant-Microbe Interactions* 28, no. 10 (2015): 1142–1152.

Maghuly, F, S. Pabinger, J. Krainer and M. Laimer "The pattern and distribution of induced mutations in *J. curcas* using reduced representation sequencing." *Frontiers in Plant Science* 9 (2018): 524.

Mani, M., J. Smith, K. Kandavelou, J. M. Berg and S. Chandrasegaran. "Binding of two zinc finger nuclease monomers to two specific sites is required for effective double-strand DNA cleavage." *Biochemical and Biophysical Research Communications* 334, no. 4 (2005): 1191–1197.

Mansoor, S., I. Amin, M. Hussain, Y. Zafar and R. W. Briddon. "Engineering novel traits in plants through RNA interference." *Trends in Plant Science* 11, no. 11 (2006): 559–565.

Marroni, F., S. Pinosio, E. Di Centa, et al. "Large-scale detection of rare variants via pooled multiplexed next-generation sequencing: towards next-generation Ecotilling." *The Plant Journal* 67, no. 4 (2011a): 736–745.

Marroni, F., S. Pinosio, G. Zaina, et al. "Nucleotide diversity and linkage disequilibrium in Populus nigra cinnamyl alcohol dehydrogenase (CAD4) gene." *Tree Genetics and Genomes* 7, no. 5 (2011b): 1011–1023.

McCallum, C. M., L. Comai, E. A. Greene and S. Henikoff. "Targeting induced locallesions in genomes (TILLING) for plant functional genomics." *Plant Physiology* 123, no. 2 (2000): 439–442.

Minoia, S., A. Petrozza, O. D'Onofrio, et al. "A new mutant genetic resource for tomato crop improvement by TILLING technology." *BMC Research Notes* 3, no. 1 (2010): 69.

Mitter, N., V. Koundal, S. Williams and H. Pappu. "Differential expression of tomato spotted wilt virus-derived viral small RNAs in infected commercial and experimental host plants." *PLoS One* 8, no. 10 (2013): e76276.

Moscou, M. J. and A. J. Bogdanove. "A simple cipher governs DNA recognition by TAL effectors." *Science* 326, no. 5959 (2009): 1501–1501.

Moynahan, M. E., and M. Jasin. "Mitotic homologous recombination maintains genomic stability and suppresses tumorigenesis." *Nature Reviews Molecular Cell Biology* 11, no. 3 (2010): 196–207.

Muthusamy, S. K., M. Dalal, V. Chinnusamy and K. C. Bansal. "Differential regulation of genes coding for organelle and cytosolic ClpATPases under biotic and abiotic stresses in wheat." *Frontiers in Plant Science* 7 (2016): 929.

Muthusamy, S. K., M. Dalal, V. Chinnusamy and K. C. Bansal. "Genome-wide identification and analysis of biotic and abiotic stress regulation of small heat shock protein (HSP20) family genes in bread wheat." *Journal of Plant Physiology* 211 (2017a): 100–113.

Muthusamy, S. K., P. N. Sivalingam, J. Sridhar, D. Singh, S. M. Haldhar and P. Kaushal. "Biotic stress inducible promoters in crop plants-a review." *Journal of Agriculture and Ecology*, 4 (2017b): 14–24.

Muthusamy, S. K., S. K. Lenka, A. Katiyar, V. Chinnusamy, A. K. Singh and K. C. Bansal. "Genome-wide identification and analysis of biotic and abiotic stress regulation of C4 photosynthetic pathway genes in rice." *Applied Biochemistry and Biotechnology* 187, no. 1 (2019a): 221–238.

Muthusamy, S. K., A. Marathe, M. Mahajan, et al. "Gene trap strategies to unravel and characterize novel biotic-stress inducible genes in plants." *Research Journal of Biotechnology*, 14, no. 12 (2019b): 168–175.

Nalam, V. J., S. Alam, J. Keereetaweep, B. Venables, D. Burdan, H. Lee, H. N. Trick, S. Sarowar, R. Makandar, J. Shah. "Facilitation of *Fusarium graminearum* Infection by 9-Lipoxygenases in *Arabidopsis* and Wheat Facilitation of *Fusarium graminearum* Infection by 9-Lipoxygenases in *Arabidopsis* and Wheat." *MPMI*, 28, no. 10 (2015): 1142–1152.

Napoli, C., C. Lemieux and R. Jorgensen "Introduction of a chimeric chalcone synthase gene into petunia results in reversible co-suppression of homologous genes in trans." *The Plant Cell* 2, no. 4 (1990): 279–289.

Navarro-León, E., J. M. Ruiz, A. lbacete and Blasco B. "Tolerance to cadmium toxicity and phytoremediation potential of three Brassica rapa CAX1a TILLING mutants." *Ecotoxicology and Environmental Safety* 189 (2020): 109961.

Nawaz-ul-Rehman, M. S., S. Mansoor, A. A. Khan, Y. Zafar, and R. W. Briddon. "RNAi-mediated male sterility of tobacco by silencing TA29." *Molecular Biotechnology* 36, no. 2 (2007): 159–165.

Nowara, D., A. Gay, C. Lacomme, et al. "HIGS: Host-induced gene silencing in the obligate biotrophic fungal pathogen *Blumeria graminis*." *The Plant Cell* 22, no. 9 (2010): 3130–3141.

O'Malley, R. C. and J. R. Ecker "Linking genotype to phenotype using the Arabidopsis unimutant collection." *The Plant Journal* 61, no. 6 (2010): 928–940.

O'Malley, R. C., C. C. Barragan and J. R. Ecker. "A user's guide to the Arabidopsis T-DNA insertion mutant collections." In *Plant Functional Genomics*, edited by J. Alonso and A. Stepanova, pp. 323–342. Humana Press, New York (2015).

Page, D. R. and U. Grossniklaus. "The art and design of genetic screens: *Arabidopsis thaliana*." *Nature Reviews Genetics* 3, no. 2 (2002): 124–136.

Panwar, V., B. McCallum and G. Bakkeren. "Endogenous silencing of *Puccinia triticina* pathogenicity genes through in planta-expressed sequences leads to the suppression of rust diseases on wheat." *The Plant Journal* 73, no. 3 (2013a): 521–532.

Panwar, V., B. McCallum and G. Bakkeren. "Host-induced gene silencing of wheat leaf rust fungus *Puccinia triticina* pathogenicity genes mediated by the Barley stripe mosaic virus." *Plant Molecular Biology* 81, no. 6 (2013b): 595–608.

Parry, M. A., P. J. Madgwick, C. Bayon, et al. "Mutation discovery for crop improvement." *Journal of Experimental Botany* 60, no. 10 (2009): 2817–2825.

Pérez-Martín, F., F. J. Yuste-Lisbona, B. Pineda, et al. "A collection of enhancer trap insertional mutants for functional genomics in tomato." *Plant Biotechnology Journal* 15, no. 11 (2017): 1439–1452.

Pradhan, A., N. Naik and K. K. Sahoo. "RNAi mediated drought and salinity stress tolerance in plants." *American Journal of Plant Sciences* 6, no. 12 (2015): 1990.

Provart, N. J., J. Alonso, S. M. Assmann, et al. "50 years of Arabidopsis research: Highlights and future directions." *New Phytologist* 209, no. 3 (2016): 921–944.

Qiao, Y., W. Jiang, J. Lee, et al. "SPL28 encodes a clathrin-associated adaptor protein complex 1, medium subunit μ1 (AP1M1) and is responsible for spotted leaf and early senescence in rice (Oryza sativa)." *New Phytologist* 185, no. 1 (2010): 258–274.

Ran, Y., N. Patron, P. Kay, et al. "Zinc finger nuclease-mediated precision genome editing of an endogenous gene in hexaploid bread wheat (*Triticum aestivum*) using a DNA repair template." *Plant Biotechnology Journal* 16, no. 12 (2018): 2088–2101.

Rawat, N., S. K. Sehgal, A. Joshi, et al. "A diploid wheat TILLING resource for wheat functional genomics." *BMC Plant Biology* 12, no. 1 (2012): 205.

Richaud, D., C. Stange, A. Gadaleta, P. Colasuonno, R. Parada, A. R. Schwember, "Identification of Lycopene epsilon cyclase (LCYE) gene mutants to potentially increase β-carotene content in durum wheat (Triticum turgidum L.ssp. durum) through TILLING." *PLoS ONE* 13, no. 12 (2018): e0208948.

Sander, J. D., E. J. Dahlborg, M. J. Goodwin, et al. "Selection-free zinc-finger-nuclease engineering by context-dependent assembly (CoDA)." *Nature Methods* 8, no. 1 (2011): 67–69.

Sandhu, A. P. S., R. V. Abdelnoor and S. A. Mackenzie. "Transgenic induction of mitochondrial rearrangements for cytoplasmic male sterility in crop plants." *Proceedings of the National Academy of Sciences of the United States of America* 104, no. 6 (2007): 1766–1770.

Sankaranarayanan, R., S. N. Palani, A. Kumar, A. S. Punitha Selvakumar and J. Tennyson "Prediction and experimental confirmation of banana bract mosaic virus encoding miRNAs and their targets." *ExRNA* 2, no. 1 (2020): 5.

Schnepf, V., A. C. Vlot, K. Kugler, and R. Hückelhoven "Barley susceptibility factor RACB modulates transcript levels of signalling protein genes in compatible interaction with *Blumeria graminis f. sp. hordei.*" *Molecular Plant Pathology* 19, no. 2 (2018): 393–404.

Segal, G., R. Song and J. Messing. "A new opaque variant of maize by a single dominant RNA-interference-inducing transgene." *Genetics* 165, no. 1 (2003): 387–397.

Shan, Q., Y. Wang, J. Li, et al. "Targeted genome modification of crop plants using a CRISPR-Cas system." *Nature Biotechnology* 31, no. 8 (2013): 686–688.

Shan, Q., Y. Zhang, K. Chen, K. Zhang and C. Gao "Creation of fragrant rice by targeted knockout of the *OsBADH2* gene using TALEN technology." *Plant Biotechnology Journal* 13, no. 6 (2015): 791–800.

Sharma, P., B. Pandey, S. K. Muthusamy, et al. "Development and validation of microsatellite markers for Karnal bunt (*Tilletia indica*) and loose smut (*Ustilago segetum tritici*) of wheat from related fungal species." *Journal of Phytopathology* 166, no. 10 (2018): 729–738.

Shukla, V. K., Y. Doyon, J. C. Miller, et al. "Precise genome modification in the crop species Zea mays using zinc-finger nucleases." *Nature* 459, no. 7245 (2009): 437–441.

Suprasanna, P., S. M. Jain, S. J. Ochatt, V. M. Kulkarni and S. Predieri, "Applications of in vitro techniques in mutation breeding of vegetatively propagated crops." In Plant Mutation Breeding and Biotechnology, edited by Q. Y. Shu, B. P. Forster and H. Nakagawa, pp. 371–385. CABI, Wallingford (2012).

Tang, X., Q. Ren, L. Yang, et al. "Single transcript unit CRISPR 2.0 systems for robust Cas9 and Cas12a mediated plant genome editing." *Plant Biotechnology Journal* 17, no. 7 (2019): 1431–1445.

Tijsterman, M., and R. H. Plasterk, "Dicers at RISC: The mechanism of RNAi." *Cell* 117, no. 1 (2004): 1–3.

Till, B. J., S. H. Reynolds, E. A. Greene, et al. "Large-scale discovery of induced point mutations with high-throughput TILLING." *Genome Research* 13, no. 3 (2003): 524–530.

Tomari, Y., and P. D. Zamore. "Perspective: Machines for RNAi." *Genes and Development* 19, no. 5 (2005): 517–529.

Uauy, C., F. Paraiso, P. Colasuonno, et al. "A modified TILLING approach to detect induced mutations in tetraploid and hexaploid wheat." *BMC Plant Biology* 9, no. 1 (2009): 115.

Valarmathi, P., G. Kumar, S. Robin, S. Manonmani, I. Dasgupta and R. Rabindran. "Evaluation of virus resistance and agronomic performance of rice cultivar ASD 16 after transfer of transgene against rice tungro bacilliform virus by backcross breeding." *Virus Genes* 52, no. 4 (2016): 521–529.

Valton, J., A. Dupuy, F. Daboussi, et al. "Overcoming transcription activator-like effector (TALE) DNA binding domain sensitivity to cytosine methylation." *Journal of Biological Chemistry* 287, no. 46 (2012): 38427–38432.

Viana, V. E., C. Pegoraro, C. Busanello and A. C. de Oliveira "Mutagenesis in rice: The basis for breeding a new super plant." *Frontiers in Plant Science* 10 (2019): 1–18.

Wang, M., A. Weiberg, F. M. Lin, B. P. Thomma, H. D. Huang and H. Jin, "Bidirectional cross-kingdom RNAi and fungal uptake of external RNAs confer plant protection." *Nature Plants* 2, no. 10 (2016): 1–10.

Weber, E., C. Engler, R. Gruetzner, S. Werner and S. Marillonnet "A modular cloning system for standardized assembly of multigene constructs." *PLoS One* 6, no. 2 (2011): e16765.

Woodward, A. W., and B. Bartel. "Biology in bloom: a primer on the *Arabidopsis thaliana* model system." *Genetics* 208, no. 4 (2018): 1337–1349.

Wu, H., C. Sparks, B. Amoah, et al. "Factors influencing successful *Agrobacterium*-mediated genetic transformation of wheat." *Plant Cell Reports* 21, (2003): 659–668.

Wu, J. L., C. Wu, C. Lei, et al. "Chemical-and irradiation-induced mutants of indica rice IR64 for forward and reverse genetics." *Plant Molecular Biology* 59, no. 1 (2005): 85–97.

Wu, C., C. You, C. Li, et al. "RID1, encoding a Cys2/His2-type zinc finger transcription factor, acts as a master switch from vegetative to floral development in rice." *Proceedings of the National Academy of Sciences of the United States of America* 105, no. 35 (2008): 12915–12920.

Yang, L., W. Huang, F. Xiong, et al. "Silencing of Sl PL, which encodes a pectate lyase in tomato, confers enhanced fruit firmness, prolonged shelf-life and reduced susceptibility to grey mould." *Plant Biotechnology Journal* 15, no. 12 (2017): 1544–1555.

Yin, C., J. E. Jurgenson and S. H. Hulbert "Development of a host-induced RNAi system in the wheat stripe rust fungus *Puccinia striiformis f. sp. tritici.*" *Molecular Plant-Microbe Interactions* 24, no. 5 (2011): 554–561.

Zhang, A., Y. Liu, F. Wang "Enhanced rice salinity tolerance via CRISPR/Cas9-targeted mutagenesis of the *OsRR22* gene." *Molecular Breeding* 39, no. 3 (2019): 47.

Zhou, Q., Z. Zhang, T. Liu, B. Gao and X. Xiong "Identification and map-based cloning of the Light-Induced Lesion Mimic Mutant 1 (LIL1) gene in rice." *Frontiers in Plant Science* 8 (2017): 2122.

Zhu, X., S. Guo, Z. Wang, et al. "Map-based cloning and functional analysis of YGL8, which controls leaf colour in rice (*Oryza sativa*)." *BMC Plant Biology* 16, no. 1 (2016): 134.

Zluhan-Martínez, E., V. Pérez-Koldenkova, M. V. Ponce-Castañeda, et al. "Beyond what your retina can see: Similarities of retinoblastoma function between plants and animals, from developmental processes to epigenetic regulation." *International Journal of Molecular Sciences* 21, no. 14 (2020): 4925.

Zong, Y., Y. Wang, C. Li, et al. "Precise base editing in rice, wheat and maize with a Cas9-cytidine deaminase fusion." *Nature Biotechnology* 35, no. 5 (2017): 438–440.

14 Genome Editing in Plant: Challenges and Opportunities

Gaurav Kumar
ICAR-National Rice Research Institute

Bhupendra Singh Panwar
Karnavati Seeds Pvt Ltd

Nabaneeta Basak
ICAR-National Rice Research Institute

CONTENTS

14.1 INTRODUCTION

The world relies upon agriculture, and agriculture itself faces stupendous difficulties, for example addressing the availability of sufficient food for consistently expanding human population and enough fodder for our livestock. It is important to address issues like utilizing the shrinking-accessible arable land while additionally adjusting to the changing climatic conditions which cause frequent flooding, long dry periods and continuous high temperatures. Breeding crop plants that can withstand various biotic and environmental stresses with little utilization of environmentally harmful agrochemicals will help in maintaining a sustainable and favourable habitat.

One of the ways out is to modulate the adaptability of crop plant against these stresses through genome editing. Genome editing allows to make exact genome adjustments, for example deletion and substitution, insertion and epiallelic modifications of the genomic material in live organism. Basically, genome engineering is creating DNA double-stranded breaks (DSBs) *in vivo* through modified endonucleases that are designed to target the predefined genomic sequence and ultimately restoration of modified sequence by exploiting the endogenous DSB repair mechanism

(Carroll, 2014). The customized nucleases are broadly categorized into meganucleases (Gong and Golic, 2003), ZFNs (zinc finger nucleases) (Urnov et al., 2005), TALENs (transcription activator-like effector nucleases) (Christian et al., 2010; Li et al., 2011) and CRISPR/Cas9 (clustered regulatory interspaced short palindromic repeats and CRISPR-associated nuclease) (Jinek et al., 2012; Cong et al., 2013; Mali et al., 2013). Moreover, variants of programmable nuclease like base editors can perform gene modification without DNA DSBs (Veillet et al., 2019) and epigenomic editor (Holtzman and Gersbach, 2018).

Developed in the late 1900s, genome editing technology is the most significant biotechnological toolkit that enables scientists to make changes in DNA, leading to changes in physical traits due to metabolic alteration. It assists in a comprehensive understanding of applied and basic research, and strengthens the field of molecular biology with huge commitments that have empowered fast advancement in industry, medication and agriculture (Gaj et al., 2016). In the last decade, CRISPR-based genome editing is widely accepted and used in plant genome editing in diverse fields, including crop improvement and plant functional genomics. The sequential steps involved in genome editing technologies in plants are as follows: sequence-specific modification of endonucleases, transport of editing tools, generation and selection of edited cells/events, validation followed by the exploitation of modified plants and at last, repercussion like acceptance by general public and regulatory consent of the genetically modified organisms (GMOs). In the last decade, the platform of genome editing has been established and successfully used for genome editing in ~45 different plant genera (Shan et al., 2020).

14.2 CHALLENGES IN PLANT GENOME EDITING

Though, a comprehensive insight into the genetic and molecular mechanism of genome engineering, progressive improvement and novel usages of genome editing tool in the agriculture field still requires significant research exploration. In particular, even after numerous efforts by plant biotechnologists, plant genome editing faces significant difficulties like lack of efficient transformation system, regeneration into full plant, simultaneous manipulation of more than one target at a time etc., which are sketched out underneath.

14.2.1 Defeating Species and Genotype-Dependent Transformation

Plant transformation has empowered major bits of knowledge in crop science and reformed commercial agriculture. Tragically, most crops resist cell transformation and plant regeneration and stay laborious even after over 30 years of innovative achievement. Genome alteration gives novel chances to upgrade crop profitability yet depends on successful plant transformation and regeneration, which becomes a serious obstacle in the practice. Successful transformation includes two major successive steps, i.e. DNA delivery into the plant cell and stable integration of introducing DNA into cell genome. Each of the progression is helpful in fundamental as well as applied research; however, the second step that involves a stable integration of transgene into the cell genome is important to produce the inheritable trait of interest in transgenic plant. For most crops, the ease of regeneration of transformed tissue is a prerequisite for the production of transgenic plant (National Research Council, 1987). Hence, except few crops, the step of transformation and plant regeneration is still a bottleneck during the process of genetic engineering.

Agrobacterium-mediated transformation and particle bombardment transformation have been drilled for over 30 years for the delivery of genetic material through a carrier. Nonetheless, current methodologies stay inefficient for many agricultural crops. The essential challenges include low frequency; stable integration; prolonged tissue culture cycle to regenerate transgenic plant from transformed tissue; low transgene titre that is insufficient for homologous recombination, conveyed by agrobacterium; and low precision of the particle bombardment method for gene transfer. Each challenge speaks to scientific research issues that, when settled, will essentially decrease time and labour constraints for crop engineering. For instance, the floral dip transformation method of *Arabidopsis thaliana* has been a game-changing innovation for that species, not on the grounds that it increases the transformation efficacy since it wiped out the requirement for tissue culture to regenerate plants (Hwang et al., 2017). Besides, this strategy is so in fact easy to perform that even entry-level researcher can effectively transform Arabidopsis. The development of gene transfer protocol that doesn't require tissue culture would be the best way out and could be used in numerous laboratories. Notwithstanding, there is no clear path for such advances, and therefore, it is essential to keep on improving tissue culture-based methods that are generally used to design crop plants (Harwood, 2012).

14.2.2 Genome Editing for Targets Expansion

The genome editing in plants/crops is difficult due to the complex crop genome and lack of complete information regarding genes and other regulatory elements. In spite of these hurdles, it can be overwhelmed through fine-tuning the in-depth knowledge of concerned genes and other genetic regulatory elements of crop plants (Kwon et al., 2020; Oliva et al., 2019). The intricate biological processes of plants like genes and pathways regulation and their interactions with external factors limit the precise targets for crop improvement via genome editing. Fortunately, the CRISPR/Cas technology emerges as a tool, which can perform high-throughput functional genomics and genome-wide screening and traits/gene discovery in model plants or agriculturally important crops (Lu et al., 2017; Meng et al., 2017). The solution relies in coordinating multidisciplinary studies for candidate gene identification suitable for genome editing with a remarkable potential for superior agronomic traits (Araus et al., 2018).

14.2.3 Base Editing with High Efficiency

Most genome editing applications and improvements are centred around targeted mutagenesis. Interestingly, the strategies and applications for the base substitution stay restricted, especially in plant genome modification because of the exceptionally low efficiency in certain crops, whereas most of the crops resist for genome editing. In crop plants, often SNPs (single-nucleotide polymorphisms) governed the genetic variability linked to superior traits; the main challenge is to recognize and incorporate them in elite germplasm/variety in a simplified way. This gives an extraordinary chance to restate the target alleles in the selected genomes through base substitution. Homology-directed repair (HDR)-mediated precise gene inclusion or base substitution gives hope in this. The recently announced RNA and protein-based technology that uses Cas9 and guide RNA variant to modify a pre-targeted DNA sequence additionally holds promise in base substitution whenever exploited and validated in plants (Anzalone et al., 2019). Cas9-based 'prime editing' technology is better in certain examples over the base editors, which translate C to T and A to G in a designated manner of a guide RNA-specified DNA site since it facilitates the substitution of a small genomic region (Rees and Liu, 2018). It's high time for researchers to work for the improvement of methodologies for better efficiency and explore these technologies in a wide range of crops.

14.2.4 Public Perspective and Regulatory Supervision

Recent genome editing technologies are competent enough to imitate the genetic makeup and physiological variations in plants that are undetectable and indistinguishable from those got through spontaneous mutagenesis or chemical mutagenesis methods; hence, the genome-edited plants don't promptly fit the current definition of GMOs as described in the most regulatory authority (Wolt et al., 2016). These technologies facilitate the crop breeding, which is comparatively less time-consuming and more accurate; augments the hereditary capability of crop yield; and creates a more tolerant germplasm. Hence, the challenge is to exploit genome editing technique over the utilization of transgenic approaches by regulatory bodies and their adaptation by general public. It is important to influence the regulatory framework and encourage the public trust, which in turn will encourage the utility of genome modification in crop breeding, and also strengthen the knowledge exchange regarding methods involved in genome editing and its application (Yang, 2020).

14.3 OPPORTUNITIES

Genome editing and the related strategies possibly overcome any barrier between the superior allele's discovery and their quick and proficient use in plant breeding for the direct crop improvement. In spite of the fact that genome editing for crop improvement is in its beginning phase, some crop species have been generated successfully and hence make use of this innovation. Here, we overview the achievement and opportunity of genome editing in various domains of agriculture.

14.3.1 Targeted Mutagenesis or Gene Knockout

ZFNs, TALENs and CRISPR/Cas9 nucleases all are able to introduce highly precise DSBs in chromosomal DNA. These DSBs are repaired by cell DNA repair mechanism mostly by NHEJ (non-homologous end-joining), leading to mutation or sequence replacement (Voytas and Gao, 2014). These nucleases are being utilized to characterize genes through down-regulation, up-regulation or disruption in many economically important organisms, including plants.

Crop yield is considered as one of the most significant characteristics in agriculture. It is controlled by several genes and hence is a quantitative characteristic (Xing and Zhang, 2010; Bai et al., 2012); also, a significant research work has been done for the identification of quantitative trait loci (QTLs) governing yield in different crop plants (Bai et al., 2012; Jianru and Jiayang, 2014). Classical breeding approaches for yield enhancement are time-consuming processes since they rely on stacking different combinations of QTL to identify the most promising ones (Shen et al., 2018). Furthermore, QTL introgression between the varieties isn't in every case simple, particularly with linked loci. Genome editing gives a promising system to quickly and explicitly alter any genomic region in a definite manner. The most direct method of enhancing yield is to knock out genes negatively influencing the yield (Ma et al., 2016; Song et al., 2016). Recently, in rice cultivar Zhonghua 11, four negative regulators of yield, namely the genes IPA1, GS3, DEP1 and Gn1a, were individually knocked out using CRISPR/Cas9 genome editing system. Three of the four genes, gs3, dep1 and gn1a mutants, showed the improved yield parameters, *viz* larger grain size, dense and erect panicle and improved grain number, respectively (Li et al., 2016a). Likewise, three homoalleles of GASR7 gene were targeted by Zhang et al. (2016) in bread wheat using the CRISPR/Cas9 system. GASR7 gene is a negative regulator of kernel weight and width. GASR7 gene knockout causes an increase in the thousand-kernel weight. Rodríguez-Leal et al. (2017) used the CRISPR/Cas9 system to manipulate cis-regulatory element of CLAVATA-WUSCHEL genes in stem cells to produce tomato of large fruit size as well as an increase in the number of locules. Similarly, several genome editing approaches have been deployed by various group of scientists across the globe to knock out genes, which influences the yield characteristics of the crop (Lawrenson et al., 2015; Soyk et al., 2017; Braatz et al., 2017; Li and Yang, 2017; Ma et al., 2018; Zhang et al., 2018a).

Plant diseases are major and significant limitations that compromise the agricultural development and threat to global food security (Taylor et al., 2004). Diseases can considerably decrease the crop quality as well as yield (Savary et al., 2012). Substantial knowledge of physiological and genetic basis of plant–pathogen interaction has been accumulated for economically important plant species (Michelmore, 1995; Kosack and Jones, 1996), which can be exploited further. Genome editing technologies can be utilized to target genes/regulatory elements, which are significant for pathogen infection, thus creating an immunized plant. The first report of utilization of TALEN genome editing for crop improvement was in rice for bacterial blight. Li et al. (2012) utilized TALENs to disrupt cis-regulating element precisely, effector binding site in the promoter region of OsSWEET14 gene (sucrose-efflux transporter gene), which resulted in bacterial blight-resistant rice. Likewise, Wang et al. (2014) successfully deployed TALEN and CRISPR/Cas9 to knock out three TaMLO homoeologs (TaMLO-A, TaMLO-B and TaMLO-D) responsible for wheat susceptibility towards powdery mildew and created a resistant mutant. In other effort, Zhang et al. (2017) targeted TaEDR1 gene homologs to enhance powdery mildew disease resistance. Likewise, in Tomelo (transgene-free tomato), Nekrasov et al. (2017) developed a powdery mildew-resistant tomato through CRISPR/Cas9 system by targeting SlMlo1 gene responsible for susceptibility.

Genome editing technologies can also be harnessed to design DNA viruses such as bean yellow dwarf virus (BeYDV), tomato yellow leaf curl virus (TYLCV), Merremia mosaic virus (MeMV), beet severe curly top virus (BSCTV) in immunized plants (Ali et al., 2015, 2016; Baltes and Voytas, 2015; Ji et al., 2015). Recently, an RNA targeting CRISPR/Cas system known as CRISPR/LshCas13a has been deployed to create RNA virus interference against turnip mosaic virus (TuMV) in plant (Aman et al., 2018). Despite the fact that Cas13a demonstrated an unobtrusive action against TuMV infection, this finding featured the presence of CRISPR/Cas13 system as a potential antivirus strategy. Further development of RNA targeting through CRISPR encourages research not only on RNA virus interference but also on plant engineering strategy focused on RNA targeting (Mahas et al., 2018; Ali et al., 2018). These examples show the extraordinary opportunity for genome editing technologies to create immunized plants.

Like biotic stresses, abiotic stresses comprising drought, extreme temperature and salinity also play a significant role in reducing the crop yield across the globe by hindering in plant physiology (Pandey et al., 2017). Because of the complex attribute of abiotic stresses, fewer genome editing researches have so far been done in this domain. Recently, DuPoint scientists utilized HDR DNA repair mechanism to introduce maize GOS2 promoter into the 5′ UTR (untranslated region) of ARGOS8 gene using CRISPR/Cas9 system. ARGOS8 gene encodes maize ethylene-responses negative regulator protein. This genome manipulation results in the maize line that can survive drought and has yield advantage under a water-deficit condition (Shi et al., 2017). In other effort, Osakabe et al. (2016) mutated Arabidopsis OST2 gene, which resulted in plant showing an environmental factor-responsive altered stomata closing pattern. These studies confirmed that genome editing has a wide opportunity to mitigate abiotic stress for the future crop development.

Engineered plant for herbicide resistance is also one of the potential areas where genome editing technology can play a remarkable role. Weeds in farm field compete for resources such as nutrients, water, space and light, thus resulting in a reduction in the crop yield. Weeds are usually managed by utilizing chemical herbicides which have a negative impact on environment as well as human health. Introgression of herbicide-tolerant gene(s) in the plant has been revolutionized agronomic practices in many crops. Classical methods involved the transfer of herbicide resistance gene from different sources into target crop to generate transgenic (Lombardo et al., 2016). Recently, scientist initiated to use genome editing approaches to produce an herbicide-tolerant plant by manipulating or target mutagenesis of endogenous genes such as EPSPS and ALS (acetolactate synthase) (Lombardo et al., 2016). ALS is a key enzyme in the metabolism of branched-chain amino acids such as leucine, valine and isoleucine (Lee et al., 1988; Chipman et al., 1998). The enzyme activity of ALS is inhibited by certain groups of herbicide, including imidazolinones, pyrimidinylthio benzoates, sulphonylureas, triazolopyrimidines and sulphonylamino-carbonyl-triazolinones (Mazur et al., 1987; Zhou et al., 2007). Genome editing has been utilized to target the conserved region of ALS responsible for susceptibility towards herbicides through a precise DNA alteration. The initial report of ALS gene alteration using genome editing tools was appeared in 2009 when group of researchers utilized ZFNs to introduce target mutagenesis in tobacco ALS genes to deliberate plant resistance against the herbicide sulphonylurea (Cai et al., 2009; Shukla et al., 2009; Townsend et al., 2009). Later on, the same gene has been edited in other crops such as soya bean, maize, rice and potato for herbicide tolerance by exploiting TALENs and CRISPR/Cas9 (Butler et al., 2015; Svitashev et al., 2015; Li et al., 2016b; Sun et al., 2016). EPSPS (5-enolpyruvylshikimate-3-phosphate synthase) is an enzyme involved in the biosynthesis of aromatic amino acids tryptophan, tyrosine and phenylalanine through the shikimate pathway. Since EPSPS is absent in animal, it is a potential target for herbicide such as glyphosate (Ganesh and Dilip, 1988; Schönbrunn et al., 2001). To manipulate EPSPS, genome editing technique was first deployed by Sauer et al. (2016) in *Linum Usitatissimum* (flax). They use the CRISPR/Cas9 system to substitute two nucleotides in the sequence that correspond to the glyphosate-binding site using HDR. Similar approach has been utilized by Li et al. (2016b) to produce glyphosate-resistant rice. This example displayed a plethora of opportunities intended for genome editing tools in agriculture to generate a broad-spectrum herbicide-tolerant crops.

Genome editing can likewise improve the crop nutritional properties to deliver healthier food sources. For example, Liang et al. (2014) exploited genome editing tools (TALENs and CRISPR/Cas9) to reduce the phytate content in maize by targeting ZmIPK gene. Phytate is considered as an antinutrient because of its tendency to bind proteins and mineral, which results in a decreased bioavailability of nutrients during digestion. In potato, during storage, reducing sugars were accumulated, which reacted with the available free amino acid and produced potential carcinogens. Vacuolar invertase (VInv) is an enzyme responsible for reducing starch into glucose and fructose. Clasen et al. (2016) used TALEN genome editing to knock out *VInv* gene, thereby producing carcinogen-free potato. Likewise, CASPR/cas9 system is also deployed to generate wheat with hypo-immunogenic gluten and rice with higher amylopectin through targeting knockout of alpha-gliadin genes and SBEIIb, respectively (Sun et al., 2017; Sánchez et al., 2018). Development of a waxy maize by DuPont Pioneer through target mutagenesis of Wx1 gene (amylose biosynthesis gene) using CRISPR/Cas9 (Waltz, 2016) is another example of successful genome editing for improving the food quality.

The present status of this innovative technology permits numerous applications suitable for improving plant disease resistance, mitigating climate change and enhancing crop productivity.

14.3.2 Multiplex Genome Editing (MGE) and Trait Staking

Multiplex genome editing-based genome engineering permits the sophisticated applications comprising chromosome deletion, gene activation, gene knockout, gene knock-in, gene repression and epigenomic modification etc. Current methodology used to provide the multiple gRNA target includes gRNAs expressing multigene cassettes, arrays of crRNAs, Csy4-based excision, ribozyme-flanked gRNAs, tRNA-based cleavage of gRNAs and direct entry of Cas9 protein loaded with predefined gRNAs. In this section, we divided multiplex genome editing applications and opportunities into the three broad categories, viz multiplex genome editing in metabolic engineering, modulating plant development and molecular farming (Ansari et al., 2020).

Metabolic engineering is referring to the improving genetic and regulatory pathways within cells in order to enhance/diminish the cell production of certain metabolites. One of the best examples reflecting opportunities of multiplex genome editing is the manipulation of gamma-aminobutyric acid (GABA) metabolic pathway in tomato. GABA is a four-carbon-containing non-protein amino acid having a health-promoting function. Recently, Li et al. used the multiplex genome editing to mutate simultaneously five genes, viz. SSADH, CAT9, GABA-TP1, GABA-TP2 and GABA-TP3 regulating the GABA shunt metabolic pathway. By using multiplex gRNA expression cassette, the author was successfully able to edit four genes except GABA-TP2. The GABA accumulation in the edited plant was found to be significantly higher than in the wild type (Li et al., 2018a). In a same manner, they also develop the tomato plant with an enhanced lycopene content by targeting five genes, *viz* SGR1, Blc, LCY-E, LCY-B1 and LCY-B2, involved in the carotenoid pathway (Li et al., 2018b). In rice, three genes (GW2, GW5 and TGW6) negatively regulating the grain weight have been simultaneously knocked out using a CRISPR/Cas9-mediated multiplexing genome editing system (Xu et al., 2016). As a result, a significant increase in thousand-grain weight has been observed in modified plants. Similarly, Andersson et al. (2017) edited the potato plant for only amylopectin starch by simultaneously knocking out four alleles of GBSS (granule-bound starch synthase) gene in potato.

On the other hand, Sun et al. (2017) generated high amylose-containing rice by knocking out SBEI and SBEIIb (starch-branching enzyme genes) deploying the CRISPR/Cas9 system. Multiplex genome editing is also proved to be significant to manipulate the fatty acid composition in food oils. In order to manipulate the fatty acid composition in soya beans, Haun et al. (2014) knocked out two genes, *viz* FAD2-1A and FAD2-1B, which resulted in an increase in oleic acid level. By using the same strategy but different genome editing tool, i.e. CRISPR/Cas9, Jiang et al. (2017) simultaneously knocked out FAD2 homologs in allohexaploid crop *Camelina sativa*. Gene knockout reduced the level of undesirable polyunsaturated fatty acids and increased the oleic acid level.

In rice, to understand the unique function of redundant gene, Minkenberg et al. (2017) used CRISPR/Cas9 genome editing to mutate four genes (MPK1, MPK2, MPK5 and MPK6) involved in map kinase signalling pathway. Other example of multiplex gene editing to study the plant development is again in rice. Simultaneously, 11 of the 13 members of floral regulator FT-like family were targeted through one construct by Ma et al. (2015b). Highest number of simultaneous mutations identified in one line was seven, causing an early senescence phenotype. Furthermore, Li et al. (2017) utilized MGE to simultaneously mutate the three floral repressor genes, *viz* Hd2, Hd4 and Hd5, to generate early-maturing varieties. The authors reported 40 days' early flowering in mutant as compared to the wild type.

Apart from economically important agronomic traits manipulation, genome editing also has applications in molecular farming. In molecular farming, genome editing is used to modify the nature of recombinant protein expressed in the plant system specifically through the adjustment of plant metabolism to elude the synthesis of glycans. The key enzyme targets are XylT (b(1,2)xylosyltransferase) and FucT (a(1,3)-fucosyltransferase), which are accountable for the plant protein glycan signatures that differ from mammals. Recently, two groups of scientists independently create BY2 mutant tobacco cell lines lacking the plant-specific glycoprotein signature through MGE of genes encoding XylT and FucT enzymes (Mercx et al., 2017; Hanania et al., 2017).

14.3.3 Gene Replacement

Gene or DNA fragments replacement with the desired sequence allows a quick way to introduce elite allele into economically important crops. Genome editing-mediated HDR of DSBs possibly gives us such an opportunity to bringing the elite alleles into crops within two to three generations. In rice, single-nucleotide polymorphism in the nitrate transporter gene *NRT1.1B* is responsible for difference in nitrogen-use efficiency (NUE) between japonica and indica cultivars. The *NRT1.1B* gene from indica cultivar is more efficient in NUE compared to the gene from japonica cultivar. By using genome editing tool CRISPR/Cas system, japonica *NRT1.1B* gene was replaced by indica counterpart in just one generation (Li et al., 2018c). Although a plethora of studies are available describing the gene replacement using genome editing tools in *Arabidopsis*, maize, tobacco, tomato, potato, soya bean, rice and wheat, it still remains challenging to execute a precise gene replacement of agronomically important genes in most of crop plants with high efficiency and frequency.

14.3.4 Gene Regulation

Genome editing tools can be utilized for numerous other applications besides genome sequence manipulation; e.g., ectopic gene expression regulation and epigenetic modification provide an insight into gene function and can also be used to design novel metabolic networks for the synthetic biological application. Either TAL effector or dead Cas9 can be assembled to an activator-like VP64 or a gene repressor like SRDX or epigenetic acetylation or methylation modifier (Lo and Qi, 2017). Among the genome editing techniques, TAL effector and Cas9 are known to be more effective in activating/repressing/epigenetic modification of gene expression than the other. But until this point, no comparative study has been completed to make an exact comparison of the two genome editing systems for gene regulation in plants.

TAL effector proteins are natural plant transcription activator. The endogenous activation domain of TAL effectors is a potent activator but larger in size, which can be swapped with VP64 domain to make smaller proteins. In an experiment conducted by Mahfouz et al. (2011), they fused TAL effectors with SRDX repressor domain and transformed Arabidopsis. The result shows the twofold repression of gene expression. Despite the established concept, no reports describing the manipulation of gene expression in crop plants are available. This could be due to difficulties in *de novo* engineering of TAL effectors and their expression regulation in the plant system.

Theoretically, CRISPR/Cas9 system is more suitable for manipulating gene regulation due to the simplest multiplexing and engineering. Furthermore, successful gene activation and repression were already demonstrated in tobacco and Arabidopsis using CRISPR-dCas9-based system. Lowder et al. (2015) fused dCas9 with VP64 activator domain to make CRISPR-dCas9-VP64 activator, which increased the mRNA expression of Arabidopsis endogenous imprinted gene AtFIS2 up to 400-fold. Despite the fact that these outcomes are revitalizing, they only represent the first generation of such activators/repressors. Further development of CRISPR-dCas9-based transcriptional regulation for superior productivity in crop plants is envisioned.

14.3.5 Visualization of Loci in Large Genomes

For decades, the visualization of sequence-specific chromatin predominantly depends on fluorescence in situ hybridization (FISH)-based techniques. In FISH, the target DNA sequence is identified through the complementary base pairing of fluorophore-labelled DNA probe with denatured genomic DNA. However, this fluorophore-labelled DNA probe may meddle with chromatin elements; this strategy just gives a roundabout information about the native locus (Shakoori, 2017). Thus, live-cell approaches to deal with the spatiotemporal dynamic of endogenous loci depend on DNA-interacting proteins, which show the target binding in a sequence-specific manner.

User-defined sequence recognition and labelling were first comprehended by exploiting zinc finger (ZFN) proteins. To recognize the specific genomic loci, zinc finger motifs were tandemly arranged into polydactyl ZFN (PZF) protein and fused to a green fluorescent protein. GFP-tagged PZF was the first DNA-binding protein module, which was exploited to visualize and characterize the repetitive DNA sequence *in vivo* (Kim and Kini, 2017). However, there is a limitation of target sequence prediction; hence, newly engineered PZFs need to be subjected for a rigorous selection process, rendering this strategy difficult and expensive.

To conquer these impediments, transcription activator-like effectors (TALEs) fused with the fluorescent protein have been utilized to substitute PZFs. Regardless of precision and efficiency over PZFs, dTALEs have their constraints that prevent their efficient binding to the target DNA, including the formation of aggregates due to self-assembly because of highly repetitive motifs (Zhang et al., 2018b).

Contrary to PZF- or dTALE-based strategy, CRISPR/Cas system is considered as a next-generation method to understand the chromatin dynamic and genome since in the CRISPR system, the target specificity is exclusively determined by the guide RNA. Since the target specificity in CRISPR imaging system is completely driven by sgRNA, it can be customized simply by swapping sgRNA to visualize the new target sequence without the need to substitute protein itself. In humans, GFP-tagged dCas9 has been exploited to understand telomere dynamics by the direct visualization and imaging of telomere repetitive sequence (Ma et al., 2015a). Recently, CRISPR imaging has been utilized in the plant to study telomere dynamic of cells in interphase and DNA–protein interaction *in vivo*. Nevertheless, the true potential of CRISPR imaging in the molecular plant cytology still needs to be investigated (Khosravi et al., 2020).

14.3.6 Creating Marker-Free Transgenic

Transgenic crops are known for good agronomic traits but they still are not acknowledged around the world. An unseen concern associated with GM crop is the horizontal transfer of selectable marker genes from GM crops to non-GM crops of same species or even other related species.

Also, the incorporated DNA very frequently alters the endogenous genes or neighbouring genes expression. To resolve such complications, a modified genome editing technique is required, which can manipulate the DNA sequence without leaving the traditional selectable marker genes or any external nucleic acid in GM plants. In this aspect, the CRISPR-Cas9 platform provides a relevant method of genome modification, which can generate GM crops deprived of foreign DNA. Such plants are categorized under 'cisgenic' and ought to be exempted from GMO regulations. Two broad methodologies of CRISPR-Cas9 platform are being utilized for generating marker or foreign DNA-free plants: the first one is DNA-based strategy and the second is ribonucleoprotein (RNP)-based technique (Soda et al., 2018).

In DNA-based technique, at first the CRISPR-Cas9 platform is used to insert the selectable marker or foreign DNA in plants by using the Agrobacterium-mediated or biolistic transformation protocol. In the subsequent generation, the T-DNA null segregants are selected, which will be marker-free genetically modified plants. However, in RNP-based technique, a sgRNA specific to the predefined target and modified Cas9 protein complex is introduced in plant in a similar manner as explained above. In this method, Cas9 is guided by sgRNA to the predefined target gene where Cas9 creates precise required mutations without any integration in genomic DNA. The RNA–protein complex is then degenerated in the growing tissue without any carry forward to the next generation. Woo et al. (2015) were the first who successfully utilized this method for the plant genome modification. Recently, CRISPR-Cas9 DNA has been successfully used to raise the marker-free transgenic wheat (Zhang et al., 2016). The same group further grew transgene-free wheat by utilizing the RNP technology. RNP platform is also known to be effective against off-target mutations while using customized sequence-specific nucleases for genome editing. Since RNP-based technique doesn't require any foreign DNA, we get transgene-free genetically modified plants for commercial purposes. Most of the crop plants show very less efficiency of transformation and hence resist the genome editing. To overcome this problem, the advanced technology of CRISPR-Cas9 RNPs approach has been successfully utilized for genome editing in maize protoplast (Svitashev et al., 2015). In the end, these progressive advanced techniques in genome editing would help to override all the persisting flaws of genetic engineering in plants. These studies open many doors for genome editing applications in economically important crops.

14.4 CONCLUDING REMARKS

The unique ability of sequence-defined targeted genome modification causing genome-wide genetic diversity in plants initiated a complex but innovative era of basic and applied research in genetics and crop breeding. The intelligible, robust and versatile nature of this genome editing system makes it a highly effective and efficient tool for the precise crop breeding through point mutation, gene knockdown or fine tuning of gene regulation and other desired modifications. It is also reliable for some advanced challenging research works like high-throughput mutant library construction and breeding against virus-resistant plants. However, an efficient technology transfer from the laboratory to the field requires a quick and influential discovery of traits and their linked genetic background, precise gene targeting, efficient delivery systems with the standardized protocol for plant cell and subsequent regeneration, and simple screening method for transformed cells/plants. Fortunately, the available genome editing technologies have all these potentials, which can be exploited meticulously for the production of improved crop varieties, hence ensuring the global food security. Simultaneously, by limiting the cost and time consumption on regulatory scrutiny, the genome editing technologies have fascinated the biotechnology companies and encouraged them to adopt them for a fast and adequate result.

Along with the applied research, the advanced genome editing techniques are strengthening the basic research by revealing the gene function, the gene regulatory mechanism and the associated pathways. The multidisciplinary approach of exploiting the advanced platforms like system and synthetic biology, the next-generation high-throughput sequencing, functional genomics and genome editing technologies will facilitate the engineered crop production with improved qualities.

As a result, ground-breaking changes in plant genetics and crop breeding can be expected in coming years, which will satisfy the growing demand of quality food and hence fulfil the global food security concern.

REFERENCES

Ali, Z., A. Abulfaraj, A. Idris, S. Ali, M. Tashkandi, and M. M. Mahfouz. "CRISPR/Cas9-mediated viral interference in plants." *Genome Biology* 16, no. 1 (2015): 238.

Ali, Z., S. Ali, M. Tashkandi, S. S. E. A. Zaidi, and M. M. Mahfouz. "CRISPR/Cas9-mediated immunity to geminiviruses: Differential interference and evasion." *Scientific Reports* 6, no. 1 (2016): 1–13.

Ali, Z., A. Eid, S. Ali, and M. M. Mahfouz. "Pea early-browning virus-mediated genome editing via the CRISPR/Cas9 system in *Nicotiana benthamiana* and Arabidopsis." *Virus Research* 244 (2018): 333–337.

Aman, R., Z. Ali, H. Butt, et al. "RNA virus interference via CRISPR/Cas13a system in plants." *Genome Biology* 19, no. 1 (2018): 1.

Andersson, M., H. Turesson, A. Nicolia, A. S. Fält, M. Samuelsson, and P. Hofvander. "Efficient targeted multiallelic mutagenesis in tetraploid potato (*Solanum tuberosum*) by transient CRISPR-Cas9 expression in protoplasts." *Plant Cell Reports* 36, no. 1 (2017): 117–128.

Ansari, W. A., S. U. Chandanshive, V. Bhatt, et al. "Genome editing in cereals: Approaches, applications and challenges." *International Journal of Molecular Sciences* 21, no. 11 (2020): 4040.

Anzalone, A. V., P. B. Randolph, J. R. Davis, et al. "Search-and-replace genome editing without double-strand breaks or donor DNA". *Nature* 576 (2019): 149–157.

Araus, J. L., S. C. Kefauver, M. Zaman-Allah, M. S. Olsen, and J. E. Cairns. "Translating high-throughput phenotyping into genetic gain." *Trends in Plant Science* 23, no. 5 (2018): 451–466.

Bai, X., B. Wu, and Y. Xing. "Yield-related QTLs and their applications in rice genetic improvement F." *Journal of Integrative Plant Biology* 54, no. 5 (2012): 300–311.

Baltes, N. J., and D. F. Voytas. "Enabling plant synthetic biology through genome engineering." *Trends in Biotechnology* 33, no. 2 (2015): 120–131.

Braatz, J., H. J. Harloff, M. Mascher, N. Stein, A. Himmelbach, and C. Jung. "CRISPR-Cas9 targeted mutagenesis leads to simultaneous modification of different homoeologous gene copies in polyploid oilseed rape (Brassica napus)." *Plant Physiology* 174, no. 2 (2017): 935–942.

Butler, N. M., P. A. Atkins, D. F. Voytas, and D. S. Douches. "Generation and inheritance of targeted mutations in potato (*Solanum tuberosum L.*) using the CRISPR/Cas system." *PLoS One* 10, no. 12 (2015): e0144591.

Cai, C. Q., Y. Doyon, W. M. Ainley, et al. "Targeted transgene integration in plant cells using designed zinc finger nucleases." *Plant Molecular Biology* 69, no. 6 (2009): 699–709.

Carroll, D. "Genome engineering with targetable nucleases." *Annual Review of Biochemistry* 83 (2014): 409–439.

Chipman, D., Z. E. Barak, and J. V. Schloss. "Biosynthesis of 2-aceto-2-hydroxy acids: Acetolactate synthases and acetohydroxyacid synthases." *Biochimica et Biophysica Acta (BBA)-Protein Structure and Molecular Enzymology* 1385, no. 2 (1998): 401–419.

Christian, M., T. Cermak, E. L. Doyle, et al. "Targeting DNA double-strand breaks with TAL effector nucleases." *Genetics* 186, no. 2 (2010): 757–761.

Clasen, B. M., T. J. Stoddard, S. Luo, et al. "Improving cold storage and processing traits in potato through targeted gene knockout." *Plant Biotechnology Journal* 14, no. 1 (2016): 169–176.

Cong, L., F. A. Ran, D. Cox, et al. "Multiplex genome engineering using CRISPR/Cas systems." *Science* 339, no. 6121 (2013): 819–823.

Gaj, T., S. J. Sirk, S. Shui, and J. Liu. "Genome-editing technologies: Principles and applications." *Cold Spring Harbor Perspectives in Biology* 8, no. 12 (2016): a023754.

Ganesh, M. K., and M. S. Dilip. "Amino acid biosynthesis inhibitors as herbicides." *Annual Review in Biochemistry* 57 (1988): 627–663.

Gong, W. J., and K. G. Golic. "Ends-out, or replacement, gene targeting in Drosophila." *Proceedings of the National Academy of Sciences of the United States of America* 100, no. 5 (2003): 2556–2561.

Hanania, U., T. Ariel, Y. Tekoah, et al. "Establishment of a tobacco BY2 cell line devoid of plant-specific xylose and fucose as a platform for the production of biotherapeutic proteins." *Plant Biotechnology Journal* 15, no. 9 (2017): 1120–1129.

Harwood, W. A. "Advances and remaining challenges in the transformation of barley and wheat." *Journal of Experimental Botany* 63, no. 5 (2012): 1791–1798.

Haun, W., A. Coffman, B. M. Clasen, Z. L. Demorest, A. Lowy, E. Ray, A. Retterath, T. Stoddard, A. Juillerat, F. Cedrone, L. Mathis, D. F. Voytas, F. Zhang. "Improved soybean oil quality by targeted mutagenesis of the fatty acid desaturase 2 gene family." *Plant Biotechnology Journal* 12, no. 7 (2014): 934–940.

Holtzman, L., and C. A. Gersbach. "Editing the epigenome: Reshaping the genomic landscape." *Annual Review Genomics and Human Genetics* 19 (2018): 43–71.

Hwang, H. H., M. Yu, and E. M. Lai. "Agrobacterium-mediated plant transformation: Biology and applications." *The Arabidopsis Book* 15 (2017). doi: 10.1199/tab.0186.

Ji, X., H. Zhang, Y. Zhang, Y. Wang, and C. Gao. "Establishing a CRISPR–Cas-like immune system conferring DNA virus resistance in plants." *Nature Plants* 1 (2015): 15144.

Jiang, W. Z., I. M. Henry, P. G. Lynagh, L. Comai, E. B. Cahoon, and D. P. Weeks. "Significant enhancement of fatty acid composition in seeds of the allohexaploid, Camelina sativa, using CRISPR/Cas9 gene editing." *Plant Biotechnology Journal* 15, no. 5 (2017): 648–657.

Jianru, Z., and L. Jiayang. "Molecular genetic dissection of quantitative trait loci regulating rice grain size." *Annual Review Genetics* 48 (2014): 99–118.

Jinek, M., K. Chylinski, I. Fonfara, M. Hauer, J. A. Doudna, and E. Charpentier. "A programmable dual-RNA-guided DNA endonuclease in adaptive bacterial immunity." *Science* 337, no. 6096 (2012): 816–821.

Khosravi, S., T. Ishii, S. Dreissig, and A. Houben. "Application and prospects of CRISPR/Cas9-based methods to trace defined genomic sequences in living and fixed plant cells." *Chromosome Research* 28, no. 4 (2020): 1–11.

Kim, M. S., and A. G. Kini. "Engineering and application of zinc finger proteins and TALEs for biomedical research." *Molecules and Cells* 40, no. 8 (2017): 533.

Kosack, H. K. E., and J. D. Jones. "Resistance gene-dependent plant defence response." *Plant Cell* 7, no. 2 (1996): 1773–1786.

Kwon, C. T., J. Heo, Z. H. Lemmon, et al. "Rapid customization of Solanaceae fruit crops for urban agriculture." *Nature Biotechnology* 38, no. 2 (2020): 182–188.

Lawrenson, T., O. Shorinola, N. Stacey, et al. "Induction of targeted, heritable mutations in barley and Brassica oleracea using RNA-guided Cas9 nuclease." *Genome Biology* 16, no. 1 (2015): 258.

Lee, K. Y., J. Townsend, J. Tepperman, et al. "The molecular basis of sulfonylurea herbicide resistance in tobacco." *The EMBO Journal* 7, no. 5 (1988): 1241–1248.

Li, W., and B. Yang. "Translational genomics of grain size regulation in wheat." *Theoretical and Applied Genetics* 130, no. 9 (2017): 1765–1771.

Li, T., S. Huang, W. Z. Jiang, et al. "TAL nucleases (TALNs): Hybrid proteins composed of TAL effectors and FokI DNA-cleavage domain." *Nucleic Acids Research* 39, no. 1 (2011): 359–372.

Li, T., B. Liu, M. H. Spalding, D. P. Weeks, and B. Yang. "High-efficiency TALEN-based gene editing produces disease-resistant rice." *Nature Biotechnology* 30, no. 5 (2012): 390.

Li, M., X. Li, Z. Zhou, et al. "Reassessment of the four yield-related genes Gn1a, DEP1, GS3, and IPA1 in rice using a CRISPR/Cas9 system." *Frontiers in Plant Science* 7 (2016a): 377.

Li, J., X. Meng, Y. Zong, et al. "Gene replacements and insertions in rice by intron targeting using CRISPR–Cas9." *Nature Plants* 2, no. 10 (2016b): 1–6.

Li, R., R. Li, X. Li, et al. "Multiplexed CRISPR/Cas9-mediated metabolic engineering of γ-aminobutyric acid levels in *Solanum lycopersicum*." *Plant Biotechnology Journal* 16, no. 2 (2018a): 415–427.

Li, X., Y. Wang, S. Chen, et al. "Lycopene is enriched in tomato fruit by CRISPR/Cas9-mediated multiplex genome editing." *Frontiers in Plant Science* 9 (2018b): 559.

Li, J., X. Zhang, Y. Sun, J. Zhang, et al. "Efficient allelic replacement in rice by gene editing: A case study of the NRT1. 1B gene." *Journal of Integrative Plant Biology* 60, no. 7 (2018c): 536–540.

Li, C., Y. Zong, Y. Wang, et al. "Expanded base editing in rice and wheat using a Cas9-adenosine deaminase fusion." *Genome Biology* 19, no. 1 (2018d): 59.

Liang, Z., K. Zhang, K. Chen, and C. Gao. "Targeted mutagenesis in Zea mays using TALENs and the CRISPR/Cas system." *Journal of Genetics and Genomics* 41, no. 2 (2014): 63–68.

Lo, A., and L. Qi. "Genetic and epigenetic control of gene expression by CRISPR–Cas systems." *F1000Research* 6 (2017): 747.

Lombardo, L., G. Coppola, and S. Zelasco. "New technologies for insect-resistant and herbicide-tolerant plants." *Trends in Biotechnology* 34, no. 1 (2016): 49–57.

Lowder, L. G., D. Zhang, N. J. Baltes, J. W. Paul, X. Tang, X. Zheng, D. F. Voytas, T.-F. Hsieh, Y. Zhang, and Y. Qi. "A CRISPR/Cas9 toolbox for multiplexed plant genome editing and transcriptional regulation." *Plant Physiology* 169, no. 2 (2015): 971–985.

Lu, Y., X. Ye, R. Guo, et al. "Genome-wide targeted mutagenesis in rice using the CRISPR/Cas9 system." *Molecular Plant* 10, no. 9 (2017): 1242–1245.

Ma, H., A. Naseri, P. Reyes-Gutierrez, S. A. Wolfe, S. Zhang, and T. Pederson. "Multicolor CRISPR labeling of chromosomal loci in human cells." *Proceedings of the National Academy of Sciences of the United States of America* 112, no. 10 (2015a): 3002–3007.

Ma, X., Q. Zhang, Q. Zhu, et al. "A robust CRISPR/Cas9 system for convenient, high-efficiency multiplex genome editing in monocot and dicot plants." *Molecular Plant* 8, no. 8 (2015b): 1274–1284.

Ma, X., Q. Zhu, Y. Chen, and Y. G. Liu. CRISPR/Cas9 platforms for genome editing in plants: Developments and applications. *Molecular Plant* 9 (2016): 961–974.

Ma, J., J. Chen, M. Wang, et al. "Disruption of OsSEC3A increases the content of salicylic acid and induces plant defense responses in rice." *Journal of Experimental Botany* 69, no. 5 (2018): 1051–1064.

Mahas, A., C. N. Stewart Jr, and M. M. Mahfouz. "Harnessing CRISPR/Cas systems for programmable transcriptional and post-transcriptional regulation." *Biotechnology Advances* 36, no. 1 (2018): 295–310.

Mahfouz, M. M., L. Li, M. Shamimuzzaman, A. Wibowo, X. Fang, and J. K. Zhu. "De novo-engineered transcription activator-like effector (TALE) hybrid nuclease with novel DNA binding specificity creates double-strand breaks." *Proceedings of the National Academy of Sciences of the United States of America* 108, no. 6 (2011): 2623–2628.

Mali, P., L. Yang, K. M. Esvelt, et al. "RNA-guided human genome engineering via Cas9." *Science* 339, no. 6121 (2013): 823–826.

Mazur, B. J., C. F. Chui, and J. K. Smith. "Isolation and characterization of plant genes coding for acetolactate synthase, the target enzyme for two classes of herbicides." *Plant Physiology* 85, no. 4 (1987): 1110–1117.

Meng, X., H. Yu, Y. Zhang, et al. "Construction of a genome-wide mutant library in rice using CRISPR/Cas9." *Molecular Plant* 10, no. 9 (2017): 1238–1241.

Mercx, S., N. Smargiasso, F. Chaumont, E. D. Pauw, M. Boutry, and C. Navarre. "Inactivation of the β (1, 2)-xylosyltransferase and the α (1, 3)-fucosyltransferase genes in Nicotiana tabacum BY-2 cells by a multiplex CRISPR/Cas9 strategy results in glycoproteins without plant-specific glycans." *Frontiers in Plant Science* 8 (2017): 403.

Michelmore, R. "Molecular approaches to manipulation of disease resistance genes." *Annual Review of Phytopathology* 33, no. 1 (1995): 393–427.

Minkenberg, B., K. Xie, and Y. Yang. "Discovery of rice essential genes by characterizing a CRISPR-edited mutation of closely related rice MAP kinase genes." *The Plant Journal* 89, no. 3 (2017): 636–648.

National Research Council. "Gene transfer methods applicable to agricultural organisms." In *Agricultural Biotechnology: Strategies for National Competitiveness*. National Academies Press, Washington, DC (1987).

Nekrasov, V., C. Wang, J. Win, C. Lanz, D. Weigel, and S. Kamoun. "Rapid generation of a transgene-free powdery mildew resistant tomato by genome deletion." *Scientific Reports* 7, no. 1 (2017): 1–6.

Oliva, R., C. Ji, G. Atienza-Grande, et al. "Broad-spectrum resistance to bacterial blight in rice using genome editing." *Nature Biotechnology* 37, no. 11 (2019): 1344–1350.

Osakabe, Y., T. Watanabe, S. S. Sugano, et al. "Optimization of CRISPR/Cas9 genome editing to modify abiotic stress responses in plants." *Scientific Reports* 6 (2016): 26685.

Pandey, P., V. Irulappan, M. V. Bagavathiannan, and M. Senthil-Kumar. "Impact of combined abiotic and biotic stresses on plant growth and avenues for crop improvement by exploiting physio-morphological traits." *Frontiers in Plant Science* 8 (2017): 537.

Rees, H. A., and D. R. Liu. "Base editing: Precision chemistry on the genome and transcriptome of living cells." *Nature Reviews Genetics* 19 (2018): 770–788.

Rodríguez-Leal, D., Z. H. Lemmon, J. Man, M. E. Bartlett, and Z. B. Lippman. "Engineering quantitative trait variation for crop improvement by genome editing." *Cell* 171, no. 2 (2017): 470–480.

Sánchez, L. S., J. G. Humanes, C. V. Ozuna, et al. "Low-gluten, nontransgenic wheat engineered with CRISPR/Cas9." *Plant Biotechnology Journal* 16, no. 4 (2018): 902–910.

Sauer, N. J., J. Mozoruk, R. B. Miller, et al. "Oligonucleotide-directed mutagenesis for precision gene editing." *Plant Biotechnology Journal* 14, no. 2 (2016): 496–502.

Savary, S., A. Ficke, J. N. Aubertot, and C. Hollier. "Crop losses due to diseases and their implications for global food production losses and food security." *Food Security* 4, no. 4 (2012): 519–537.

Schönbrunn, E., S. Eschenburg, W. A. Shuttleworth, et al. "Interaction of the herbicide glyphosate with its target enzyme 5-enolpyruvylshikimate 3-phosphate synthase in atomic detail." *Proceedings of the National Academy of Sciences of the United States of America* 98, no. 4 (2001): 1376–1380.

Shakoori, A. R. "Fluorescence In Situ Hybridization (FISH) and its applications." In Bhat, T. and Wani, A. (eds). *Chromosome Structure and Aberrations*, pp. 343–367. Springer, New Delhi (2017). https://doi.org/10.1007/978-81-322-3673-3_16

Shan, S., P. S. Soltis, D. E. Soltis, and B. Yang. "Considerations in adapting CRISPR/Cas9 in nongenetic model plant systems." *Applications in Plant Sciences* 8, no. 1 (2020): e11314.

Shen, L., C. Wang, Y. Fu, et al. "QTL editing confers opposing yield performance in different rice varieties." *Journal of Integrative Plant Biology* 60, no. 2 (2018): 89–93.

Shi, J., G. Huirong, W. Hongyu, et al. "ARGOS8 variants generated by CRISPR-Cas9 improve maize grain yield under field drought stress conditions." *Plant Biotechnology Journal* 15, no. 2 (2017): 207–216.

Shukla, V. K., Y. Doyon, J. C. Miller, et al. "Precise genome modification in the crop species Zea mays using zinc-finger nucleases." *Nature* 459, no. 7245 (2009): 437–441.

Soda, N., L. Verma, and J. Giri. "CRISPR-Cas9 based plant genome editing: Significance, opportunities and recent advances." *Plant Physiology and Biochemistry* 131 (2018): 2–11.

Song, G., M. Jia, K. Chen, et al. "CRISPR/Cas9: A powerful tool for crop genome editing." *The Crop Journal* 4, no. 2 (2016): 75–82.

Soyk, S., N. A. Müller, S. J. Park, et al. "Variation in the flowering gene SELF PRUNING 5G promotes day-neutrality and early yield in tomato." *Nature Genetics* 49, no. 1 (2017): 162–168.

Sun, Y., X. Zhang, C. Wu, et al. "Engineering herbicide-resistant rice plants through CRISPR/Cas9-mediated homologous recombination of acetolactate synthase." *Molecular Plant* 9, no. 4 (2016): 628–631.

Sun, Y., G. Jiao, Z. Liu, et al. "Generation of high-amylose rice through CRISPR/Cas9-mediated targeted mutagenesis of starch branching enzymes." *Frontiers in Plant Science* 8 (2017): 298.

Svitashev, S., J. Young, C. Schwartz, H. Gao, S. C. Falco, and A. M. Cigan. "Targeted mutagenesis, precise gene editing and site-specific gene insertion in maize using Cas9 and guide RNA." *Plant Physiology* 169, no. 2 (2015): 931–945.

Taylor, J. E., P. E. Hatcher, and N. D. Paul. "Crosstalk between plant responses to pathogens and herbivores: A view from the outside in." *Journal of Experimental Botany* 55, no. 395 (2004): 159–168.

Townsend, J. A., D. A. Wright, R. J. Winfrey, et al. "High-frequency modification of plant genes using engineered zinc-finger nucleases." *Nature* 459, no. 7245 (2009): 442–445.

Urnov, F. D., J. C. Miller, Y. L. Lee, et al. "Highly efficient endogenous human gene correction using designed zinc-finger nucleases." *Nature* 435, no. 7042 (2005): 646–651.

Veillet, F., L. Perrot, L. Chauvin, et al. "Transgene-free genome editing in tomato and potato plants using agrobacterium-mediated delivery of a CRISPR/Cas9 cytidine base editor." *International Journal of Molecular Sciences* 20, no. 2 (2019): 402.

Voytas, D. F., and C. Gao. "Precision genome engineering and agriculture: Opportunities and regulatory challenges." *PLoS Biology* 12, no. 6 (2014): e1001877.

Waltz, E. "CRISPR-edited crops free to enter market, skip regulation." *Nature Biotechnology* 34, no. (6) (2016): 582.

Wang, Y., X. Cheng, Q. Shan, et al. "Simultaneous editing of three homoeoalleles in hexaploid bread wheat confers heritable resistance to powdery mildew." *Nature Biotechnology* 32, no. 9 (2014): 947–951.

Wolt, J. D., K. Wang, and B. Yang. "The Regulatory Status of Genome-edited Crops." *Plant Biotechnology Journal* 14, no. 2 (2016): 510–518.

Woo, J. W., K. Jungeun, I. K. Soon, et al. "DNA-free genome editing in plants with preassembled CRISPR-Cas9 ribonucleoproteins." *Nature Biotechnology* 33, no. 11 (2015): 1162–1164.

Xing, Y. and Q. Zhang. "Genetic and molecular bases of rice yield." *Annual Review of Plant Biology* 61 (2010): 421–442.

Xu, R., Y. Yang, R. Qin, et al. "Rapid improvement of grain weight via highly efficient CRISPR/Cas9-mediated multiplex genome editing in rice." *Journal of Genetics and Genomics Yi chuan xue bao* 43, no. 8 (2016): 529.

Yang, B. "Grand challenges in genome editing in plants." *Frontiers in Genome Editing* 2 (2020): 2.

Zhang, Y., Z. Liang, Y. Zong, et al. "Efficient and transgene-free genome editing in wheat through transient expression of CRISPR/Cas9 DNA or RNA." *Nature Communications* 7, no. 1 (2016): 1–8.

Zhang, K., N. Raboanatahiry, B. Zhu, and M. Li. "Progress in genome editing technology and its application in plants." *Frontiers in Plant science* 8 (2017): 177.

Zhang, Y., S. Li, S. Xue, S. Yang, J. Huang, and L. Wang. "Phylogenetic and CRISPR/Cas9 studies in deciphering the evolutionary trajectory and phenotypic impacts of rice ERECTA genes." *Frontiers in Plant Science* 9 (2018a): 473.

Zhang, Y., K. Massel, I. D. Godwin, and C. Gao. "Applications and potential of genome editing in crop improvement." *Genome Biology* 19, no. 1 (2018b): 210.

Zhou, Q., W. Liu, Y. Zhang, and K. K. Liu. "Action mechanisms of acetolactate synthase-inhibiting herbicides." *Pesticide Biochemistry and Physiology* 89, no. 2 (2007): 89–96.

15 Regulatory Approaches for Genome-Edited Crops

Awadhesh Kumar and Lopamudra Nayak
ICAR-National Rice Research Institute

Chandrasekhar Sahu
Centurion University of Technology and Management

CONTENTS

15.1 INTRODUCTION

Genome or gene editing is a genetic engineering tool where the host genome is modified through insertion, deletion, modification or replacement of DNA (RNA) segment at the targeted locations (Friedrichs et al., 2019). At present, it is the most advanced and a promising technique used in applied biological science as well as in industrial innovations. Genome edition empowers researchers to precisely alter some of the billions of nucleotides in the genome of the living organisms. Researchers are using genome editing to get to know the function of various genes and utilizing the information in developing different types of crops with improved nutritional value, insect pest resistance and capability to grow in dry and arid lands. Crops give nourishment, feed and fibre for humans, with the best 60 species grown in 96% of the world's arable land (Abdallah et al., 2015). The endurance and prosperity of humans fundamentally rely upon the yield of these crops. The ever-growing human population faces huge challenges, which include climate change, degradation of arable land and ecological debasement, thus making agriculture sustainable and climate-resilient. These issues are multifaceted yet breeding of different crops without a doubt has a role in meeting the objectives of agriculture and production of food. For addressing these challenges, it is necessary to explore the latest development in research techniques with due emphasis on genome editing tools

(Huang et al., 2016). Genome editing begins with the insertion of a double-stranded DNA segment at a targeted locus using a sequence-specific nuclease and hence forms a powerful technique for crop improvement. In principle, it is a clear straightforward technique to erase an undesirable trait or to introduce a favourable one through knockout mutations in the responsible genes (Sikandar, 2019).

15.2 DIFFERENT TYPES OF GENOME EDITING TECHNIQUES

Genome editing consists of a broad range of tools like a site-specific nuclease (SSN) or a site-specific recombinase (SSR), which require a known sequence recognition. The SSN system creates single- or double-stranded DNA breaks followed by the activation of repair mechanism of endogenous DNA. SSR technology, like Flp/FRT and Cre/loxP-mediated tools, knocks down or knocks in the targeted genes of the genome, depending upon the sites (FLP, loxP etc.) and the specific orientation of the flanking target site. Four different types of SSN have been developed to break genome sequences, which include meganucleases (homing endonuclease), transcriptional activator-like effector nucleases (TALENs), zinc finger nucleases (ZFNs) and the latest clustered regularly interspaced short palindromic repeat/CRISPR-associated protein (CRISPR/Cas) system. Genome engineering (mediated by recombinase) depends on high frequencies of homologous recombination by recombinase enzymes at the target site (Abdallah et al., 2015).

15.2.1 Meganuclease-Based Engineering

Meganucleases are long base pair structures (naturally found endonucleases in the genome) discovered during the 1980s and also known as molecular DNA scissors (Abdallah et al., 2015). They generally excise large DNA sequences which recognize recently as a genetic tool for DNA modification in the laboratory through the modification of the recognition sites. This technique plays around some basic steps: firstly, recognizing the cleavage site and secondly, the use of endonucleases that splice out the region (McMurrough et al., 2018). The positive aspect of this tool is that it provides highly site-specific cleavage and shows only little toxicity, as they are naturally occurring nucleases. The disadvantage of meganuclease is that the construction of sequence-specific enzymes is time-consuming and costly (Sikandar, 2019).

15.2.2 Engineering Approach Based on Zinc Finger Nuclease

ZFNs are artificial structures that give rise to a combinatorial approach where zinc finger-binding protein domain is combined with restriction endonucleases (Sikandar, 2019). When DNA-binding protein domain reaches the specific splice site, special restriction endonucleases (FokI) cut at a specific codon. This technology has restriction in terms of attachment with three codons on both sides of the DNA chain. Due to its simplicity and specificity, this technique is being popularized on a large scale in the recent years. It has found a utility in certain clinical applications (Ochiai and Yamamoto, 2017; Ji et al., 2018). There are few disadvantages of the zinc finger-based technology, which include the complexity of procedure with higher protein domain construction cost for the specific genome locus. An inaccurate cleavage (off-target break) of target DNA due to substitutions of single nucleotide or inappropriate interaction between domains may also be noticed (Puchta and Hohn, 2010).

15.2.3 TALEN-Based Approach of Genome Editing

The TALEN system was developed from the phytopathogenic bacteria of the *Xanthomonas* genus that produces transcription activator-like effectors (TALEs) (Boch and Bonas, 2010). TALENs are alike to ZFNs in terms of production along with a mode of action. They are designed on the same principle where a restriction nuclease is attached to a domain of DNA-binding protein called as

TAL effector (Hensel and Kumlehn, 2019). The vital difference between TALENs and ZFNs is that the former can target three nucleotides at a time, while the latter can address only one, which makes TALENs more precise with fewer off-target effects (Chandrasegaran and Carroll, 2016).

15.2.4 CRISPR/CAS-Mediated Genome Edition Technique

CRISPR is a novel genome editing tool, developed in 2013. CRISPRs are natural genetic elements that impart defence to bacteria against virus attack. Three main stages are involved in the protection mechanism: first adaptation followed by transcription, then interference. During adaptation, foreign DNA enters into bacterial cells; then a small DNA fragment is integrated into the CRISPR locus of host genome, which is a specific repeat structure separated by short palindromic repeats (Abdallah et al., 2015). The Cas genes are found nearby the CRISPR cassette expression having helicase and nuclease activity (Haft et al., 2005). During the transcriptional stage, a long polyspacer precursor crRNA (pre-crRNA) is transcribed by the CRISPR locus, which further synthesizes to short CRISPR RNA (39–45 nucleotides) with one spacer sequence (Figure 15.1). In the recent years, CRISPR/Cas9 technique has gained worldwide popularity due to its specificity and simplicity. Several experiments and advancements have redefined the technology into three specific types of CRISPR-Cas techniques based on crRNA processing and other related steps. The first among them is Type 1 CRISPR/Cas system, which utilizes Cas5 or Cas6 protein for pre-processing of crRNA, followed by cleavage involving Cas3, cascade and crRNA for interference.

CRISPR-Cas system is the simplest endonuclease activity needed for the interference of gene function where the foreign DNA is concentrated in a single Cas9 multidomain, which is guided by a co-processed dual tracrRNA: crRNA. Though Cas9 protein mainly functions under the guidance of crRNA to target the DNA, RNase III, trans-activating RNA (tracrRNA) and some unknown proteins are involved in the 5′ end trimming. Type III is the first system that provides some key aspects of the molecular mechanisms behind the CRISPR-Cas immunity system. Similar to Type 1 system, this category uses Cas6 protein for processing of crRNA at 3′ end trimming. This technique uniquely targets RNA, which forms a specific complex termed as type III Csm/Cmr complex (Rath et al., 2015).

15.3 APPLICATION OF GENOME EDITING TECHNIQUE TO DIFFERENT CROPS

In the present condition, genome editing is in blossoming stage for the purpose of crop improvement; still some improved crop species have been developed using genome editing.

Gene-knockout approach utilizing SSNs has been implemented to interpret some significant and valuable genes, which can utilized for the purpose of improvement of diverse crop species, including rice, maize, sorghum *etc.* (Voytas and Gao, 2014; Curtin et al., 2013; Qi et al., 2013; Shan et al., 2013; Gao et al., 2010; Shukla et al., 2009; Liang et al., 2013; Wang et al., 2013; Mao et al., 2013; Miao et al., 2013; Xie and Yang 2013).

Among the achievements of genetic engineering, drought- and salinity-tolerant soya bean (*Glycine max*), high amylopectin-containing waxy corn (*Zea mays*) and high omega-3 fatty acid-containing oilseed (*Camelina sativa*) are few examples.

EMN (engineered meganucleases) was first exhibited for the purpose of genome editing by utilizing a native endonuclease manipulated through a structure-based protein-designing strategy to recognize and induce the specific double-stranded break at a highly specific locus in maize, resulting in the disruption of gene through the insertion or deletion of small base segment by non-homologous end-joining (Gao et al., 2010). In the recent past, a cotton plant was developed by EMN engineering with the use of an yeast endonuclease by inserting the target gene precisely into plants to develop tolerance to herbicide at a specific site where insect resistance gene was previously inserted (D'Halluin et al., 2013). A low phytase maize along with herbicide tolerance has been

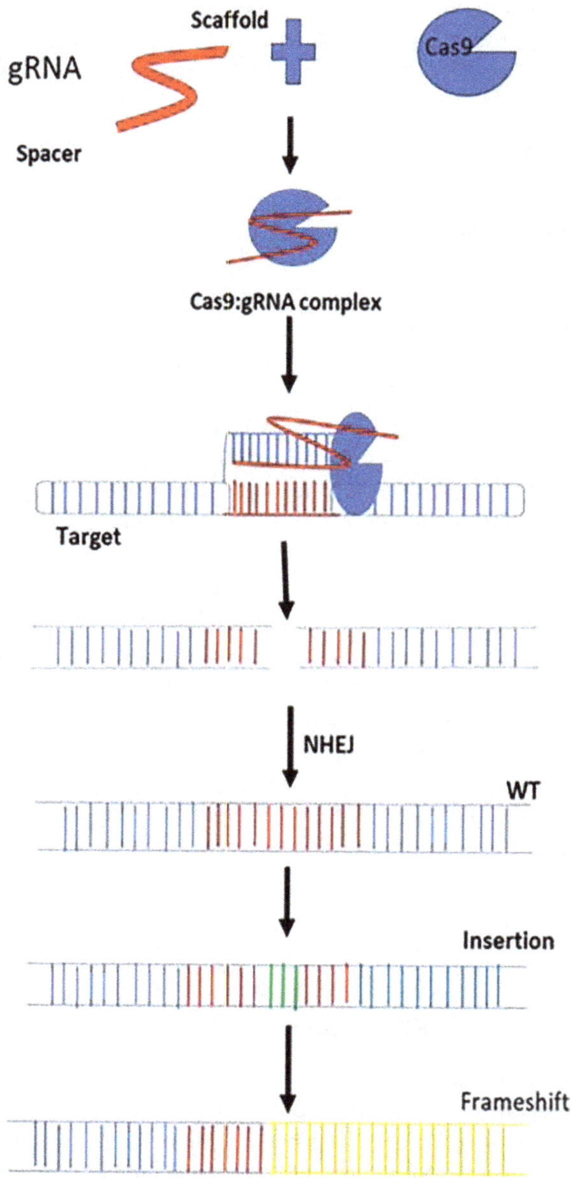

FIGURE 15.1 Schematic diagram of CRISPR/Cas technology (gRNA, guide RNA; NHEJ, non-homologous end-joining; WT, wild type).

developed by using site-specific ZFN by the disruption of the *inositol phosphate kinase I* (*IPK1*) gene (Shukla et al., 2009). Further, a site-specific trait stacking approach using a ZFN was achieved in maize by initially producing a line with a herbicide resistance gene linked to the synthetic zinc finger target site, followed by the integration of a second herbicide resistance to zinc finger target site; thus, multiple trait stacking was accomplished at a specific locus (Ainley et al., 2013). For the first time, TALENs were used for the purpose of crop improvement in rice where *OsSWEET14*, a disease-susceptible gene, was modified by site-directed mutagenesis ranging from 3 to 55 base pairs. Subsequently, disease-resistant null segregant rice was reported with the absence of TALEN genes and selection marker. Further, TALEN technology was applied in economically important plants like rice, maize and barley (Char et al., 2015; Gurushidze et al., 2014; Shan et al., 2013).

A Cas9-mediated single guide RNA (sgRNA) system from *Streptococcus pyogenes*, representing the type II CRISPR/Cas system, has been successfully utilized for genome editing of monocot as well as dicot crop species. Cas9/sgRNA encoding genes and a mutant green fluorescence protein (GFP) were delivered into sorghum and *Arabidopsis* through *Agrobacterium tumefaciens*-mediated transformation. In a second case of Cas9-sgRNA demonstration, constructs for targeting the promoter region of *OsSWEET11* and *OsSWEET14* bacterial blight susceptibility genes with mutated DNA sequences were transformed into rice protoplast cells at the selected target sites. Further, fragrant rice has been developed by knocking down of *OsBADH2* gene (controlled primarily by recessive alleles) using TALEN technology.

15.4 REGULATORY APPROACHES FOLLOWED IN DIFFERENT COUNTRIES

There are regulators of different countries handling the assessment and evaluation process of genetically modified crops, and they are in support of scientific view of development of plant phenotype by the use of genetic engineering and focus on safety aspects for people.

However in the different countries of the world, the guidelines for GM crops are different; the methodology and the process used usually determine the way to its approval and regulatory assessment (Wolt et al., 2016). The progressive effect of genome editing has just been shown on a wide assortment of agricultural commodities. Its fruitful applications vary from the improvement of the effectiveness of plant towards quality, yield and resistance to diseases and pests. However, a rapidly growing genome editing application creates a concern for policy implications pertaining to this tool (Figure 15.2).

15.4.1 Regulation of GMOs and New Breeding Techniques (NBTs) in Argentina

All NBTs include recombinant DNA strategies, which result in GMOs. In case of NBT non-involvement, a non-GM regulatory classification is generally applied for a new genetic material combination, which does not use a transgene or escape in the final product. The above-mentioned procedure can be applied to speculative and real items where it seeks essential data on the total breeding procedure, changes in heredity, attributes, removal of helper transgenes and so forth. In the event that the NBT has a new genetic combination (e.g. which utilizes a transgene that stays in the final product), the regulatory classification considers the final product under GM category. Argentina was one of the 'six founder' nations as it has allowed the commercial cultivation of GM crops since 1996. Argentina is the third largest producer of both GE and GM crops under 23 M-ha area coverage. It is likewise the top exporter of soya oil, the second highest exporter of corn and also the third largest exporter of soya. From the commencement of NBT resolution launched, 12 cases had been checked, mainly at the hypothetical design stage. Products that sought NBT classification basically differed from the conventional GM classification (Lema, 2019).

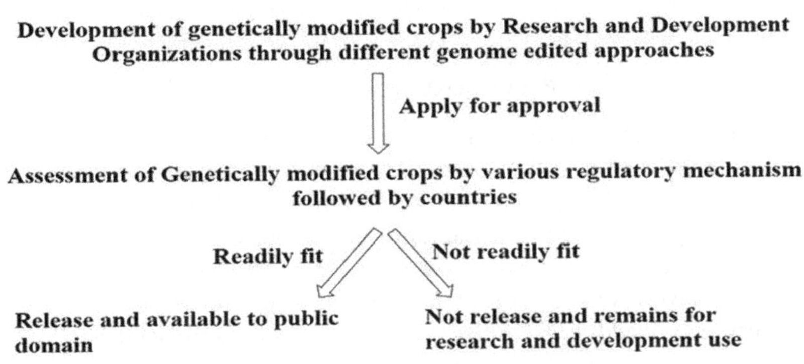

FIGURE 15.2 General regulatory mechanism for genome-edited crops.

15.4.2 Genome Editing and Regulation of GE/GM in Canada

In Canada, the product-oriented and risk-based regulatory approach is generally followed. In Canada, the regulatory control mechanism related to genetic engineering is somewhat complex. It means that various GM products in Canada market are managed by eight agencies considering eight acts and strategies.

These are ISED (Innovation, Science and Economic Development, Canada), AAFC (Agriculture and Agri-Food, Canada), ECCC (Environment and Climate Change, Canada) GAC (Global Affairs, Canada) and DFO (Fisheries and Oceans Canada) (Kleter et al., 2019; Friedrichs et al., 2019). Canada needs a pre-market safety evaluation for agricultural GM items, which include items that were created through gene editing, but only if the novelty is there (for example if they express new features) and could therefore a concern of risk. But for the genome-edited crop having a novel trait, a pre-market safety assessment is not required.

15.4.3 Regulations on Genome Editing and GE/GM in USA

In United States for all GM products, a product-oriented regulatory guideline is applied under their existing law, giving a system of agency jurisdiction (Araki and Ishii, 2015). Modernization of the regulatory mechanism was initiated in 2015 for GM products aiming at trust of public and transparency in the regulatory framework, which enhanced coordination along with the effective regulatory system. GM products are regulated by The USDA Animal and Plant Health Inspection Service (APHIS) through control mechanism (i.e. permits for environmental release, import notifications and interstate movement) of the regulated commodities (Ishii and Araki, 2017).

Under the 'Am I regulated' (AIR) procedure, APHIS encouraging submission of letters of inquiry by the developers of the product comes under the relevant regulation (https://www.aphis.usda.gov/aphis/ourfocus/biotechnology/am-i-regulated). A written statement was issued by USA Agriculture Secretary in 2018 to clarify 'USDA oversight of plants produced through innovative novel breeding technologies that includes genome editing'.

15.4.4 Regulatory Mechanism of Australian GMO (Genetically Modified Organisms)

The Gene Technology Act (GT Act, 2000) was introduced in Australia in 2000.

The following year, 2001, Gene Technology Regulations was introduced concerning mutagenesis, nuclear transfer of somatic cell, protoplast fusion and a natural process without involving the genetically modified component (i.e. an organism formed by exchange of DNA from donor host species where the vector DNA has no heterogeneous DNA). This was the intergovernmental gene technique agreement between the Federal State and the Territory governments of Australia (http://www.ogtr.gov.au/internet/ogtr/publishing.nsf/Content/ reviewregulations-1).

15.4.5 European Union (EU) Regulatory Framework for GMOs

The EU legislation on GMOs intends to ensure an effective protection for humans, animals and environmental well-being and a well-working EU internal market (Kleter et al., 2019). For the release of any authorized GMOs into the market, a pre-market authorization and postmarket environmental monitoring are required as per the European Parliament and the Council for release of the GMOs directive (Directive 90/220/EEC; dated 12 March 2001) (Scheben and Edwards, 2018). The market of the EU involves more than 500 million consumers where seed market is worth 7 billion Euros for every year; it adds to a positive trade balance in the agro-segment. As indicated by a representative from the EU Commission, though the EU approved the release of GMOs each year; their marketing faces criticism at the higher level.

15.4.6 GMOs and GE Regulation in India

India introduced the rules for manufacture, use, export, import and storage of hazardous microorganisms/GMOs or cells in 1989. In 2006, the country introduced the 'Food Safety and Standards Act, 2006', where FSSAI (Food Safety and Standards Authority of India) regulates the GM foods. It defines GM foods as 'the ultimate form of food and their ingredients as a consequence of bioengineering with an altered and/or modified genetic constitutions obtained through modern biotechnology'. At present, more than 85 crop species are under various stages of research and development in India. Currently, the three crop species have been either approved or are awaiting for approval. The BT-cotton was the alone crop that has been fully approved, whereas BT-brinjal (eggplant) was subjected to a moratorium, during the approval process due to an adverse public reaction. The moratorium on the approval of BT-brinjal could well be a political issue. India's neighbour country Bangladesh, however, goes ahead and approved the cultivation and commercialization of the plant. The genetically modified mustard in India is also awaiting for approval (Friedrichs et al., 2019).

15.5 SAFETY ASSESSMENT MEASURE OF GM CROPS

The present difference in national regulations on new plant breeding techniques (NPBTs) has its root in the debate on whether NPBTs establish procedures of genetic modification and whether the resulting organisms fall under GMO legislation and whether it warrants guideline and pre-market security appraisal. Though a comparative disparity between nations in their regularity mechanism applies to GM crops too, the pre-market safety evaluation of such crops in countries worldwide is performed by the globally agreed methodology. Any safety assessment's common features include, *inter alia,* the probable toxicity, allergenicity of novel expressed proteins and assessment of any potential effects of a genetic modification of GM crops. In genome editing through CRISPR-Cas9 technique, a special attention should be paid towards the possible off-target edit effects, especially single locus-targeted sequence, which is reported at a low frequency in GM crops (Zhang et al., 2014). The questionable safeness of using GM crops can pass through the safety assessment procedures through more convincing NPBTs where an alteration in the genetic makeup can be obtained through natural selection rather than an induced change in the genes. The scientific community has experience of more than 20 years of regulatory safety assessment related to GM crops. Now it appears that the possible adverse consequences of the GM crops have been overstated.

15.6 CONCLUSION

Genome editing techniques help researchers to precisely modify any gene sequences. These advanced techniques enable us to modify the gene expression patterns in a desired manner. Engineered nucleases in genome editing represent a highly precise and efficient tool for crop improvement with novel traits. Government regulations in many countries are major bottlenecks for the large-scale adoption of genetically modified crops. To achieve an appropriate regulatory standard, a system should be developed where integrating crop data can be shared, which accelerate the prioritization for strong engagement from government side and also address environmental and public health concerns.

REFERENCES

Abdallah, N.A., Channapatna, S.P., and Alan, G.M. "Genome editing for crop improvement: Challenges and opportunities." *GM Crops & Food* 6, no. 4(2015): 183–205.

Ainley, W.M., Sastry-Dent, L., Welter, M.E., et al. "Trait stacking via targeted genome editing." *Plant Biotechnology Journal* 11(2013): 1126–1134.

Araki, M., and Ishii, T. "Towards social acceptance of plant breeding by genome editing." *Trends in Plant Science* 20, no. 3(2015): 145–149.

Boch, J., and Bonas, U. "Xanthomonas AvrBs3 family-type III effectors: Discovery and function." *Annual Review of Phytopathology* 48(2010): 419–436.

Chandrasegaran, S., and Carroll, D. "Origins of programmable nucleases for genome engineering." *Journal of Molecular Biology* 428(2016): 963–989.

Char, S.N., Unger-Wallace, E., Frame, B., et al. "Heritable site-specific mutagenesis using TALENs in maize." *Plant Biotechnology Journal* 13(2015): 1002–1010.

Curtin, S.J., Anderson, J.E., Starke, C.G., et al. "Targeted mutagenesis for functional analysis of gene duplication in legumes." *Methods in Molecular Biology* 1069(2013): 25–42.

D'Halluin, K., Vanderstraeten, C., Van Hulle, J., et al. "Targeted molecular trait stacking in cotton through targeted double-strand break induction." *Plant Biotechnology Journal* 11(2013): 933–941.

Friedrichs, S., Yoko, T., Peter, K., et al. "An overview of regulatory approaches to genome editing in agriculture." *Biotechnology Research and Innovation* 3, no. 2(2019): 208–220.

Gao, H., Smith, J., Yang, M., et al. "Heritable targeted mutagenesis in maize using a designed endonuclease." *Plant Journal* 61(2010): 176–187.

Gurushidze, M., Hensel, G., Hiekel, S., et al. "True-breeding targeted gene knock-out in barley using designer TALE nuclease in haploid cells." *PLoS One* 9(2014): e92046.

Haft, D.H., Selengut, J., Mongodin, E.F., et al. "A guild of 45 CRISPR-associated (Cas) protein families and multiple CRISPR/Cas subtypes exist in prokaryotic genomes." *PLoS Computational Biology* 1, no. 6(2005): e60.

Hensel, G., and Kumlehn, J. "Genome engineering using TALENs." *Methods in Molecular Biology* 1900(2019): 195–215.

Huang, S., Detlef, W., Roger, N.B., et al. "A proposed regulatory framework for genome-edited crops." *Nature Genetics* 48, no. 2 (2016): 109–111.

Ishii, T., and Araki, M. "A future scenario of the global regulatory landscape regarding genome-edited crops." *GM Crops & Food* 8, no. 1 (2017): 44–56.

Ji, H., Lu, P., Liu, B., et al. "Zinc-finger nucleases induced by HIV-1 Tat excise HIV-1 from the host genome in infected and latently infected cells." *Molecular Therapy: Nucleic Acids* 12(2018): 67–74.

Kleter, G.A., Harry, A.K., and Esther, J.K. "Gene-edited crops: Towards a harmonized safety assessment." *Trends in Biotechnology* 37, no. 5(2019): 443–447.

Lema, M.A. "Regulatory aspects of gene editing in argentina." *Transgenic Research* 28(2019): 147–150.

Liang, Z., Zhang, K., Chen, K., and Gao, C. "Targeted mutagenesis in *Zea mays* using TALENs and the CRISPR/Cas system." *Journal of Genetics and Genomics* 41(2013): 63–68.

Mao, Y., Zhang, H., Xu, N., et al. "Application of the CRISPR-Cas system for efficient genome engineering in plants." *Molecular Plant* 6(2013): 2008–2011.

McMurrough, T.A., Brown, C.M., Zhang, K., et al. "Active site residue identity regulates cleavage preference of LAGLIDADG homing endonucleases." *Nucleic Acids Research* 46(2018): 11990–12007.

Miao, J., Guo, D., Zhang, J., et al. "Targeted mutagenesis in rice using CRISPR-Cas system." *Cell Research* 23(2013): 1233–1236.

Ochiai, H., and Yamamoto, T. "Construction and evaluation of zinc finger nucleases." *Methods in Molecular Biology* 1630(2017): 1–24.

Puchta, H., and Hohn, B. "Breaking news: Plants mutate right on target." *Proceedings of the National Academy of Sciences of the United States of America* 107, no. 26(2010): 11657–11658.

Qi, Y., Li, X., Zhang, Y., et al. "Targeted deletion and inversion of tandemly arrayed genes in Arabidopsis thaliana using zinc finger nucleases." *G3-Genes Genomes Genetics* 3(2013):1707–1715.

Rath, D., Amlinger, L., Rath, A., et al. "The CRISPR-Cas immune system: Biology, mechanisms and applications". *Biochimie* 117(2015): 119–128.

Scheben, A, and Edwards, D. "Bottlenecks for genome-edited crops on the road from lab to farm." *Genome Biology* 19, no. 1(2018): 1–7.

Shan, Q., Wang, Y., Chen, K., et al. Rapid and efficient gene modification in rice and Brachypodium using TALENs." *Molecular Plant* 6(2013): 1365–1368.

Shukla, V.K., Doyon, Y., Miller, J.C., et al. "Precise genome modification in the crop species *Zea mays* using zinc-finger nucleases." *Nature* 459(2009): 437–441.

Sikandar, H.K. "Genome-editing technologies: Concept, pros, and cons of various genome-editing techniques and bioethical concerns for clinical application." *Molecular Therapy Nucleic Acids* 16(2019): 326–334.

Voytas, D.F., and Gao, C. "Precision genome engineering and agriculture: Opportunities and regulatory challenges," *PLOS Biology* 12, no. 6(2014): 1–6.

Wang, C.L., Qin, T.F., Yu, H.M., et al. "The broad bacterial blight resistance of rice line CBB23 is triggered by a novel TAL effector of *Xanthomonas oryzae* pv. oryzae." *Molecular Plant Pathology* 15(2013): 333–341.

Wolt, J.D., Kan, W., and Bing, Y. "The regulatory status of genome-edited crops." *Plant Biotechnology Journal* 14, no. 2(2016): 510–518.

Xie, K., and Yang, Y. "RNA-guided genome editing in plants using a CRISPR-Cas system." *Molecular Plant* 6(2013):1975–1983.

Zhang, H., Zhang, J., Wei, P., et al. "The CRISPR/Cas9 system produces specific and homozygous targeted gene editing in rice in one generation." *Plant Biotechnology Journal* 12(2014): 797–807.

16 Advances in Genome Editing

Shweta Singh
University of Minho (UMINHO)

Rupesh Kumar Singh
Universidade *de* Trás-os-Montes e Alto Douro (UTAD)

Sivalingam Anandhan
ICAR-Directorate of Onion and Garlic Research

Jyoti Priya
ICAR-Indian Agricultural Research Institute

Sandeep Jaiswal
ICAR-Research Complex for North-Eastern Hill Region

Ashok Kumar
ICAR-Directorate of Onion and Garlic Research

CONTENTS

16.1 INTRODUCTION

The world population is increasing exponentially and is expected to reach 9 billion by 2050. The limited arable land and changing environmental conditions are the serious threats to produce food for the planet. In order to fill the gap between food supply and its demand, the major focus of plant biologist is to generate crops with higher efficiency and higher nutritive values. In the past century, the domestication of high yielding varieties of crop plants relied on conventional breeding approach that depended upon the natural and mutant-induced genetic variations and 'cross the elite with the elite and wait for the best' strategy (van Ginkel and Ortiz, 2018). The mutation breeding programme was recognized as a tool in the 1940s and assisted to improve the yield of wheat during the Green Revolution. This classic strategy for the improvement of germplasm is a very laborious and long-term process. On the other hand, genome editing has emerged as a process of targeted genome manipulation especially for the crops with complex genome, which are hard to manipulate by the

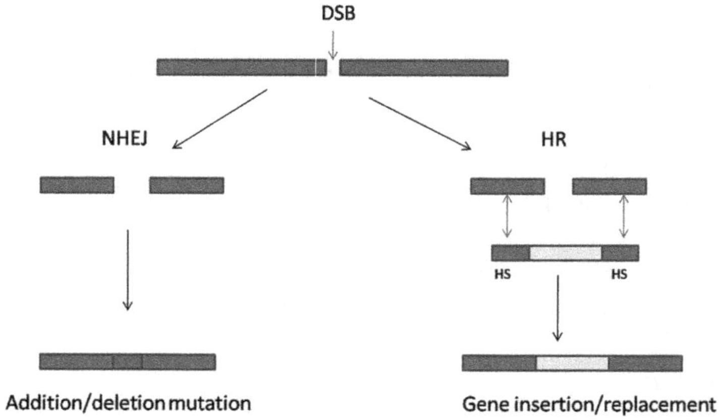

FIGURE 16.1 Double-stranded break and repair of DNA. Joining of double-strand in non-precise way by non-homologous end-joining (NHEJ) method, which causes the addition or deletion mutation. Joining of double-strand can be in a precise way by homologous recombination (HR) method that leads to gene insertion or replacement.

traditional breeding approach. The genome editing allows modifying the specific genome in a precise manner to increase the yield and nutritive value of crop (Feng et al., 2013; Brooks et al., 2014).

The concept of genome editing started with the double-stranded breaks (DSBs) of DNA (Capecchi, 1980), and further development was dependent on site-directed DSBs by sequence-specific nucleases (SSNs) (Rouet et al., 1994). The machinery of the cell can repair the DSBs in the precise manner (homology-directed repair (HDR)) (Paszkowski et al., 1988) or an imprecise way (non-homologous end-joining (NHEJ)) (Puchta, 2004) (Figure 16.1). Site-specific break in the double-strand of nucleic acid is possible by the bimodular fusion proteins. In these proteins, one has a DNA-binding domain to bind to the specific site of DNA and the other domain that can cleave that specific DNA sequence (Puchta, 2016).

The DSB gives an opportunity to edit the genome by harnessing the cell's own DNA repair machinery. The limitation was to create the site-specific DSBs by the SSNs, which were designed to bind DNA and cleave it in a site-specific manner (Puchta, 2016). The first-generation SSNs, which can cleave the DNA at predefined sites, include meganucleases (MNs), zinc finger nuclease (ZFN), transcription activator-like effector nucleases (TALENs) and clustered interspaced short palindromic repeats/CRISPR-associated protein-9 (CRISPR/Cas9) (Mohanta et al., 2017). Among these SSNs, ZFN and TALENs are protein-based and require the specific protein designing for every target sequence, while CRISPR/Cas9 is a RNA-guided system which can be easily designed to edit DNA (Arora and Narula, 2017).

16.2 IMPORTANCE AND NECESSITY OF GENOME EDITING

Scientific community across the globe has been manipulating the genome of plants by breeding within related and compatible families to obtain a superior phenotype via artificial selection methods. However, the major limitations were compatibility and naturally existing variations. Genome engineering technology has allowed the researchers to cross this barrier and manipulate with genomes across the compatibility of genus and species. Subsequent generations may express the desired characteristic, which can be further used in breeding programmes to develop the improved varieties.

The improvement of agricultural crops by genetic manipulation for various objectives has gained a major attention of biologists in order to provide food security for exponentially increasing world population. Production of crop varieties showing disease resistance and higher level of tolerance for various abiotic stresses such as salinity, high temperature, low temperature and drought is required to increase the crop production in changing climatic conditions (Takeda and Matsuoka, 2008). Cost-effective production with improved quality has also been a major objective. Gene modification by using modern molecular technologies has shown a potential to fulfil these objectives.

Advances in the field of genetic engineering had allowed to generate genetically modified varieties of plants, using the *Agrobacterium tumefaciens*-mediated transformation of desired traits from intergeneric species. One of the most popular examples is 'golden rice', which is a genetically modified variety of rice showing the enhanced production of β-carotene, a precursor of vitamin A (Paine et al., 2005). Herbicide resistance in soya bean was obtained by transferring *enzyme5-enolpyruvylshikimate-3-phosphate synthase* (*EPSPS*) gene from *Agrobacterium tumefaciens* to soya bean; insecticidal protein from *Bacillus thuringiensis* (Cry1Ab) was introduced in corn to provide pest resistance; canola was engineered to develop high fatty acid composition; plum was developed using coat protein (CP) gene from the virus for resistance against diseases (Phillips, 2008). Salmon fish eggs were engineered to obtain faster growth with significant improvement (Devlin et al., 1994) and in cattle for improved resistance to mad cow disease (Phillips, 2008). Additionally, a series of genome engineering has been achieved for a diverse range of objectives for humankind as for human growth hormone (Barta et al., 1986) and antibody production (Hiatt et al., 1989).

Transgenic technology has major limitations due to the conflict between intended and unintended effects of genetically modified organism/plant in comparison with its substantial equivalence counterpart (Millstone et al., 1999) in terms of safety issues and regulations. However, international organizations have been involved with plenty of expert advices to evaluate and establish the methods on the concept of substantial equivalence for genetically modified products for food (Goodman et al., 2008). The concept was recommended for newly introduced DNA as a target where no further modifications were expected in metabolic pathways (OCED, 2000). Bioinformatics tools and computational analysis have also been implemented to counter the unintended phenotyping occurred at the RNA level, amino acids or in metabolites (Stiekema and Nap, 2004).

Genome information has increased tremendously over the last decade for various crops, from model plants to cultivated crops, and increasing bioinformatics tools with computational analytical methods provided opportunities to design the genome engineering in a more precise manner. Additionally, emergence of nucleases to cleave and target the specific site by incorporating the own repair mechanism became an advantage. These new genome-editing tools may be divided into two different groups. The first one is the protein-based method where two different mannered proteins along with a DNA-binding protein domain and a C-terminal nuclease domain were utilized to edit a specific genome site; it is very time-consuming and complex mechanism as it involves protein engineering each time for a new target (Pabo et al., 2001; Boch et al., 2009; Moscou and Bogdanove, 2009). These protein-based approaches emerged as ZFN- and TALEN-mediated editing.

Cong et al. (2013) had described RNA-guided nucleases from CRISPRs and CRISPR-associated (Cas) proteins to facilitate genome-editing methods, which have been incorporated for knock-in and knock-out approaches for model plants as well as in agricultural crops successfully. These developments altogether facilitated the site-specific genome editing in crops, and thus, the developed crops are easy to screen in first generation for unintended effects. These approaches potentially facilitate the concept of substantial equivalence comparison for any changes and may escape the long and strict regulations for genetically modified organisms prior to commercialization.

16.3 ADVANCEMENT IN GENOME EDITING

In recent years, genome editing has become very popular because of precision and efficiency of technique that induce heritable changes in various plant species without altering the other desirable traits (Brooks et al., 2014; Alagoz et al., 2016). The SSNs played a major role in this technology, are bimodular fusion proteins with a DNA-binding domain and a protein with endonuclease activity. MNs were the first nucleases used for genome editing in plants (Stoddard, 2011; D'Halluin et al., 2013).

Because of the non-modular property of the specific protein, MNs were hard to redesign for different targets and their use in plant was restricted only to the natural occurring MNs (Prieto et al., 2007). ZFNs were used for the targeted transgene insertion and for precise genome editing in plants (Cai et al., 2009; Zhang et al., 2010). However, ZFNs are not popular now because of low target specificity and high off-target genome changes (Zhang et al., 2013). The other important system of nuclease is TALEN, which has been used for modifying various traits in *Arabidopsis*, rice, *Glycine max*, tobacco and other crops (Sedeek et al., 2019). The discovery of CRISPR/Cas has revolutionized the genome editing by making it less time-consuming and more efficient.

16.3.1 Meganuclease

MNs are homing endonucleases present in most of the microbes. They recognize relatively long stretch of DNA sequences and introduce DSBs, which are repaired by host repair pathways leading to insertion, deletion and point mutation. Journey of targeted genome editing started with MNs: the first MN *I-SecI* was characterized in yeast. *I-SecI* is an enzyme encoded by an intronic region of large ribosomal RNA subunit (LsrRNA) of yeast mitochondria. Later, *I-SecI* is used for the landmark demonstration of precise genome editing in tobacco plant by Puchta et al. (1993). D'Halluin et al. (2008) reported a positive HR event for restoring *bar* gene expression providing phosphinothricin resistance in maize. The same group reported gene stacking of two herbicide tolerance transgenes, 4-hydroxyphenyl pyruvate dioxygenase (hppd) from *Pseudomonas fluorescence* and enol-pyruvyl shikimate-3-phosphate synthase gene (*epsps*) from *Zea mays* in cotton providing resistance against hppd inhibitors and glyphosate, respectively (D'Halluin et al., 2013). Engineering MNs for the desired sequence specificity has been the major bottleneck mainly due to overlapping DNA-binding and cleavage domain (Stoddard, 2011) and the lack of efficient methods for protein engineering; even substitution of one amino acid requisite for sequence specificity leads to compromised catalytic activity. Several engineered MNs were developed on *I-creI* scaffold through a large-scale random mutagenesis in protein subunit and screening for site specificity and cleavage activity but attention remained diverted due to the development of other site-directed nucleases such as ZFNs, TALENs and CRISPR/Cas9 system.

16.3.2 Zinc Finger Nucleases

ZFNs involve a combination of FokI nuclease and pre-engineered zinc finger DNA-binding domain, which contains up to 30-amino acid domain for specific binding capacity. This amino acid binding domain was fabricated to allow recruiting the FokI domain to the desired DNA region by dimerization. DNA-binding domain represents the domain of folded proteins or may include nucleic acids as well with a motif to read the specific nucleotide sequence called as the recognition sequence. These domains were characterized as larger proteins with further protein domains with different functional roles or sometimes overlapping roles for a specific phenomenon, for example replication and modifications. The nuclease FokI used in this method belongs to a bacterial type IIS restriction endonuclease from *Flavobacterium okeanokoites.* This endonuclease contains a non-specific cleavage domain for DNA at C-terminal and DNA-binding domain at its N-terminal. The binding domain of endonuclease first recognizes the binding site and attaches with that site; the sooner activates the cleavage domain to chop down first strand by 9 nucleotides (nt) downstream and the second strand by 13 nt upstream of the recognition site at the nearest nucleotides. After the cleavage, host cell machinery itself performs the DNA repair mechanism, which may be homologous or non-homologous. Moreover, a donor DNA may be used for this repair mechanism for desired modification in the sequence (Porteus and Carroll, 2005; Weinthal et al., 2010, 2013).

ZFNs were used for the targeted transgene insertion and for the precise genome editing in plants for the past two decades (Cai et al., 2009; Zhang et al., 2010). ZFNs are useful to manipulate the genomes of many plants, including *Arabidopsis* for targeting *ADH1* and *TT4* because of their specific phenotypes (Zhang et al., 2010), *where Osakabe* et al. *(2010) performed genome editing of* stress-response regulator, the *ABA-INSENSITIVE4* (*ABI4*) gene. Cai et al. (2009) designed a dimer of 4-finger ZFNs that enabled intrachromosomal reconstitution of reporter gene and site-specific transgene integration, and tobacco cells were used for co-transformation with a donor construct comprising sequences necessary to complement a non-functional pat herbicide resistance gene, thereby suggesting ZFNs as a novel approach for predetermined gene addition, replacement or stacking in plants. Another study in tobacco with specific mutations to confer resistance to imidazolinone and sulphonylurea herbicides at SuR loci by ZFN-mediated gene targeting was performed, and the results demonstrated the possibility of targeted sequence modification in endogenous genes of plants (Townsend et al., 2009). Engineered ZFNs targeting many endogenous gene(s) of RNA silencing, including *DICER-LIKE* (*DCL*), were used for targeted genome manipulation in soya bean to investigate the efficiency of ZFN. ZFN-induced mutations in the paralogous genes *DCL4a* and *DCL4b* showed efficient heritable transmission across the subsequent generation. The results observed in this study provided the evidence of ZFN-induced mutagenesis as an efficient method for soya bean genome editing (Curtin et al., 2011). Corn has also been used to check the efficiency of ZFNs and provided a significant potential of this method (Shukla et al., 2009).

ZFNs have emerged as a very significant tool for genome editing in the last decade due to the specific targeted approach which can be fabricated for most of the genes. EXZACT™ is brand from Dow AgroSciences's genome-editing platform based on ZFNs and has validated the technology in monocots and dicots for addition or deletion with a targeted approach (Cai et al., 2009; Shukla et al., 2009, 2013; Kumar et al., 2015, 2016; Ainley, 2013a,b, 2014) ZFNs have certain advantages to bind and cut on the most positions with high fidelity in comparison with other methods (Paschon et al., 2019). CRISPR method is more advanced but needs a protospacer-adjacent motif (PAM) sequence at the target, which limits its application in comparison with ZFNs (Jinek et al., 2012; Zetsche et al., 2015). Engineering of zinc finger protein having a requisite specificity and affinity to the target sites needs a lot of efforts and scientific skills to define this domain for diverse DNA-targeted sites (Greisman and Pabo, 1997; Mani et al., 2005). Moreover, ZFNs are multifinger protein which should be optimized because the context-dependent and cooperative contacts can affect DNA recognition. Optimization process has also been improved by directed domain shuffling and rapid cell-based selections (Hurt et al., 2003). However, there are new tools available online for the technical support to define and select the best zinc finger domain for experiments; the present method could not become cost-effective and easier for users, which limited its popularity and gave a space to emerge other more advanced methods.

16.3.2 Transcription Activator-Like Effector Nuclease

The other system of nuclease TALENs was first derived from *Xanthomonas oryzae*, which has emerged as a popular method and succeeded over ZFNs (Zhang et al., 2013; Doyle et al., 2013). In nature, when this pathogen infects the plant cell, it produces transcription activator-like (TAL) effectors, which utilize the host cellular machinery and activate gene expression by binding their promoters. These effector proteins comprise a specific DNA-binding domain and an effector domain. DNA-binding domain is the pre-fabricated region having an affinity to bind specifically to the nucleotide sequence in the host genome. These pre-fabricated DNA-binding domains were linked to an activator/repressor/nuclease according to the desired purpose towards the targeted genetic manipulation and have been tested in a range of organisms. GeneArt TALs are commercially available tools from 'ThermoFisher Scientific', also derived from *Xanthomonas* TAL effectors. This product contains a DNA-binding domain, carrying a variable number of amino acid repeats. Each repeat contains 33–35 amino acids and recognizes a single DNA base pair. Repeat-variable diresidues (RVDs) are 2 hypervariable amino acid residues at positions 12 and 13, which define the DNA recognition pattern. TAL effector repeats having the requisite specificity and affinity for a target DNA sequence can be assembled module by module by varying these RVDs. Even long

TAL effectors can be designed to specifically target any locus in the genome by linking the repeats having RVDs. Engineering the DNA-binding domain of TALE and fusion of it with nucleases has generated a site-specific DNA cleavage (Christian et al., 2010; Li et al., 2012). TALEN-mediated gene editing has been used to modify various traits in model plants for the functional analysis of novel genes as well as in crop species to modify the overall performance of plant (Sedeek et al., 2019). TALEN has been proven as a gene editing technique with high precision but it needs to engineer the specific binding protein for each target that makes it a time-consuming and tedious process (Nemudryi et al., 2014).

TALEN technique has been accepted for eukaryotes, including model plants as well as agricultural crops for the targeted genome manipulation. Several plants, including rice, tobacco and *Arabidopsis,* have been tested by using this method for efficiency and specificity (Sanagala et al., 2017). Emerald–GATEWAY TALEN system was developed by Kusano et al. (2016) recently, which was a very simple GATEWAY-mediated vector for TALEN genes and effective for plant genome-based objectives. This system was developed by using a destination plasmid (pDual35SGw1301) and a vector (pPlat plasmids), engineered for concerned TALEN genes. Granule-bound starch synthase (*GBSS*) gene was targeted for site-specific mutation in potato cells by using this system, and the results demonstrated the high efficiency and specificity for plant genome.

These two targeted editing methods (ZFNs and TALENs) include FokI endonuclease and fabricated DNA-binding domain for site-specific recognition on the host genome to edit with precision. Both these methods require a DSB, which has the possibility for off-target cleavage and cannot be used for multiple editing at a single time (Bortesi and Fischer, 2015). Several bioinformatics tools become available to design appropriate TALENs for the predetermined target and need computational skills (Cermak et al., 2011). Together all these new methods with certain limitations gave a motivation to researchers for further improvement in targeted genome engineering; ZFNs and TALENs have been improved and replaced by clustered regularly interspaced short palindromic repeats (CRISPR)/CRISPR-associated (Cas) type II prokaryotic system (Sorek et al., 2013).

16.3.3 Clustered Regulatory Interspaced Short Palindromic Repeats/Cas

After 2 years of discovery of TALEN, new genome editing system, namely CRISPR/Cas, was discovered, which is based on non-coding RNA along with Cas protein (Haft et al., 2005; Nemudryi et al., 2014). The discovery of CRISPR/Cas made the genome editing less time-consuming and more efficient as it requires only the synthesis of target-specific single guide RNA of 20 nucleotides cloned with PAM sequence, which guides Cas endonuclease to break the double-strand at specific sites near the target.

The CRISPR/Cas9 is an adaptive immunity mechanism of prokaryotes, which is involved in cleaving the foreign DNA in bacterial cell in a sequence-dependent manner (Makarova et al., 2011). The CRISPR/Cas9 system is dependent on two components: a single guide RNA (sgRNA) and a non-specific endonuclease Cas9 that creates the DSBs at specific sites (Jinek et al., 2012). sgRNA recognizes the 20-bp complimentary sequence in DNA where PAM sequence is present downstream of the target sequence (Figure 16.2). The last 80 nucleotides of sgRNA form a secondary structure that provides a site for the attachment of Cas9. In the guidance of first 20 nucleotide sequences, Cas9 can cut the DNA in a site-specific manner (Jinek et al., 2012). The cleavage of double-stranded DNA subsequently triggers the DNA repair pathways. The repair in plant is generally by NHEJ, which causes the insertion or deletion mutation (Li et al., 2018a).

The major concern of genome editing by CRISPR/Cas9 is the off-target cleavage of DNA by Cas9 that can be avoided by selecting unique target sites. CRISPR-Plant (https://www.genome.arizona.edu/crispr/) (Xie et al., 2014), GT-scan (https://gt-scan.csiro.au/) (O'Brien and Bailey, 2014), CRISPRdirect (https://crispr.dbcls.jp/) (Naito et al., 2014) and many other tools are developed to select the unique targets and avoid the off-targets.

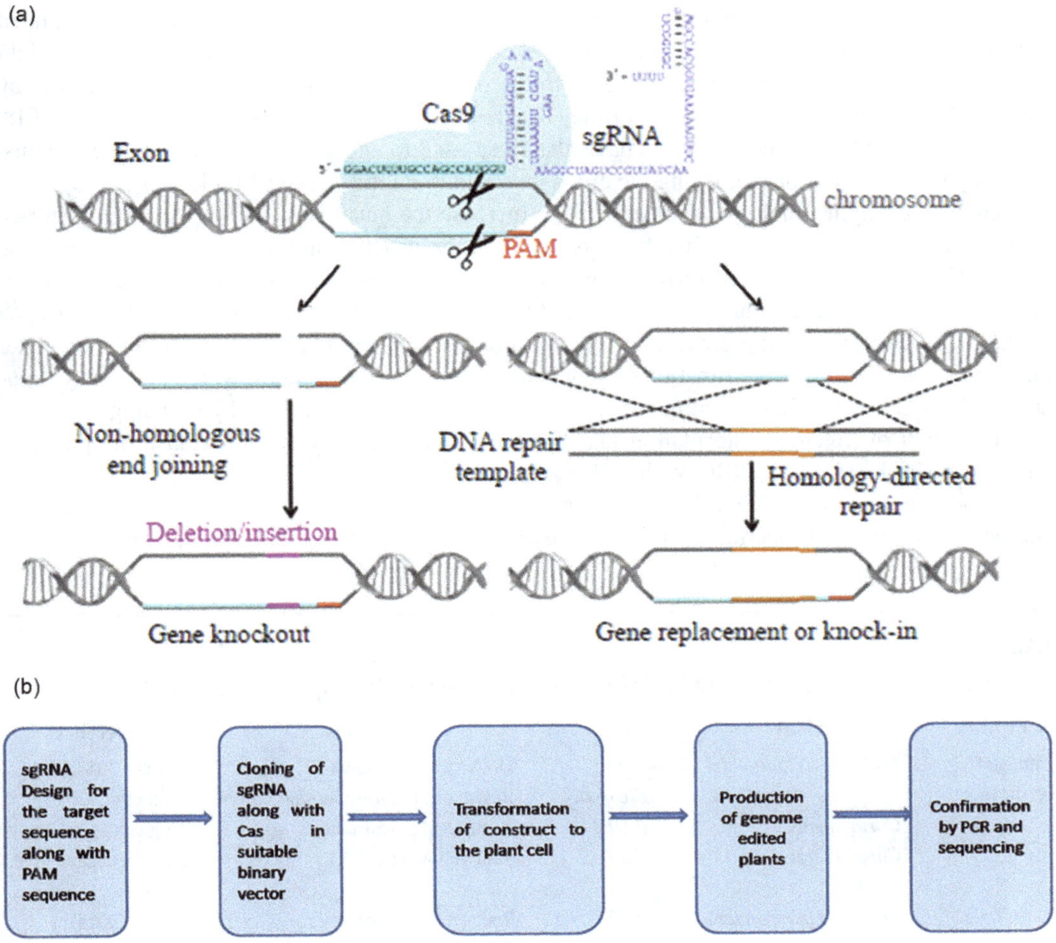

FIGURE 16.2 Targeted gene manipulation by CRISPR/Cas9 in plants. (a) A single guide RNA (sgRNA) having 20 nucleotides at 5′ end and PAM motif, which guides Cas9 nuclease to cleave the double-stranded DNA at a specific site. (b) Stepwise representation of CRISPR/Cas9-mediated plant genome editing. (Adopted from Zhang et al. (2016).)

Because of the simplicity and high efficiency of CRISPR/Cas9, it has been proven as a potential tool for the crop improvement. This method was initially utilized for the functional characterization of novel genes in model plants like *Arabidopsis thaliana*, *Nicotiana benthamiana* and also in cereal crops like *Oryza sativa* (Zhang et al., 2016). In recent years, CRISPR/Cas9 system has been used to generate heritable mutations without affecting the other valuable traits. The knock-out of some genes has encouraged superior traits, including abiotic stress tolerance, disease resistance and increased nutritive values (Arora and Narula, 2017). Thus, CRISPR/Cas9-mediated genome editing has a great potential to improve the nutritive quality and the yield of crops.

16.4 TARGETED GENOME EDITING AND IMPROVEMENT OF CROP TRAITS

Genome editing is relatively new technology but it has been adapted to improve the yield, abiotic and biotic stress tolerance, and herbicide resistance in a wide range of crops (Sedeek et al., 2019). The identification of genes responsible for desirable phenotype, tolerance to various stress and biosynthesis of various metabolites offers the possibilities for the use of genome editing for targeted trait improvement.

Knocking out of three genes GW2, GW5 and TGW6 that negatively regulated the weight of grains in rice, by CRISPR/Cas9, had resulted in mutants with a significant increase in grain weight (Xu et al., 2016). Various other traits related to the overall plant yields were also knocked out by CRISPR/Cas9, during the small span of time (Braatz et al., 2017; Soyk et al., 2017; Ma et al., 2018; Zhang et al., 2018). Gene editing tools have also been used to enhance the plant resistance against various pathogens and for modifying the crops for various abiotic stresses (Table 16.1).

Genome editing technology has been used to increase the nutritive value of plants. Golden rice is an important achievement in this direction through the genetic modification technology (Paine et al., 2005). Discovery of CRISPR/Cas9 has extensively boosted the research in the field of metabolic engineering where plant cells are targeted for the production of specific metabolites (Alagoz et al., 2016). Genome editing provides a tool to modify the genes in a very specific manner. After knowing the role and specific function of genes, we can mutate these genes, which negatively regulate the biosynthesis of important metabolites. Classic work has been done by reducing the acrylamide content of potato (Halterman et al., 2016) and reducing polyphenol oxidase in apple and mushroom (Nishitani et al., 2016; Waltz, 2016).

Further application of CRISPR/Cas9 includes intensive research in the field of metabolic engineering to produce specific metabolites in plant cells. In *Papaver somniferum*, by knocking out

TABLE 16.1
Application of Genome Editing in Different Plant Species to Improve Various Traits

Plant Traits	Plant	Gene	Functions	References
Pathogen resistance	*Arabidopsis thaliana*	*BSCTV*	DNA virus resistance	Ji et al. (2015)
		elF(iso)4E	Resistance to turnip mosaic virus	Pyott et al. (2016)
	Citrus sinensis	*CsLOB1*	Citrus canker resistance	Peng et al. (2017)
	Cucumis sativus	*eLF4E*	Vein yellow virus	Chandrasekaran et al. (2016)
	Nicotiana benthamiana	*BSCTV*	Beet severe curly top virus resistance	Ji et al. (2015)
		TYLCV	Yellow leaf curl virus resistance	Ali et al. (2015)
		AGO2	Virus resistance	Ludman et al. (2017)
	Oryza sativa	*OsSWEET14*	Bacterial blight resistance	Li et al. (2012)
		OsSWEET13	Bacterial blight resistance	Zhou et al. (2015)
		OsERF922.	Fungal blast resistance	Wang et al. (2016)
	Solanum lycopersicum	*SlMlo*	Powdery mildew resistance	Nekrasov et al. (2017)
	Triticum aestivum	*TaMLO-A1*	Powdery mildew resistance	Wang et al. (2014)
Abiotic stress tolerance	*Arabidopsis thaliana*	*Mir169a*	Drought resistance tolerance	Zhao et al. (2016)
	Oryza sativa	*OsRAV2*	Salt stress tolerance	Duan et al. (2016)
	Zea mays	*ARGOS8*	Drought tolerance	Shi et al. (2017)
Herbicide resistance	*Linum usitatissimum*	*EPSPS*	Glyphosate resistance	Sauer et al. (2016)
	Nicotiana tabacum	*MEL1*	Herbicide resistance	Cai et al. (2009)
		ALS	Imidazolinone and sulphonylurea resistance	Zhang et al. (2013)
	Oryza sativa	*ALS*	Clorsulfuron and bispyribac sodium tolerance	Sun et al. (2016)
		EPSPS	Glyphosate resistance	Li et al. (2016)
	Solanum tuberosum	*ALS1*	Clorsulfuron and bispyribac sodium tolerance	Butler et al. (2016)
	Zea mays	*IPK1*	Herbicide resistance	Shukla et al. (2009)

TABLE 16.2
List of Crops That Are Edited for Various Biosynthetic Genes

S. No.	Plant	Gene	Functions	References
1	*Solenum lycopersicum*	*ANT1*	Anthocyanin biosynthesis	Čermák et al. (2015)
		Phytoene desaturase (SlPDS)	Carotenoid biosynthesis	Pan et al. (2016)
		Phytoene synthase (PSY1)	Fruit colour	Filler Hayut et al. (2017)
		GABA-TP1, GABA-TP2, GABA-TP3, CAT9 and SSADH	γ-Aminobutyric acid (GABA) biosynthesis	Li et al. (2018a)
2	*Solenum tuberosum*	*Granule-bound starch synthase (GBSS)*	Starch quality	Andersson et al. (2017)
3	*Vitis vinifera*	*Phytoene desaturase (VvPDS)*	Carotenoid biosynthesis	Nakajima et al. (2017)
4	*Citrus sinensis*	*Phytoene desaturase (CsPDS)*	Carotenoid biosynthesis	Jia and Wang (2014)
5	*Citrullus lanatus*	*Phytoene desaturase (ClPDS)*	Carotenoid biosynthesis	Tian et al. (2017)
6	*Salvia miltiorrhiza*	*Diterpene synthase (SmCPS1)*	Tanshinone biosynthesis	Li et al. (2017)
7	*Zea mays*	*ZmIPK*	Phytic acid synthesis	Liang et al. (2014)
8	*Camelina sativa*	*FAD2*	Seed oil biosynthesis	Jiang et al. (2017)
9	*Oryza sativa*	*OsBADH2* *SBE1* and *SBEIIb*	Fragrance content Starch-branching enzyme	Shan et al. (2015) and Sun et al. (2017)

the 3′ *OMT2* gene that is involved in the biosynthesis of benzylisoquinoline alkaloid pathway, the synthesis of various alkaloids like noscapine, papaverine and morphine was altered (Alagoz et al., 2016). CRISPR/Cas9 has also been used to develop hypoimmunic gluten wheat and tomato with enhanced lycopene content (Li et al., 2018b; Sanchez-Leon et al., 2018). A list of some crops edited for various biosynthetic pathways are mentioned in Table 16.2.

16.5 FUTURE PERSPECTIVE

Genome editing has revolutionized the fundamental research as well as crop improvement. The functional validation of various genes has become faster and easier in many plant species by developing knock-out mutant through gene editing tools. It has also given a pace to conventional breeding approach and can be extended to the crops that are difficult to breed. CRISPR/Cas9 system has allowed incorporating stable genetic changes in plants without altering the other plant traits in a fast and easy way. The genome editing has many potential applications for improving the qualitative and quantitative traits of crops, which can potentially reshape the face of agriculture. Various technological improvements are still needed to improve the genome editing. The process of genome editing is dependent upon the *in vitro* regeneration and transformation of crop plant. To bypass this need, an efficient system is required to deliver the genome-editing machinery to the germline cells of a wide range of plant species. Further, a conscious application of this technology will be helpful for improving plant productivity and its nutritive values that will reshape the face of agriculture and ensure the food security for drastically increasing world population.

REFERENCES

Ainley, W. M., Bing, J. W., Corbin, D. R., et al. "Precision gene targeting to a particular locusin maize". *U.S. Patent* 14, no. 105 (2013a): 569.

Ainley, W. M., Sastry-Dent, L., Welter, M. E., et al. "Trait stacking via targeted genome editing". *Plant Biotechnology Journal* 11, no. 9 (2013b): 1126–1134.

Ainley, W. M., Guschin, D. Y., Hayden, M., et al. "Methods and compositions for integration of an exogenous sequence within the genome of plants". *US Patent Application U. S.* 14, no. 243 (2014): 553.

Alagoz, Y., Gurkok, T., Zhang, B. & Unver, T. "Manipulating the biosynthesis of bioactive compound alkaloids for next-generation metabolic engineering in opium poppy using CRISPR-Cas 9 genome editing technology". *Scientific Reports* 6, no. 1 (2016): 30910.

Ali, Z., Abul-faraj, A., Li, L., et al. "Efficient virus-mediated genome editing in plants using the CRISPR/Cas9 system". *Molecular Plant* 8, no. 8 (2015): 1288–1291.

Andersson, M., Turesson, H., Nicolia, A., Fält, A.-S., Samuelsson, M. & Hofvander, P. "Efficient targeted multiallelic mutagenesis in tetraploid potato (*Solanum tuberosum*) by transient CRISPR-Cas9 expression in protoplasts". *Plant Cell Reports* 36, no. 1 (2017): 117–128.

Arora, L. & Narula, A. "Gene editing and crop improvement using CRISPR-Cas9 system". *Frontiers in Plant Science* 8 (2017): 1932.

Barta, A., Sommergruber, K., Thompson, D., et al. "The expression of a nopaline synthase: Human growth hormone chimaeric gene in transformed tobacco and sunflower callus tissue". *Plant Molecular Biology* 6, no. 5 (1986): 347–357.

Boch, J., Scholze, H., Schornack, S., Landgraf, A., Hahn, S., Kay, S., Lahaye, T., Nickstadt, A. & Bonas, U. "Breaking the code of DNA binding specificity of TAL-type III effectors". *Science* 326, no. 5959 (2009): 1509–1512.

Bortesi, L. & Fischer, R. "The CRISPR/Cas9 system for plant genome editing and beyond". *Biotechnology Advances* 33, no. 1 (2015): 41–52.

Braatz, J., Harloff, H.-J., Mascher, M., et al. "CRISPR-Cas9 targeted mutagenesis leads to simultaneous modification of different homoeologous gene copies in polyploid oilseed rape (*Brassica napus*)". *Plant Physiology* 174, no. 2 (2017): 935–942.

Brooks, C., Nekrasov, V., Lippman, Z. B. & Van Eck, J. "Efficient gene editing in tomato in the first generation using the clustered regularly interspaced short palindromic repeats/CRISPR-Associated9 system." *Plant Physiology* 166, no. 3 (2014): 1292–1297.

Butler, N. M., Baltes, N. J., Voytas, D. F. & Douches, D. S. "Geminivirus-mediated genome editing in potato (*Solanum tuberosum L.*) using sequence-specific nucleases". *Frontiers in Plant Science* 7 (2016): 1045.

Cai, C. Q., Doyon, Y., Ainley, et al. "Targeted transgene integration in plant cells using designed zinc finger nucleases". *Plant Molecular Biology* 69, no. 6 (2009): 699–709.

Capecchi, M. R. "High efficiency transformation by direct microinjection of DNA into cultured mammalian cells". *Cell* 22, no. 2 (1980): 479–488.

Cermak, T., Doyle, E. L., Christian, M., et al. "Efficient design and assembly of custom TALEN and other TAL effector-based constructs for DNA targeting". *Nucleic Acids Research* 39, no. 12 (2011): e82.

Čermák, T., Baltes, N. J., Čegan, R., et al. "High-frequency, precise modification of the tomato genome." *Genome Biology* 16, no. 1 (2015): 232.

Chandrasekaran, J., Brumin, M., Wolf, D., et al. "Development of broad virus resistance in non-transgenic cucumber using CRISPR/Cas9 technology". *Molecular Plant Pathology* 17, no. 7 (2016): 1140–1153.

Christian, M., Cermak, T., Doyle, E. L., et al. "Targeting DNA double-strand breaks with TAL effector nucleases". *Genetics* 186, no. 2 (2010): 757–761.

Cong, L., Ran, F. A., Cox, D., et al. "Multiplex genome engineering using CRISPR/Cas systems." *Science* 339, no. 6121 (2013): 819–823.

Curtin, S. J., Zhang, F., Sander, J. D., et al. "Targeted mutagenesis of duplicated genes in soybean with zinc-finger nucleases". *Plant Physiology* 156, no. 2 (2011): 466–473.

D'Halluin, K., Vanderstraeten, C., Stals, E., Cornelissen, M. & Ruiter, R. "Homologous recombination: a basis for targeted genome optimization in crop species such as maize." *Plant Biotechnology Journal* 6, (2008): 93–102.

D'Halluin, K., Vanderstraeten, C., Van Hulle, J., et al. "Targeted molecular trait stacking in cotton through targeted double-strand break induction." *Plant Biotechnology Journal* 11 (8) (2013): 933–941.

Devlin, R. H., Yesaki, T. Y., Biagi, C. A., et al. "Extraordinary salmon growth." *Nature* 371, no. 6494 (1994): 209–210.

Doyle, E. L., Stoddard, B. L., Voytas, D. F. & Bogdanove, A. J. "TAL effectors: Highly adaptable phytobacterial virulence factors and readily engineered DNA-targeting proteins." *Trends Cell Biology* 23, no. 8 (2013): 390–398.

Duan, Y. B., Li, J., Qin, R. Y., et al. "Identification of a regulatory element responsible for salt induction of rice OsRAV2 through ex situ and in situ promoter analysis". *Plant Molecular Biology* 90, no. 1 (2016): 49–62.

Feng, Z., Zhang, B., Ding, W., et al. "Efficient genome editing in plants using a CRISPR/Cas system". *Cell Research* 23, no. 10 (2013): 1229–1232.

Filler Hayut, S., Melamed Bessudo, C. & Levy, A. A. "Targeted recombination between homologous chromosomes for precise breeding in tomato". *Nature Communications* 8 (2017): 15605.

Goodman, R. E., Vieths, S., Sampson, H. A., et al. "Allergenicity assessment of genetically modified crops: What makes sense?" *Nature Biotechnology* 26, no. 1 (2008): 73–81.

Greisman, H. A. & Pabo, C. O. "A general strategy for selecting high-affinity zinc finger proteins for diverse DNA target sites." *Science* 275, no. 5300 (1997): 657–661.

Haft, D. H., Selengut, J., Mongodin, E. F. & Nelson, K. E. "A guild of 45 CRISPR-associated (Cas) protein families and multiple CRISPR/Cas subtypes exist in prokaryotic genomes". *PLoS Computational Biology* 1, no. 6 (2005): e60.

Halterman, D., Guenthner, J., Collinge, S., et al. "Biotech potatoes in the 21st century: 20 years since the first biotech potato". *American Journal of Potato Research* 93, no. 1 (2016): 1–20.

Hiatt, A., Cafferkey, R. & Bowdish, K. "Production of antibodies in transgenic plants". *Nature* 342, no. 6245 (1989): 76–78.

Hurt, J. A., Thibodeau, S. A., Hirsh, A. S., et al. "Highly specific zinc finger proteins obtained by directed domain shuffling and cell-based selection". *Proceedings of the National Academy of Sciences of the United States of America* 100, no. 21 (2003): 12271–12276.

Ji, X., Zhang, H., Zhang, Y., Wang, Y. & Gao, C. "Establishing a CRISPR-Cas-like immune system conferring DNA virus resistance in plants. *Nature Plants* 1 (2015): 15144.

Jia, H. & Wang, N. "Targeted genome editing of sweet orange using Cas9/sgRNA. *PLoS One* 9, no. 4 (2014): e93806.

Jiang, W. Z., Henry, I. M., Lynagh, P. G., et al. "Significant enhancement of fatty acid composition in seeds of the allohexaploid, *Camelina sativa*, using CRISPR/Cas9 gene editing". *Plant Biotechnology Journal* 15, no. 5 (2017): 648–657.

Jinek, M., Chylinski, K., Fonfara, I., et al. "A programmable dual-RNA-guided DNA endonuclease in adaptive bacterial immunity". *Science* 337, no. 6096 (2012): 816–821.

Kumar, S., AlAbed, D., Worden, A., et al. "A modular gene targeting system for sequential transgene stacking in plants". *Journal of Biotechnology* 207 (2015): 12–20.

Kumar, S., Barone, P. & Smith, M. "Gene targeting and transgene stacking using intra genomic homologous recombination in plants" *Plant Methods* 12 (2016): 11.

Kusano, H., Onodera, H., Kihira, M., et al. "A simple Gateway-assisted construction system of TALEN genes for plant genome editing". *Scientific Reports* 6, no. 1 (2016): 30234.

Li, L., Piatek, M. J., Atef, A., et al. "Rapid and highly efficient construction of TALE-based transcriptional regulators and nucleases for genome modification". *Plant Molecular Biology* 78, no. 4–5 (2012): 407–416.

Li, J., Meng, X., Zong, Y., Chen, K., Zhang, H., Liu, J., Li, J. & Gao, C. "Gene replacements and insertions in rice by intron targeting using CRISPR-Cas9". *Nature Plants* 2 (2016): 16139.

Li, B., Cui, G., Shen, G., Zhan, Z., Huang, L., Chen, J. & Qi, X. "Targeted mutagenesis in the medicinal plant Salvia miltiorrhiza". *Scientific Reports* 7 (2017): 43320.

Li, R., Li, R., Li, X., et al. "Multiplexed CRISPR/Cas9-mediated metabolic engineering of gamma-aminobutyric acid levels in *Solanum lycopersicum*". *Plant Biotechnology Journal* 16, no. 2 (2018a): 415–427.

Li, X., Wang, Y., Chen, S., et al. "Lycopene is enriched in tomato fruit by CRISPR/Cas9-mediated multiplex genome editing". *Frontiers in Plant Science* 9 (2018b): 559.

Liang, Z., Zhang, K., Chen, K. & Gao, C. "Targeted mutagenesis in Zea mays using TALENs and the CRISPR/Cas system". *Journal of Genetics and Genomics* 41, no. 2 (2014): 63–68.

Ludman, M., Burgyán, J. & Fátyol, K. "Crispr/Cas9 mediated inactivation of argonaute 2 reveals its differential involvement in antiviral responses. *Scientific Reports* 7, no. 1 (2017): 1010.

Ma, J., Chen, J., Wang, M., Ren, Y., Wang, S., Lei, C., Cheng, Z. & Sodmergen. "Disruption of OsSEC3A increases the content of salicylic acid and induces plant defense responses in rice". *Journal of Experimental Botany* 69, no. 5 (2018): 1051–1064.

Makarova, K. S., Aravind, L., Wolf, Y. I. & Koonin, E. V. "Unification of Cas protein families and a simple scenario for the origin and evolution of CRISPR-Cas systems". *Biology Direct* 6 (2011): 38.

Mani, M., Kandavelou, K., Dy, F. J., Durai, S. & Chandrasegaran, S. "Design, engineering, and characterization of zinc finger nucleases". *Biochemical and Biophysical Research Communications* 335, no. 2 (2005): 447–457.

Millstone, E., Brunner, E. & Mayer, S. "Beyond substantial equivalence". *Nature* 401, no. 6753 (1999): 525–526.

Mohanta, T. K., Bashir, T., Hashem, A., Abd Allah, E. F. & Bae, H. "Genome editing tools in plants". *Genes (Basel)* 8 (2017): 12.

Moscou, M. J. & Bogdanove, A. J. "A simple cipher governs DNA recognition by TAL effectors". *Science* 326, no. 5959 (2009): 1501.

Naito, Y., Hino, K., Bono, H. & Ui-Tei, K. "CRISPRdirect: Software for designing CRISPR/Cas guide RNA with reduced off-target sites". *Bioinformatics* 31, no. 7 (2014): 1120–1123.

Nakajima, I., Ban, Y., Azuma, A., et al. "CRISPR/Cas9-mediated targeted mutagenesis in grape". *PLoS One* 12, no. 5 (2017): e0177966.

Nekrasov, V., Wang, C., Win, J., et al. "Rapid generation of a transgene-free powdery mildew resistant tomato by genome deletion". *Scientific Reports* 7, no. 1 (2017): 482.

Nemudryi, A. A., Valetdinova, K. R., Medvedev, S. P., et al. "TALEN and CRISPR/Cas genome editing systems: Tools of discovery". *Acta Naturae* 6, no. 3 (2014): 19–40.

Nishitani, C., Hirai, N., Komori, S., et al. "Efficient genome editing in apple using a CRISPR/Cas9 system". *Scientific Reports* 6, no. 1 (2016): 31481.

O'Brien, A. & Bailey, T. L. (2014). "GT-Scan: Identifying unique genomic targets". *Bioinformatics* 30 (18): 2673–2675.

OCED. "Report of the task force for the safety of novel foods and feeds". Reports no. C(2000)86/ADD1 (2000).

Osakabe, K., Osakabe, Y. & Toki, S. "Site-directed mutagenesis in Arabidopsis using custom-designed zinc finger nucleases". *Proceedings of the National Academy of Sciences of the United States of America* 107, no. 26 (2010): 12034–12039.

Pabo, C. O., Peisach, E. & Grant, R. A. "Design and selection of novel Cys2His2 zinc finger proteins". *Annual Review of Biochemistry* 70 (2001): 313–340.

Paine, J. A., Shipton, C. A., Chaggar, S., et al. "Improving the nutritional value of golden rice through increased pro-vitamin A content". *Nature Biotechnology* 23, no. 4 (2005): 482–487.

Pan, C., Ye, L., Qin, L., et al. "CRISPR/Cas9-mediated efficient and heritable targeted mutagenesis in tomato plants in the first and later generations". *Scientific Reports* 6 (2016): 24765.

Paschon, D. E., Lussier, S., Wangzor, T., et al. "Diversifying the structure of zinc finger nucleases for high-precision genome editing". *Nature Communications* 10, no. 1 (2019): 1133.

Paszkowski, J., Baur, M., Bogucki, A. & Potrykus, I. "Gene targeting in plants". *EMBO Journal* 7, no. 13 (1988): 4021–4026.

Peng, A., Chen, S., Lei, T., Xu, L., He, Y., et al. "Engineering canker-resistant plants through CRISPR/Cas9-targeted editing of the susceptibility gene CsLOB1 promoter in citrus". *Plant Biotechnology Journal* 15, no. 12 (2017): 1509–1519.

Phillips, T. "Genetically modified organisms (GMOs): Transgenic crops and recombinant DNA technology" *Nature Education* 1, no. 1 (2008): 213.

Porteus, M. H. & Carroll, D. "Gene targeting using zinc finger nucleases". *Nature Biotechnology* 23, no. 8 (2005): 967–973.

Prieto, J., Redondo, P., Padro, D., et al. "The C-terminal loop of the homing endonuclease I-CreI is essential for site recognition, DNA binding and cleavage". *Nucleic Acids Research* 35, no. 10 (2007): 3262–3271.

Puchta, H. "The repair of double-strand breaks in plants: Mechanisms and consequences for genome evolution" *Journal of Experimental Botany* 56, no. 409 (2004): 1–14.

Puchta, H. "Using CRISPR/Cas in three dimensions: Towards synthetic plant genomes, transcriptomes and epigenomes". *The Plant Journal* 87, no. 1 (2016): 5–15.

Puchta, H., Dujon, B., & Hohn, B. "Homologous recombination in plant cells is enhanced by in vivo induction of double strand breaks into DNA by a site-specific endonuclease." *Nucleic Acids Research* 21, (1993): 5034–5040.

Pyott, D. E., Sheehan, E. & Molnar, A. "Engineering of CRISPR/Cas9-mediated potyvirus resistance in transgene-free Arabidopsis plants". *Molecular Plant Pathology* 17, no. 8 (2016): 1276–1288.

Rouet, P., Smih, F. & Jasin, M. "Introduction of double-strand breaks into the genome of mouse cells by expression of a rare-cutting endonuclease". *Molecular Cell Biology* 14, no. 12 (1994): 8096–8106.

Sanagala, R., Moola, A. K. & Bollipo Diana, R. K. "A review on advanced methods in plant gene targeting". *Journal of Genetic Engineering and Biotechnology* 15, no. 2 (2017): 317–321.

Sanchez-Leon, S., Gil-Humanes, J., Ozuna, C. V., et al. "Low-gluten, nontransgenic wheat engineered with CRISPR/Cas9". *Plant Biotechnology Journal* 16, no. 4 (2018): 902–910.

Sauer, N. J., Mozoruk, J., Miller, R. B., et al. "Oligonucleotide-directed mutagenesis for precision gene editing". *Plant Biotechnology Journal* 14, no. 2 (2016): 496–502.

Sedeek, K. E. M., Mahas, A. & Mahfouz, M. "Plant genome engineering for targeted improvement of crop traits" *Frontiers in Plant Science* 10 (2019): 114.

Shan, Q., Zhang, Y., Chen, K., Zhang, K. & Gao, C. "Creation of fragrant rice by targeted knockout of the OsBADH2 gene using TALEN technology". *Plant Biotechnology Journal* 13, no. 6 (2015): 791–800.

Shi, J., Gao, H., Wang, H., et al. "ARGOS8 variants generated by CRISPR-Cas9 improve maize grain yield under field drought stress conditions". *Plant Biotechnology Journal* 15, no. 2 (2017): 207–216.

Shukla, V. K., Doyon, Y., Miller, J. C., et al. "Precise genome modification in the crop species Zea mays using zinc-finger nucleases". *Nature* 459, no. 7245 (2009): 437–441.

Shukla, V., Gupta, M., Urnov, F., Guschin, D., De Both, M. J., Bundock, P. "Targeted modification of malate dehydrogenase." *US Patent Application* 9, no. 523 (2013): 098.

Sorek, R., Lawrence, C. M. & Wiedenheft, B. "CRISPR-mediated adaptive immune systems in bacteria and archaea." *Annual Review of Biochemistry* 82 (2013): 237–266.

Soyk, S., Muller, N. A., Park, S. J., et al. "Variation in the flowering gene SELF PRUNING 5G promotes day-neutrality and early yield in tomato". *Nature Genetics* 49, no. 1 (2017): 162–168.

Stiekema, W. J. & Nap, J. P. H. "Bioinformatics for biosafety: Predicting the potential allergenicity of GM food". In *Genomics for Biosafety in Plant Biotechnology*, NATO Science Series No. 359, edited by J. P. Nap, A. Atanassov & W. J. Stiekema, pp. 98–114. Amsterdam: IOS Press (2004).

Stoddard, B. L. "Homing endonucleases: From microbial genetic invaders to reagents for targeted DNA modification". *Structure* 19, no. 1 (2011): 7–15.

Sun, Y., Zhang, X., Wu, C., et al. "Engineering herbicide-resistant rice plants through CRISPR/Cas9-mediated homologous recombination of Acetolactate synthase". *Molecular Plant* 9, no. 4 (2016): 628–631.

Sun, Y., Jiao, G., Liu, Z., et al. "Generation of high-amylose rice through CRISPR/Cas9-mediated targeted mutagenesis of starch branching enzymes". *Frontiers in Plant Science* 8 (2017): 298.

Takeda, S. & Matsuoka, M. "Genetic approaches to crop improvement: Responding to environmental and population changes." *Nature Reviews Genetics* 9, no. 6 (2008): 444–457.

Tian, S., Jiang, L., Gao, Q., et al. "Efficient CRISPR/Cas9-based gene knockout in watermelon." *Plant Cell Reports* 36, no. 3 (2017): 399–406.

Townsend, J. A., Wright, D. A., Winfrey, R. J., et al. "High-frequency modification of plant genes using engineered zinc-finger nucleases". *Nature* 459, no. 7245 (2009): 442–445.

van Ginkel, M. & Ortiz, R. "Cross the best with the best, and select the best: HELP in breeding selfing crops". *Crop Science* 58, no. 1 (2018): 17–30.

Waltz, E. "Gene-edited CRISPR mushroom escapes US regulation". *Nature* 532, no. 7599 (2016): 293.

Wang, Y., Cheng, X., Shan, Q., et al. "Simultaneous editing of three homoeoalleles in hexaploid bread wheat confers heritable resistance to powdery mildew". *Nature Biotechnology* 32, no 9 (2014): 947–951.

Wang, F., Wang, C., Liu, P., et al. "Enhanced rice blast resistance by CRISPR/Cas9-targeted mutagenesis of the ERF transcription factor gene OsERF922". *PLoS One* 11, no. 4 (2016): e0154027.

Weinthal, D., Tovkach, A., Zeevi, V. & Tzfira, T. "Genome editing in plant cells by zinc finger nucleases". *Trends Plant Science* 15, no. 6 (2010): 308–321.

Weinthal, D. M., Taylor, R. A. & Tzfira, T. "Nonhomologous end joining-mediated gene replacement in plant cells". *Plant Physiology* 162, no. 1 (2013): 390–400.

Xie, K., Zhang, J. & Yang, Y. "Genome-wide prediction of highly specific guide RNA spacers for CRISPR-Cas9-mediated genome editing in model plants and major crops". *Molecular Plant* 7, no. 5 (2014): 923–926.

Xu, R., Yang, Y., Qin, R., Li, H., Qiu, C., Li, L., Wei, P. & Yang, J. "Rapid improvement of grain weight via highly efficient CRISPR/Cas9-mediated multiplex genome editing in rice". *Journal of Genetics and Genomics* 43, no. 8 (2016): 529–532.

Zetsche, B., Gootenberg, J. S., Abudayyeh, O. O., et al. "Cpf1 is a single RNA-guided endonuclease of a class 2 CRISPR-Cas system". *Cell* 163, no. 3 (2015): 759–771.

Zhang, F., Maeder, M. L., Unger-Wallace, E., et al. "High frequency targeted mutagenesis in Arabidopsis thaliana using zinc finger nucleases". *Proceedings of the National Academy of Sciences of the United States of America* 107, no. 26 (2010): 12028–12033.

Zhang, Y., Zhang, F., Li, X., et al. "Transcription activator-like effector nucleases enable efficient plant genome engineering". *Plant Physiology* 161, no. 1 (2013): 20–27.

Zhang, D., Li, Z. & Li, J.-F. "Targeted gene manipulation in plants using the CRISPR/Cas technology". *Journal of Genetics and Genomics* 43, no. 5 (2016): 251–262.

Zhang, Y., Li, S., Xue, S., Yang, S., Huang, J. & Wang, L. "Phylogenetic and CRISPR/Cas9 studies in deciphering the evolutionary trajectory and phenotypic impacts of rice ERECTA genes". *Frontiers in Plant Science* 9 (2018): 473.

Zhao, Y., Zhang, C., Liu, W., et al. "An alternative strategy for targeted gene replacement in plants using a dual-sgRNA/Cas9 design". *Scientific Reports* 6, no. 1 (2016): 23890.

Zhou, J., Peng, Z., Long, J., et al. "Gene targeting by the TAL effector PthXo2 reveals cryptic resistance gene for bacterial blight of rice." *Plant Journal* 82, no. 4 (2015): 632–643.

Index

Note: **Bold** page numbers refer to tables and *Italic* page numbers refer to figures.

For Product Safety Concerns and Information please contact our EU
representative GPSR@taylorandfrancis.com
Taylor & Francis Verlag GmbH, Kaufingerstraße 24, 80331 München, Germany

www.ingramcontent.com/pod-product-compliance
Lightning Source LLC
LaVergne TN
LVHW080802111025
823277LV00005B/308

* 9 7 8 0 3 6 7 4 1 5 9 0 7 *